Fundamentals of Chemical Reactor Engineering

Fundamentals of Chemical Reactor Engineering

A Multi-Scale Approach

Timur Doğu
Professor of Chemical Engineering
Middle East Technical University
Çankaya, Ankara

Gülşen Doğu
Professor of Chemical Engineering
Gazi University
Maltepe, Ankara

This edition first published 2022.

Registered Office(s)
John Wiley & Sons, Inc., 111 River Street, Hoboken, NJ 07030, USA

Editorial Office
111 River Street, Hoboken, NJ 07030, USA

For details of our global editorial offices, customer services, and more information about Wiley products visit us at www.wiley.com.

Wiley also publishes its books in a variety of electronic formats and by print-on-demand. Some content that appears in standard print versions of this book may not be available in other formats.

Library of Congress Cataloging-in-Publication Data:

Names: Doğu, Timur, author. | Doğu, Gülşen, author.
Title: Fundamentals of chemical reactor engineering : a multi-scale approach / Timur Doğu & Gülşen Doğu, Professors of Chemical Engineering, Middle East Technical University & Gazi University.
Description: Hoboken, NJ, USA : Wiley, 2022. | Includes index.
Identifiers: LCCN 2021032013 (print) | LCCN 2021032014 (ebook) | ISBN 9781119755890 (cloth) | ISBN 9781119755906 (adobe pdf) | ISBN 9781119755913 (epub)
Subjects: LCSH: Chemical reactors. | Chemical plants.
Classification: LCC TP157 .D595 2022 (print) | LCC TP157 (ebook) | DDC 660/.2832–dc23
LC record available at https://lccn.loc.gov/2021032013
LC ebook record available at https://lccn.loc.gov/2021032014

Cover Design: Wiley
Cover Image: Courtesy of Timur Doğu and Gülşen Doğu

Set in 9.5/12.5pt STIXTwoText by Straive, Pondicherry, India

10 9 8 7 6 5 4 3 2 1

We are grateful to our mothers Halide Duralı and Mübdia Doğu, our fathers Hüseyin Duralı and Latif Doğu. Their memories are still alive and supportive.

Our endless thanks to our children Burcu Balam and Doruk, and our daughter-in-law Serap for their continuing support and affection.

Our deepest love is for our grandchildren İpek, Emre, and Demir.

This book is dedicated to all of you.

Contents

Preface

The multi-scale nature of chemical reactor engineering makes it quite distinct from most of the other courses in a chemical engineering curriculum. Its diverse nature is quite unique as compared to the courses in many of the engineering fields. Analysis and design of chemical reactors are generally taught in the undergraduate-level Chemical Reaction Engineering, Chemical Engineering Kinetics, or Reactor Design courses, as well as in the graduate-level Advanced Chemical Reaction Engineering and Catalysis courses. Any chemical engineering graduate should acquire a strong background in the fundamentals of kinetics and thermodynamics of chemical conversions, as well as hydrodynamics and transport processes taking place in a chemical reactor. The interaction of the reacting molecules on the molecular scale determines the mechanism and the kinetics of chemical reactions. Comprehension of the kinetics and thermodynamics of chemical conversions is required but not enough to analyze and design chemical reactors. Solid catalytic materials catalyze many of the chemical conversions. Most of these catalysts are porous, having very high surface area values, reaching 1000 m^2/g or even higher. In the scale of a catalyst pellet, knowledge on the adsorption and surface reaction mechanisms is needed to develop the rate expressions. Understanding pore diffusion and heat transfer effects on the observed rates of the reactions catalyzed by porous catalytic materials are crucial for analyzing and designing catalytic reactors.

Observed rates of chemical conversions strongly depend upon the temperature and concentration distributions, as well as the mixing characteristics within a reactor. Design and scale-up of a chemical reactor from bench-scale to industrial-scale require a strong knowledge of hydrodynamics and transport processes within these units. Understanding heat and mass transfer resistances within such units is required for the realistic design of chemical reactors. Transport processes between the phases should be considered in the analysis and design of multiphase reaction vessels, such as slurry, trickle-bed, fluidized-bed, and fixed-bed reactors. Pseudo-homogeneous transport equations are used to design multiphase reaction systems. These equations involve interphase and intraphase transport rate parameters in solid-catalyzed reactions.

Design and analysis of a chemical reaction vessel require the synthesis of basic knowledge of the fundamentals at each scale of the processes taking place in a reactor. This multi-scale nature of chemical reactor engineering and the fundamentals of each scale are emphasized throughout this book. Some new approaches to process intensification are also included in this text. This book is intended as a textbook for undergraduate- and graduate-level chemical reaction engineering and reactor design courses taught in the chemical engineering programs. Although Chapters 1–8 are designed mainly for an undergraduate-level chemical reaction engineering course, the choice of the chapters to be covered in the undergraduate- or graduate-level courses depends upon the background of the students taking these courses.

Chemical reactor engineering is a continuously evolving area in the chemical engineering curriculum. The first book, which treats kinetics and catalysis, is part III of the "Chemical Process

Principles" book series of Hougen and Watson (1947). The classical "Chemical Engineering Kinetics" book of J.M. Smith (1956), and the "Chemical Reaction Engineering: An Introduction" book of O. Levenspiel (1962) made lasting impacts on the understanding of principles and applications of chemical reaction engineering. Some of the other excellent books published in this area are "Chemical Reactor Theory" book by K.G. Denbigh and J.C.R. Turner (1965), "Chemical Reaction Analysis" book by E.E. Petersen (1965), "Elementary Chemical Reactor Analysis" book of R. Aris (1969), "Chemical and Catalytic Reactor Engineering" book of J.J. Carberry (1976), "An Introduction to Chemical Kinetics and Reactor Design" book by C. G. Hill (1977), "Chemical Reactor Analysis and Design" book of G.F. Froment and K.B. Bischoff (1979), "Elements of Chemical Reaction Engineering" book of H.S. Fogler (1986), and "Chemical Reactions and Chemical Reactors" book of G.W. Roberts (2009). Multiple editions of some of these books were published and used in the chemical reaction engineering courses in a number of chemical engineering programs. The authors gratefully acknowledge all of the former contributors who made their mark in chemical reaction engineering and catalysis.

Recent advances in process intensification comprise the development of more efficient, smaller, cleaner processes than the conventional ones used in the chemical industry. Cost and waste minimization, improved production yield of the desired product, less energy and reactant consumption per unit mass of the product, improved safety, and environmental cleanliness are some of the concepts, which should be considered in the design of new chemical processes and reactors. From the reactor engineering point of view, the idea of integration of reaction and separation processes into a single unit initiated the development of multifunctional reactors. Also, the use of alternative energy sources, like microwave, solar energy, ultrasound, etc. in chemical processing and design of novel reactors, such as microchannel reactors, monolithic reactors, supercritical reactors, opened new avenues in reaction engineering applications.

Process intensification concepts and modeling of multifunctional reactors, such as membrane reactors, equilibrium-stage and continuous packed-bed reactive distillation units, and sorption enhanced reactors, are covered in Chapter 13. Some other chemical reactor engineering applications in material processing (chemical vapor deposition reactors) are also illustrated with examples. Effects of the pore-size distribution of the catalysts having micro-, meso-, and macropores on the observed rates of reactions, criteria for the neglection of transport effects on the observed rates, diffusion mechanisms within porous catalysts, dynamic methods used for the evaluation of rate parameters, modeling of gas–solid non-catalytic reactions and catalyst deactivation are some of the other topics covered in this book. Several problems are included at the end of each chapter, and a list of open-ended case studies is given in Chapter 7. Different software is available in the universities for the numerical solution and simulation of engineering problems. Students are expected to write down their own codes or use any one of the available simulation software and design packages to solve some of the complex reactor design problems and improve their problem-solving abilities without being side-tracked from the basics of chemical reactor engineering. Hence, the emphasis is given to the deep understanding of the basic principles of chemical reactor analysis and heterogeneous catalysis at every scale of this subject, and these basic principles are illustrated with examples. Online Appendices are also available, covering a review of mathematical tools needed in chemical reactor engineering applications and some case study examples.

Timur Doğu
Gülşen Doğu
Ankara Turkey

Foreword

Chemical reactor engineering is one of the core subjects of the chemical engineering curriculum. It is often said that the reactor is the heart of a chemical plant. After all, this is where the magic occurs, turning reactants into more valuable products – and this ought to happen selectively, reliably, safely, and economically at the pace and scale relevant to practical applications. Hence, proper design and analysis of chemical reactors are of the utmost importance. A properly designed and operated reactor can limit or even avoid very expensive separation steps downstream, which might severely affect the environmental footprint as well. Chemical reactor engineering connects chemistry (molecular transformations, at nanoscales) with the macroscopic world, bridging a gap of up to 9, 10, sometimes even more orders of magnitude in length scales. To make this possible, transport phenomena at intermediate mesoscales need to be harnessed, which is where mass, heat, and momentum transfer come in – concepts like diffusion and convection. Too frequently, this is neglected or oversimplified, at a cost. Furthermore, a catalyst, and most often a heterogeneous porous one, is typically employed to accelerate the desired reactions, while undesired reactions are suppressed to realize a particular product or product distribution. Catalytic and reactor engineering are intimately connected, and an integral approach is, therefore, required.

This textbook introduces the readers to chemical reactor engineering, emphasizing ways to bridge the micro- and the macroworld. Its authors point out the need for a rigorous, multi-scale approach. Only then can one move beyond idealized pictures of reactors that serve a purpose as limiting cases (and thus should be learnt first, such as the classic batch, plug-flow, and the perfectly stirred-tank reactor), but do not sufficiently address the interference of transport phenomena at mesoscales. This is key to solving some of the challenges that modern chemical engineers face.

The playground for chemical engineers has been considerably expanded in recent years, from the traditional realm of refineries and petrochemicals to the production of high value-added, fine chemicals, pharmaceuticals, and specialty products, and including a broadening range of biotechnological processes, electrochemical and photocatalytic devices. Opportunities in chemical reactor engineering have increased thanks to faster computational tools, advances in materials of construction, catalysis, and progress on the fundamentals. But there is also a shift in the use of resources, design, and operational constraints imposed by sustainability targets in the face of pressing environmental concerns. Chemical reactor engineering is a fascinating subject, because it allows us to manipulate the design and operational space to address such challenges. However, to do so, a mastery of the fundamental concepts is essential. A systematic pedagogical approach is required to make the subject approachable, gradually moving from the basics to build a more complex toolbox that allows formulating the principal equations and applying these in practice. Equally important is that a solid grasp of the fundamentals makes the subject future-proof, building the foundation to study more specialized subjects, as the demands continue to shift.

Thanks to decades of experience in deeply researching and teaching the subject, world-renowned Professors Timur and Gülşen Doğu do a marvelous job at preparing the budding chemical engineer or other interested reader through the essential steps, following the multi-scale approach that makes this textbook stand out. A logical, systematic approach takes the reader from the basics of chemical kinetics and thermodynamics to reactor engineering. Here, the introduction of the basic reactor types in the first chapter provides the basis for an increasingly more detailed approach to embrace transport phenomena and heterogeneous catalysis, from simple single to multiple reactions (selectivity!), from ideal to non-ideal reactors, and from single-phase to multiphase reactors.

There is also more than the usual attention to diffusion and reaction in porous catalysts, a topic for which the authors are globally recognized. In too many reactor engineering courses, this topic is barely touched upon, which is a shame, considering that most processes employ a porous catalyst, and diffusion limitations play a key role, including in the world's (by far) most employed reactor, namely the automotive catalytic converter, and processes from ammonia synthesis to steam reforming and catalytic cracking.

A further unique aspect of this book is the inclusion of a chapter on process intensification, which is a subject of increasing interest worldwide, from processes that integrate separations and reactions to miniaturization for distributed processing, and safer, more compact, and environment-friendly production.

It is a privilege for me to write these few lines to encourage the readers, and to commend the authors for sharing their decades of experience to guide future generations of chemical engineers. As a student at the University of Ghent, under the notable chemical reactor engineering Professor Gilbert Froment in the 1990s, I had the pleasure to meet the Professors Doğu when they visited our university, and I remember reading their defining articles on diffusion and reaction in bimodal porous catalysts, which were crucial to my own starting research at the time. Years later, in 2014, I had the great honor to present the Somer Lectures at METU in Ankara and share the table with them at a special celebration. This will always remain a fond memory.

As I could be inspired by the authors already as a student, I trust the reader will enjoy the journey Professors Timur and Gülşen Doğu take us on to discover the beautiful world of chemical reactor engineering and demystify some of its magic!

Marc-Olivier Coppens
Ramsay Memorial Professor
Department of Chemical Engineering
University College London (UCL), UK

Foreword

This is a comprehensive book that can be used for both undergraduate and graduate "Chemical reaction engineering and kinetics" courses in chemical engineering curricula. While including all the fundamental concepts in a clear and concise manner, the book also encompasses many new and timely topics, such as process intensification, integration of chemical reaction and separation processes into a single unit, and the use of alternative energy sources, such as microwave, solar energy, ultrasound, in chemical processing. It successfully demonstrates the multi-scale nature of chemical reactor engineering by covering the fundamentals at every scale. It establishes the relationships among thermodynamics, transport phenomena, catalysis, and reaction engineering in a seamless manner. In addition to offering a comprehensive approach to meet the needs of instructors and students alike, it will also be an ideal reference for researchers and industrial practitioners of chemical reaction engineering.

Umit S. Ozkan
College of Engineering Distinguished Professor
Chair
William G. Lowrie Department of Chemical and Biomolecular Engineering
The Ohio State University, USA

About the Authors and Acknowledgments

Both Professor Timur Doğu and Professor Gülşen Doğu earned their BS degrees in chemical engineering from the Middle East Technical University (METU) in Ankara, Turkey, their MS degrees from Stanford University, and their PhD degrees in chemical engineering from the University of California Davis. They both started their academic careers in the Chemical Engineering Department of METU.

Professor Timur Doğu was promoted to full professorship in 1985. He also served as the Dean of the Faculty of Sciences of Ankara University. He is a member of the Turkish Academy of Sciences. Professor Gülşen Doğu was appointed as an associate professor at METU, followed by a full professorship at Gazi University in 1985. At this university, Professor Gülşen Doğu held the positions of Dean of Engineering and University Academic Vice President. Both Timur Doğu and Gülşen Doğu contributed to the teaching and research activities at McGill University in 1980–1981 as visiting professors.

Both authors of this book had the opportunity to take the heterogeneous catalysis and the chemical engineering kinetics courses taught by Professor M. Boudart and Professor D.M. Mason during their graduate studies at Stanford University. They also had the chance to participate in the chemical reaction engineering and chemical kinetics courses taught by Professor N.A. Dougharty and Professor J. Swinehart during their PhD work at UC Davis. They had the privilege of studying with the notable chemical reaction engineering Professor J.M. Smith, who was also the PhD thesis advisor of Gülşen Doğu.

Both Timur Doğu and Gülşen Doğu contributed to the advancement of diffusion and reaction in porous catalysts, environmental catalysis, multifunctional reactors, as well as to the design and analysis of chemical reactors. Contributions of Professor Gülşen Doğu and Professor Timur Doğu to chemical reaction engineering are honored by the two special issues of the *International Journal of Chemical Reactor Engineering* in 2019.

The authors are indebted to their students and colleagues for their continuing collaborations in chemical reaction engineering and heterogeneous catalysis. Prof. B. McCoy of University of California Davis, Prof. H. de Lasa of the University of Western Ontario, Prof. N. Orbey of University of Massachusetts Lowell, Prof. K. Mürtezaoğlu, Prof. N. Oktar, Prof. N. Yaşyerli, Prof. S. Yaşyerli, Prof. S. Balcı, Prof. M. Dogan, Prof. D. Varışlı of Gazi University, Prof. G. Karakaş, Prof. N.A. Sezgi of METU, Prof. T. Kopaç of Zonguldak Bülent Ecevit University, and Asst. Prof. Dr. Doruk Doğu of Atılım University are some of these continuing collaborators who influenced the approach of the authors in writing this book.

The authors are indebted to Professor Marc-Olivier Coppens of University College London and Professor Umit S. Ozkan of Ohio State University for the enlightening forewords they wrote for this book. Also, the valuable suggestions of Prof. Rachel Getman (Clemson University) and Prof. Coppens, which made this book more purposeful, are gratefully acknowledged.

Timur Doğu and Gülşen Doğu would also like to thank their friends at METU, Gazi University, and other institutions in Turkey and abroad for their support and friendship. They also thank Dr. Doruk Doğu for his valuable suggestions and their student Merve Sarıyer for her help and contributions in finalizing the manuscript of this book.

List of Symbols

a	Pore radius	m
a_b	Bubble area per unit liquid volume in a slurry	m^2/m^3
a_c	Activity factor	
a_e	External area of the catalyst pellet per unit volume in a slurry	m^{-1}
a_i	The activity of species i	
a_{mem}	Membrane surface area per reactor volume	m^2/m^3
A_c	The cross-sectional area of the reactor	m^2
A_e	External surface area of catalyst pellet	m^2
A_h	Heat transfer area	m^2
b	Half of the cell size in a monolith	m
B	Parameter defined by Eq. (8.39)	
Bi_h	Biot number for heat transfer (Eq. (10.79))	
Bi_m	Biot number for mass transfer (Eq. (10.42))	
$\bar{c}$	Specific heat of the mixture	J/kg K
C	Total concentration	mol/m^3
C_d	Darcy coefficient	
C_i	The concentration of species i in the reactor	mol/m^3
C_{i_a}, C_{i_i}	The concentration of species i in the macro- and micro-porous regions	mol/m^3
C_{ib}	Bubble phase concentration of species i in a fluidized bed	mol/m^3
C_{id}	Emulsion (dense) phase concentration of i in a fluidized bed	mol/m^3
C_{i_e}	The concentration of species i at equilibrium	
C_p	Molar heat capacity	J/mol K
d_b	Bubble diameter	m
d_p	Pellet diameter	m
d_t	Tube diameter	m
Da	Damköhler number (Eq. (10.44))	
D_a, D_i	Effective macro- and micro-pore diffusivity	m^2/s
D_{AB}	Molecular diffusion coefficient of A in B	m^2/s
D_e	Effective diffusion coefficient	m^2/s
D_{K_i}	Knudsen diffusion coefficient of species i (Eq. (12.7))	m^2/s
D_s	Surface diffusion coefficient	m^2/s
D_T	Composite diffusivity (Eq. (12.21))	m^2/s
D_{T_a}, D_{T_i}	Composite diffusivity in the macro- and micro-pore regions	m^2/s

D_z	Axial dispersion coefficient	m^2/s
e	Energy flux	$J/m^2\ s$
E_a	The activation energy of the reaction	J/mol
$E_{a_{obs}}$	Observed activation energy	J/mol
f_i	Fugacity of species i	
F_i	Molar flow rate of species i	mol/s
F_T	Total molar flow rate	mol/s
G	Gibbs free energy	J/mol
$\overline{G}$	The mean value of the molar flow rate of the vapor stream in a distillation column	mol/s
G_N	Molar flow rate of vapor stream leaving stage N in a distillation column	mol/s
h	Heat transfer coefficient	$J/m^2\ s\ K$
H_i	Enthalpy of species i	J
J_D	J factor for mass transfer (Eq. (9.24))	
J_i	Diffusion flux of species i	$mol/m^2\ s$
k	Reaction rate constant for an nth-order reaction	$(mol/m^3)^{(1-n)}/s$
k_c	Mass transfer coefficient on catalyst surface in a slurry	m/s
k_d	Deactivation rate constant	s^{-1}
k_f	Forward reaction rate constant of (nth-order reaction)	$(mol/m^3)^{(1-n)}/s$
k_b	Backward reaction rate constant (nth-order reaction)	$(mol/m^3)^{(1-n)}/s$
k_g	Gas-side mass transfer coefficient in a slurry	m/s
k_l	Liquid side mass transfer coefficient in a slurry	m/s
k_m	Mass transfer coefficient	m/s
k_{obs}	Observed rate constant (first-order reaction)	s^{-1}
k_o	Frequency factor in the Arrhenius equation	
k_w	Reaction rate constant based on catalyst mass	
K_i	Adsorption equilibrium constant of species i	
K_C	The equilibrium constant in terms of concentrations	
K_f	The equilibrium constant in terms of fugacities	
K_H	Henry's constant	
K_l	Overall mass transfer coefficient	m/s
K_P	The equilibrium constant in terms of partial pressures	
K_y	The equilibrium constant in terms of mole fractions	
L	Length of the reactor	m
L_s	Half thickness of a slab	
$\overline{L}$	Mean value of molar liquid flow rate in a distillation column	mol/s
L_N	Liquid flow rate leaving stage N in a distillation column	mol/s
m_n	The nth moment (Eq. (8.57))	
M_i	The molecular weight of species i	kg/mol
n	Unit normal vector	
n_i	Number of moles of i	mol
$n_{i,\ ads}$	Adsorbed concentration of species i per unit mass of the catalyst	mol/kg
P	Pressure	Pa, atm
P_c	Critical pressure	Pa, atm
Pe_d	Peclet number in terms of tube diameter (8.46)	

Pe_z	Axial Peclet number (Eq. (8.36))	
q	Heat transfer rate	J/s
q_G	Heat generation rate	J/s
q_R	Heat removal rate	J/s
Q	Volumetric flow rate	m^3/s
r	Radial direction	
r_c	Core radius for unreacted-core model	m
r_{cy}	The radius of the cylindrical catalyst pellet	m
r_g	Microporous particle/micro-grain radius	m
r_i	The radial direction in the microporous particles	
r_o	The radius of the tubular reactor	m
R_i	The reaction rate for species i based on reactor volume	mol/m^3 s
$R_{i,\,p}$	The reaction rate for species i based on pellet volume	mol/m^3 s
$R_{i,\,w}$	The reaction rate for species i based on catalyst mass	mol/kg s
R	Specific rate of reaction based on reactor volume	mol/m^3 s
$R_{(j),\,p}$	Rate of reaction j based on pellet volume	mol/m^3 s
R_p	Intrinsic rate of reaction based on pellet volume	mol/m^3 s
R_g	Gas constant	J/mol K
s	Laplace variable	
s_{BA}	Point selectivity of B with respect to A (Eq. (6.3))	
S_{BA}	Overall selectivity of B with respect to A (Eq. (6.4))	
Sc	Schmidt number (Eq. (8.48))	
S_g	Surface area per unit mass of a catalyst	m^2/g
S_o	Total number of sites per unit mass of catalyst	
t	Time	s
$t_{1/2}$	The half-life of a reaction	s
T	Temperature	K
T_a	Ambient temperature	K
T_c	Critical temperature	K
T_w	Wall temperature	K
U	Overall heat transfer coefficient	J/m^2 s K
U_o	Superficial velocity	m/s
v_x, v_y, v_z	Velocity components in the x, y, and z directions, respectively	m/s
v_{max}	The maximum velocity in a tubular reactor	m/s
V	Reactor volume	m^3
V_{pore}	Pore volume per unit mass	m^3/g
w	Catalyst mass in the reactor	kg
x_i	Fractional conversion of reactant i	
$x_{i,\,N}$	Liquid phase mole fraction of species i leaving the Nth stage	
$y_{i,\,N}$	Vapor phase mole fraction of species i leaving the Nth stage	
Y_i	The yield of species i (Eq. (6.6))	
z	Axial direction	
Z_{AB}	Collision rate between unlike gas molecules per unit volume	m^{-3}/s

Greek Letters

α	Parameter defined by Eq. (10.93)	
$\overline{\alpha}$	Parameter defined by Eq. (12.18)	
β	Prater group (Eq. (10.62))	
γ	Dimensionless Arrhenius group (Eq. (10.69))	
γ_i	Fugacity coefficient of species i	
Γ	Parameter defined by Eq. (8.82)	
δ	The thickness of the catalyst layer on the surfaces of a monolith	m
δ_o	Parameter defined by Eq. (12.44)	
δ_1	Parameter defined by Eq. (12.45)	
ΔH	Heat of reaction	J/mol
$\Delta H^o_{T_R}$	Standard heat of reaction at the reference temperature	J/mol
$\Delta H^o_{i,f}$	Standard heat of formation of species i	J/mol
ΔG	Gibbs free energy of the reaction	J/mol
$\Delta G^o_{i,f}$	Gibbs free energy of formation of species i at standard state	J/mol
ΔT_{LM}	Log mean temperature difference	
ε, ε_a	Porosity, macro-porosity of a pellet	
ε_b	Fixed bed void fraction	
ε_{mf}	The void fraction of a bed at minimum fluidization	
ε	Volume expansion factor defined in Eq. 3.63	
ζ	Dimensionless radial direction in a pellet (Eq. (10.4))	
ζ_i	Dimensionless radial direction in a particle (Eq. (10.90d))	
ς	Dimensionless length (Eq. (8.34))	
η	Effectiveness factor (Eq. (9.9))	
θ	The fraction of surface sites covered by the adsorbing molecules	
Θ	Dimensionless temperature (Eq. (10.68))	
λ	Mean free path (Eq. (12.6))	m
λ_b	Thermal conductivity in the reactor	J/s m K
μ	Viscosity	kg/m s
μ_1	First absolute moment (Eq. (8.58))	s
μ_2'	Second central moment (Eq. (8.59))	s^2
μ_i	Chemical potential of species i	
ν_i	Stoichiometric coefficient of species i	
ξ	Reaction extent	
ρ	Density	kg/m^3
ρ_s	Solid density	kg/m^3
τ	Space–time in the reactor ($\tau = V/Q_o$)	s
τ	Tortuosity factor of a porous material	
τ_s	Tortuosity factor for surface diffusion	
φ_n	Thiele modulus for an nth-order reaction (Eq. (10.7))	
$\widehat{\varphi}_n$	Shape and order generalized Thiele modulus (Eq. (10.31))	
φ_i	Particle Thiele modulus (Eq. (10.92))	
Φ	Observable modulus (Eq. (10.80))	
Φ_s	Observable modulus defined for a sphere (Eq. (10.108))	
Ψ	Dimensionless concentration (Eq. (8.34))	

Subscripts

e	Index denoting equilibrium
e	Index denoting effective value in transport equations
f	Index denoting exit condition of a reactor
g	Index denoting gas phase
i	Index denoting species "i"
l	Index denoting liquid phase
o	Index denoting the inlet condition of a reactor
p	Index denoting catalyst pellet
r	In the radial direction
s	At the surface condition of the catalyst pellet
z	In the axial direction

Superscripts

o	Index denoting the initial condition of a reactor
o	Index denoting standard state for thermodynamic properties
$\hat{H}$	(Over-hat) molar value of property H

About the Companion Website

This book is accompanied by a companion website

www.wiley.com/go/dogu/chemreacengin

The website includes:

1. Appendices
    - Online Appendix 1: Vector Notation
    - Online Appendix 2: Microscopic Species Conservation and Energy Equations in Different Coordinate Systems
    - Online Appendix 3: Pseudo-Homogeneous Transport Equations
    - Online Appendix 4: Some Differential Equations Relevant to Reaction Systems
    - Online Appendix 5: Some Numerical Techniques for Chemical Reactor Applications
    - Online Appendix 6: Case Study Examples
    - Online Appendix 7: Dimensionless Groups and Parameters
2. Solutions to Problems

1

Rate Concept and Species Conservation Equations in Reactors

Chemical processes developed for the production of valuable products involve both physical and chemical rate processes. Chemical reactors are usually at the heart of the process, where chemical conversion of reactants to the products occurs. Design and analysis of a chemical reactor involve information from chemical kinetics, as well as from the thermodynamics of the processes taking place in the reactor. Besides the kinetics and thermodynamics of the chemical conversions, hydrodynamics of the system, as well as the heat and mass transfer processes taking place within the reaction vessel, should be understood for the design and analysis of such units. Kinetics of a chemical conversion involves understanding the parameters affecting the rate of a reaction and the reaction mechanism. Thermodynamics gives us information, whether a chemical conversion can occur spontaneously or not and about the maximum possible extent of reaction determined by chemical equilibrium. However, the reaction rate gives us the answer to how fast the chemical conversion is going to take place at the given operating conditions of the reactor, such as temperature, the composition of the reaction medium, and pressure.

1.1 Reaction Rates of Species in Chemical Conversions

Analysis of the rates of chemical conversions using the collision theory of gases or transition-state theory shows that the reaction rate is a function of the temperature of the reaction medium and concentrations of the species involved in the chemical conversion. Experimental evidence also show that the reaction rate of any molecule in a reaction vessel can be expressed as a function of concentrations of species involved in the reaction and the reaction temperature. Thus, the reaction rate for any component i can be expressed as

$$R_i = f(\text{temperature, concentrations}).$$

The reaction rate is generally defined as the moles of species i reacted (or produced) per unit time per unit volume of the mixture in the reaction vessel. However, in some solid-catalyzed reactions, it may also be expressed based on the catalyst volume, catalyst weight, or catalyst surface area, instead of the volume of the reaction medium. For an arbitrary reaction between reactants A and B

$$a\text{A} + b\text{B} \rightleftharpoons c\text{C} + d\text{D}$$

Fundamentals of Chemical Reactor Engineering: A Multi-Scale Approach, First Edition. Timur Doğu and Gülşen Doğu.

Companion website: www.wiley.com/go/dogu/chemreacengin

the relation between the reaction rates of reactants and products can be written as follows:

$$\frac{R_A}{-a} = \frac{R_B}{-b} = \frac{R_C}{c} = \frac{R_D}{d} = R$$

Here, R is the specific rate of the reaction under consideration. More generally, the relation between the rates of different species can be expressed as

$$\frac{R_i}{\nu_i} = \frac{R_j}{\nu_j} = \cdots = \frac{R_k}{\nu_k} = R \tag{1.1}$$

where ν_i is the stoichiometric coefficient of species i in the reaction, such that

$\nu_i < 0$ for reactants

$\nu_i > 0$ for products

For instance, for the oxidation of sulfur dioxide to sulfur trioxide

$$SO_2 + \frac{1}{2}O_2 \rightleftharpoons SO_3$$

the relation between the rates of different species should be written as follows:

$$\frac{R_{SO_2}}{-1} = \frac{R_{O_2}}{-(1/2)} = \frac{R_{SO_3}}{1} \; ; \; -R_{SO_2} = -2R_{O_2} = R_{SO_3}$$

The relations between the reaction rates of different species can be obtained by writing the conservation equations for the atomic species involved in the process. For instance, for the oxidation of sulfur dioxide, the sulfur atom should be either within SO_2 or SO_3 molecules. Similarly, the oxygen atom is either in O_2 or SO_2 or SO_3 molecules. Hence, the relations for the conservation of the sulfur and oxygen atoms can be expressed as follows:

$$\text{S balance}: \; R_{SO_2} + R_{SO_3} = 0$$

$$\text{O balance}: \; 2R_{SO_2} + 2R_{O_2} + 3R_{SO_3} = 0$$

Here, we have three unknowns $(R_{SO_2}, R_{SO_3}, R_{O_2})$ and two equations for these unknowns, indicating that one of the rates can be arbitrarily selected as the independent rate. The other two unknowns (rates) can be expressed in terms of the chosen independent reaction rate. The other two rates can be expressed from the two equations written above by selecting R_{SO_3} as the independent rate and following an elimination procedure:

$$-R_{SO_2} = R_{SO_3}$$

$$-R_{O_2} = \frac{3}{2}R_{SO_3} + R_{SO_2} = \frac{3}{2}R_{SO_3} - R_{SO_3} = \frac{1}{2}R_{SO_3}$$

This analysis shows that there is only one independent rate term, indicating the presence of only one independent reaction.

In many of the industrial processes, more than one reaction may take place within the reactors. Some of these reactions may be desired reactions, while some others may be undesired parasitic reactions. For a multiple-reaction system, the relations between the rates of different species involved in the reaction cannot be directly expressed as in Eq. (1.1), using their stoichiometric coefficients. Instead, the relationships between the rates of different species can be obtained by writing atomic conservation expressions.

Let us consider the reaction of ethylene with oxygen to produce ethylene oxide over a silver-based solid catalyst. In the reactor exit stream, carbon monoxide, carbon dioxide, and water molecules are also observed, in addition to ethylene, ethylene oxide, and oxygen. Hence, there are six species, involved either as reactants or products in this reaction system. To obtain the relations between the rates of these six species, namely $R_{C_2H_4}$, $R_{C_2H_4O}$, R_{O_2}, R_{CO}, R_{CO_2}, and R_{H_2O}, we can write conservation relations for carbon, hydrogen, and oxygen atoms.

$$\text{C balance}: 2R_{C_2H_4} + 2R_{C_2H_4O} + R_{CO} + R_{CO_2} = 0$$

$$\text{H balance}: \quad 4R_{C_2H_4} + 4R_{C_2H_4O} + 2R_{H_2O} = 0$$

$$\text{O balance}: \quad R_{C_2H_4O} + 2R_{O_2} + R_{CO} + 2R_{CO_2} + R_{H_2O} = 0$$

As illustrated above, we have three relations for the rates of six different species in this system. This analysis showed three independent rates for this system, indicating the presence of three independent reactions. In general, if there are n number of species involved in a reaction system (n number of rate values) and if we can write m number of atomic conservation relations, we should expect $(n - m) = p$ number of independent rate values. Hence, we expect to have p number of independent reactions.

Returning to the ethylene oxidation example, we can arbitrarily select the reaction rate values of the three species as the independent rates. We can express the other rates in terms of these independent rates, following an elimination procedure (refer to Online Appendix O.5.2). If we arbitrarily select $R_{C_2H_4O}$, $R_{C_2H_4}$, and R_{CO_2} as the independent rates, the other three rates can be expressed as follows:

$$R_{CO} = -(2R_{C_2H_4} + 2R_{C_2H_4O} + R_{CO_2})$$

$$R_{H_2O} = -(2R_{C_2H_4} + 2R_{C_2H_4O})$$

$$R_{O_2} = 2R_{C_2H_4} + \frac{3}{2}R_{C_2H_4O} - \frac{1}{2}R_{CO_2}$$

The presence of three independent rates indicates the occurrence of three independent reactions in this system. Three possible independent reactions for this system are

$$C_2H_4 + \frac{1}{2}O_2 \rightarrow C_2H_4O \qquad \text{(R.1.1)}$$

$$C_2H_4 + 2O_2 \rightarrow 2CO + 2H_2O \qquad \text{(R.1.2)}$$

$$C_2H_4 + 3O_2 \rightarrow 2CO_2 + 2H_2O \qquad \text{(R.1.3)}$$

Of course, we could also write other reaction stoichiometries for this system. For instance, further oxidation of the product ethylene oxide is another possible reaction.

$$C_2H_4O + \frac{5}{2}O_2 \rightarrow 2CO_2 + 2H_2O \qquad \text{(R.1.4)}$$

However, this reaction is dependent on the three reactions (R.1.1)–(R.1.3) mentioned above.

1.2 Rate of a Chemical Change

The rate of any chemical conversion can be considered as the speed of that process to approach the equilibrium condition. Chemical conversion processes may take place either in a homogeneous system, where the whole process takes place within a single phase, or in a heterogeneous environment, where more than one phase is involved during the reaction. We may have liquid- or gas-phase

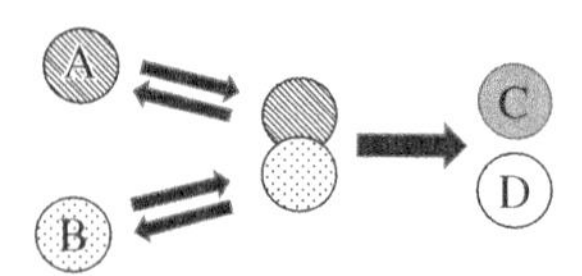

Figure 1.1 Schematic representation of an elementary reaction.

homogeneous reactions in batch or flow reaction vessels, which are called chemical reactors. In the case of heterogeneous systems, the reaction may take place at the interphase between the phases. For instance, in the case of reactions catalyzed by solid catalysts, chemical conversions may proceed on the surface of the catalyst involving the adsorbed reacting species.

One of the most commonly accepted theories for the reaction kinetics of gas-phase reactions is collision theory. During the interaction of two reactant molecules A and B to give new chemical compounds, they must first approach each other very closely. As a result of this collision process, rearrangement of chemical bonds may occur, yielding new molecules (products) if the colliding molecules have sufficient kinetic energy (Figure 1.1). However, not all of the collisions would yield product molecules. Only a fraction of collisions having the sum of the energies of the colliding molecules being higher than a critical value would give product molecules. This critical energy level is generally called the activation energy of the chemical conversion. This is illustrated in Figure 1.2.

According to the gas kinetic theory, the number of collisions between two unlike molecules per unit time and unit volume of a gas can be expressed as follows [1]:

$$Z_{AB} = \pi \mathrm{N}_A^* \mathrm{N}_B^* d_{AB}^2 \left(\frac{8R_g T}{\pi \mu_m} \right)^{1/2} \tag{1.2}$$

Here, d_{AB} is the mean molecular diameter, which is defined as $d_{AB} = \frac{(d_A + d_B)}{2}$, μ_m is the reduced mass of colliding molecules, which is defined as $\mu_m = \frac{M_A M_B}{(M_A + M_B)}$, R_g is the gas constant. N_A^* and N_B^* are the number densities of reactants A and B (number of molecules per unit volume) in the vessel. The rate of a chemical reaction as a result of this elementary process of collision of two unlike molecules is then expressed as being proportional to the product of Z_{AB} with the probability of collisions having energies equal to or higher than the critical energy level of E_a.

$$R \propto pe^{-\frac{E_a}{R_g T}} Z_{AB} = \pi d_{AB}^2 \left(\frac{8R_g T}{\pi \mu_m} \right)^{1/2} pe^{-\frac{E_a}{R_g T}} \mathrm{N}_A^* \mathrm{N}_B^* \tag{1.3}$$

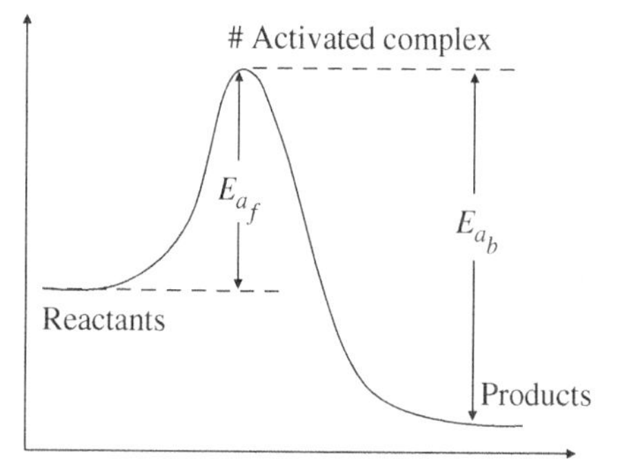

Figure 1.2 Schematic representation of the activation energies of the forward and reverse reaction rate constants (exothermic reaction).

Here, p is a steric factor, depending upon the orientation of collision. Hence, the reaction rate of this elementary reaction step can be expressed as follows:

$$R = kC_A C_B \tag{1.4}$$

Here, k is the reaction rate constant, which is generally defined as follows:

$$k = k_o e^{-\frac{E_a}{R_g T}} \ (\text{mol/m}^3)^{-1} \text{s}^{-1} \tag{1.5}$$

Here, k_o is called the frequency factor of the rate constant. This relation is called the Arrhenius equation, and it shows that the dependence of reaction rate constant on reaction temperature is exponential. Experimental data on reaction rates also supported the Arrhenius dependence of the reaction rate constant of an elementary reaction step.

The activation energy of chemical conversion E_a and the pre-exponential factor (frequency factor) of the reaction rate constant k_o are generally determined experimentally from the plot of ln(k) vs. $1/T$, as illustrated in Figure 1.3.

According to the collision theory of gas-phase reactions, the pre-exponential factor k_o is proportional to the square root of temperature. Hence, the temperature dependence of the reaction rate constant can also be expressed as follows:

$$k = k_o^* T^{1/2} e^{-\frac{E_a}{R_g T}} \tag{1.6}$$

However, the temperature dependence of the pre-exponential factor is quite weak. Hence, for many cases, it may be considered as being constant.

Another theory proposed for the mechanism of a chemical conversion is called transition-state theory (or activated complex theory). According to this theory, the reactant molecules first form an unstable activated complex, and then this activated complex decomposes to form the products.

$$\text{A} + \text{B} \rightleftharpoons (\text{AB})^{\ddagger} \rightarrow \text{Products}$$

The activated complex is assumed to be in equilibrium with the reactant molecules. Activated complex decomposes to give the products. The conversion rate of the reactants to the products is proportional to the concentration of the activated complex. The reaction rate is determined by the product of the concentration of activated complex and the frequency with which it goes over to the products. The rate expression can then be obtained by the use of statistical-thermodynamic principles. Details of both collision theory and transition-state theory can be found in physical chemistry and reaction kinetics books [1–3]. Both of these theories predict the concentration and temperature dependences of the reaction rates of elementary reactions.

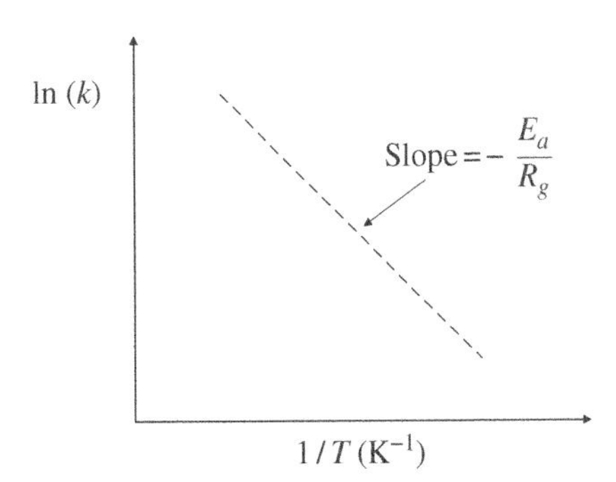

Figure 1.3 Schematic diagram of the temperature dependence of the reaction rate constant.

Reaction order is defined as the summation of exponents of the concentration terms in the reaction rate expression. Any reaction taking place through a single physicochemical process is called an elementary reaction. For the elementary reaction described above, the reaction is first order with respect to the concentration of reactant A and first order with respect to the concentration of reactant B. The overall order of this reaction is second order. In the case of an elementary reaction, reaction orders with respect to the concentrations involved in the reaction are the same as the molecularity of the species in the reaction step (stoichiometric constants in the elementary reaction step). However, many of the reactions do not proceed through a single physicochemical process, but involve a series of elementary steps to give the reaction products. Hence, for a reaction involving a number of elementary steps, the reaction orders for the concentrations of the species involved in this chemical conversion are not necessarily the same as the stoichiometric constants of the species of that overall reaction. For the general case, the rate of an irreversible chemical reaction may be written in a power-law form as follows:

$$R = kC_A^n C_B^m \cdots C_i^p \text{ (irreversible reaction)} \tag{1.7}$$

Here, the reaction orders n, m, p,... are usually determined experimentally.

In many cases, chemical conversions are reversible. In such cases, the rate expression should be expressed as the difference in the rates of the forward and the reverse reactions.

$$R = k_f C_A^n C_B^m \cdots C_i^p - k_b C_A^{n'} C_B^{m'} \cdots C_i^{p'} \tag{1.8}$$

Here, k_f and k_b are the forward and the backward rate constants, respectively. They both depend upon temperature according to the Arrhenius law. For an nth-order reaction, the unit of the reaction rate constant is expressed as $(\text{mol/l})^{(1-n)}\ \text{s}^{-1}$. In the case of a first-order reaction, the unit of the rate constant is s^{-1}.

1.3 Chemical Reactors and Conservation of Species

Chemical reactors are the vessels in which chemical conversions take place. Most reactors may be classified as batch reactors or flow reactors. There are also some reactors operating in semi-batch mode. The design of a chemical reactor involves species conservation and energy balance equations written around the reaction system. Hydrodynamics of the fluid within a reactor should also be considered during the design of a chemical reactor. Let us consider a reaction vessel in which an arbitrary reaction A + B → C + D is taking place (Figure 1.4).

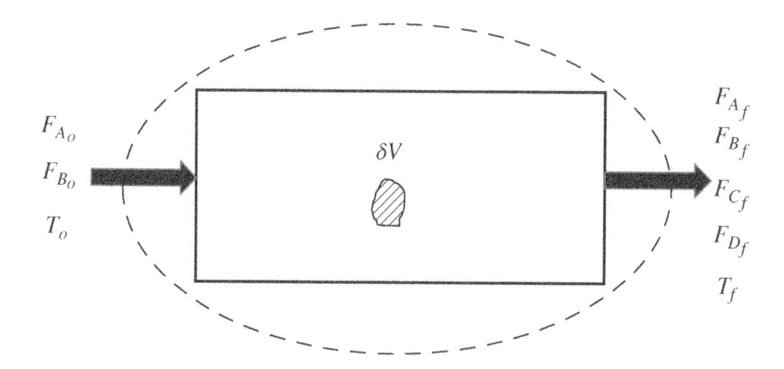

Figure 1.4 Schematic diagram of an arbitrary reaction vessel with input and output terms.

In general, reaction temperature and the concentrations of species are not necessarily uniform within the reaction vessel and may depend upon position and time. Since the reaction rate is expected to be a function of both temperature and concentrations, it is also expected to change with position and time in the reactor.

$$C_i(x, y, z, t);\ T(x, y, z, t) \ \rightarrow \ R_i(x, y, z, t)$$

If we consider an arbitrary differential volume element within a reactor, the concentration of the reacting species may be taken as uniform within this differential element.

Species conservation equation for any system can be written by selecting a control volume and applying general conservation law.

$$(\text{Input Rate}) - (\text{Output Rate}) + (\text{Generation Rate}) = (\text{Accumulation Rate}) \tag{1.9}$$

The macroscopic species conservation equation for reactant A around the reaction vessel, which is shown in Figure 1.4, can then be expressed as follows:

$$F_{A_o} - F_{A_f} + \int_V R_A dV = \frac{dn_A}{dt} \tag{1.10}$$

Here, F_{A_o} and F_{A_f} are the molar flow rates of A (mol/s) at the inlet and exit of the reactor, respectively. The generation term of Eq. (1.10) corresponds to the total moles of species A converted to the products per unit time within the reactor. Since the reaction rate may be a function of position within the reactor, the total moles of reactant A converted to the products should be evaluated by the volume integration of the reaction rate over the reactor volume. This requires information about the variation of concentrations of species and the temperature with respect to position within the reactor. Note that the generation term becomes negative for the reactants and positive for the products. The accumulation term of Eq. (1.10) corresponds to the rate of change of the total number of moles of species A in the reactor (n_A)with respect to time.

This general species conservation equation can be used to derive conservation equations for batch and flow reactors. In a batch reactor, the input rate and the output rate terms become zero, and the species conservation equation reduces to

$$\int_V R_A dV = \frac{dn_A}{dt} \tag{1.11}$$

In many cases, the mixture in a batch reactor is perfectly mixed to approach uniform temperature and concentrations of the reacting species (Figure 1.5). Although the achievement of uniform temperature and concentrations within a reactor may be difficult in most cases, the perfect mixing assumption is usually made.

With the perfect mixing assumption, Eq. (1.11) reduces to

$$R_A V = \frac{dn_A}{dt} \tag{1.12}$$

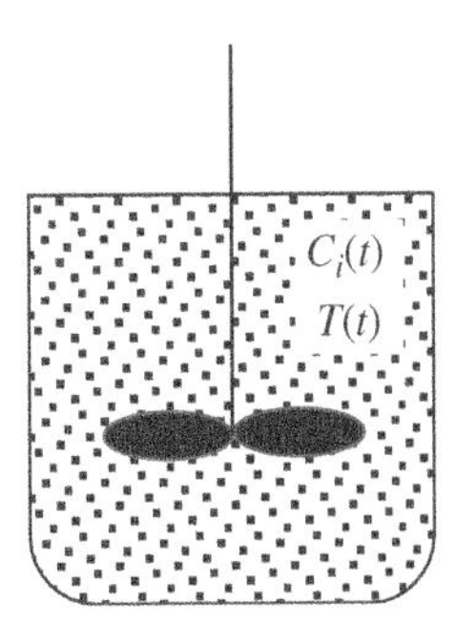

Figure 1.5 Schematic diagram of a well-mixed batch reactor.

Liquid- or gas-phase homogeneous reactions may take place in a batch reactor. In the case of liquid-phase reactions, the volume of the reaction mixture can, in general, be taken as constant. However, for gas-phase reactions taking place at constant pressure, changes in the reaction volume may occur due to changes in temperature or the total number of moles of the reaction mixture as a result of the conversion of reactants to the products. On the other hand, total pressure changes as a result of changes in the total number of moles or temperature in a constant volume batch reactor. If the volume of the reaction medium is constant, we can express Eq. (1.12) as follows:

$$R_A = \frac{dC_A}{dt} \tag{1.13}$$

In many instances, the reaction rate of species A is defined as expressed in Eq. (1.13). However, this relation can be used only with the following assumptions:

- Batch reactor
- Perfectly mixed reactor (uniform temperature and concentrations)
- Constant volume of the reaction mixture

1.4 Flow Reactors and the Reaction Rate Relations

In a flow reactor, the concentrations of reactants and the temperature may vary with respect to position and also with respect to time. If there is no change of temperature and concentrations with respect to time at a given location within the reactor, we consider a steady-state operation. At steady state, the accumulation term of the general material balance equation, which is given by Eq. (1.10), becomes zero, and this equation reduces to

$$F_{A_o} - F_{A_f} + \int_V R_A dV = 0 \tag{1.14}$$

The design of a flow reactor operating at steady state requires the integration of the rate term in Eq. (1.14). Hydrodynamics of the reaction medium in a flow reactor is the main factor in determining the variation of concentrations and temperature in the reactor. There are different types of flow reactors, which are discussed in Chapters 4–7, in this book. One of these flow reactor types is a perfectly mixed continuous stirred-tank reactor (Figure 1.6), which is usually denoted as ideal CSTR. As a result of the well-mixing assumption, all the properties are assumed as uniform within a CSTR. Therefore, the reaction rate term can be assumed to be uniform at any point in an ideal CSTR and then Eq. (1.14) reduces to

$$F_{A_o} - F_{A_f} + R_{A_f} V = 0 \tag{1.15}$$

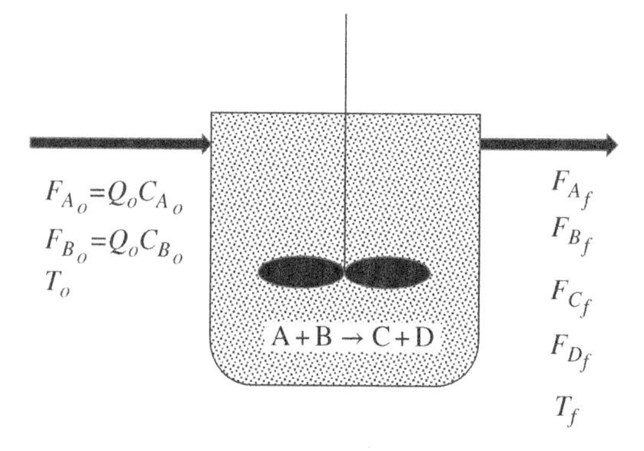

Figure 1.6 Schematic diagram of a CSTR.

Note that, in a perfectly mixed stirred-tank reactor (ideal CSTR), concentrations and temperature within the reactor are assumed to be uniform and can be taken as the same values at the reactor exit stream. Hence, for a CSTR, the following relation can be written for the reaction rate:

$$R_{A_f} = \frac{(F_{A_f} - F_{A_o})}{V} \tag{1.16}$$

At this point, we should note that for an ideal stirred tank reactor operating at steady state, one cannot use the rate relation given in Eq. (1.12), which is written for a batch reactor.

The molar flow rate of species A can be expressed as a product of volumetric flow rate (Q: m^3/s) and its concentration. By substituting $F_A = QC_A$ into Eq. (1.16), the following relation can be obtained for the reaction rate:

$$R_{A_f} = \frac{(Q_f C_{A_f} - Q_o C_{A_o})}{V} \tag{1.17}$$

Here, Q_f and Q_o are the volumetric flow rates of the reaction mixture at the exit and the inlet of the reactor, respectively. For the case of a constant flow rate ($Q_f = Q_o$), Eq. (1.17) reduces to

$$R_{A_f} = \frac{Q_o(C_{A_f} - C_{A_o})}{V} = \frac{(C_{A_f} - C_{A_o})}{\tau} \tag{1.18}$$

Here, τ is the space–time, defined as the ratio of reactor volume to the inlet volumetric flow rate ($\tau = V/Q_o$). The assumptions involved in the rate relation given in Eq. (1.18) are as follows:

- Flow reactor
- Steady-state operation
- Perfectly mixed reactor (uniform temperature and concentrations in the reactor and reactor exit values are the same as the values within the reactor)
- Constant volumetric flow rate

1.5 Comparison of Perfectly Mixed Flow and Batch Reactors

Analyses of batch and ideal stirred-tank flow reactors described in Sections 1.3 and 1.4 are based on the perfect mixing assumption. Achievement of perfect mixing and creating uniform temperature and concentrations within the reactor vessel is not easy, and this assumption may fail in the industrial-scale stirred-tank reactors. CSTR and the perfectly mixed batch reactors are ideal reactor models. Deviations from the ideal reactor performance are treated in Chapter 8.

Fractional conversion definition of reactant molecules is commonly used in the analysis of the chemical conversions and also in the design of reactors. Fractional conversion of a reactant A is defined as the ratio of its converted number of moles to the inlet/initial moles. For batch and flow reactors, fractional conversion of limiting reactant is defined as follows:

$$x_A = \frac{n_A^o - n_A}{n_A^o} \text{ (batch reactor)}; \; x_A = \frac{F_{A_o} - F_A}{F_{A_o}} \text{ (flow reactor)} \tag{1.19}$$

Here, n_A^o and F_{A_o} are the initial number of moles of A in the batch reactor and the inlet molar flow rate of A in the flow reactor, respectively. Incorporation of these fractional conversion definitions into Eqs. (1.12) and (1.16) yields the following relations for a perfectly mixed batch reactor and an ideal CSTR, respectively.

$$R_A V = -n_A^o \frac{dx_A}{dt};\ \frac{t}{C_A^o} = \int_0^{x_A} \frac{dx_A}{-R_A} \text{ (batch reactor)} \tag{1.20}$$

$$R_{A_f} = \frac{-F_{A_o} x_{A_f}}{V} = -C_{A_o} \frac{x_{A_f}}{(V/Q_o)};\ \frac{V}{F_{A_o}} = \frac{\tau}{C_{A_o}} = \frac{x_{A_f}}{-R_{A_f}} \text{ (CSTR)} \tag{1.21}$$

For a simple first-order reaction, the rate expression used in a CSTR (for the constant volumetric flow rate case) and in a batch reactor (with a constant reaction volume assumption) can be expressed in terms of fractional conversion of reactant A as follows:

$$-R_A = kC_A = k\frac{F_A}{Q} = kF_{A_o}\frac{(1-x_A)}{Q} = kC_{A_o}(1-x_A) \text{ (CSTR)} \tag{1.22}$$

$$-R_A = kC_A = k\frac{n_A}{V} = kn_A^o\frac{(1-x_A)}{V} = kC_A^o(1-x_A) \text{ (batch reactor)} \tag{1.23}$$

Here, C_{A_o} and C_A^o are the inlet and initial concentrations of reactant A in the flow and batch reactors, respectively. The species conservation equations for the batch and the perfectly mixed flow reactors (CSTR) can then be written for a first-order reaction, as follows:

$$\frac{t}{C_A^o} = \int_0^{x_A} \frac{dx_A}{kC_A^o(1-x_A)};\ kt = -\ln(1-x_A) \text{ (constant volume batch)} \tag{1.24}$$

$$\frac{\tau}{C_{A_o}} = \frac{x_{A_f}}{kC_{A_o}(1-x_{A_f})};\ k\tau = \frac{x_{A_f}}{(1-x_{A_f})} \text{ (CSTR with constant volumetric flow rate)} \tag{1.25}$$

Note that the integration shown in Eq. (1.24) is performed assuming isothermal operation in the batch reactor. The rate constant k in Eq. (1.25) should be evaluated at the reactor exit temperature of the CSTR.

1.6 Ideal Tubular Flow Reactor

Tubular flow reactors are frequently used in chemical processes (Figure 1.7). For a tubular flow reactor operating at a steady state, the concentration of reactants/products and temperature are expected to change both in the axial and radial directions. Hydrodynamics of the system is quite crucial for the development of such variations in a reactor. However, in many cases, radial variations of concentrations and temperature are neglected. Further analysis of ideal tubular reactors is

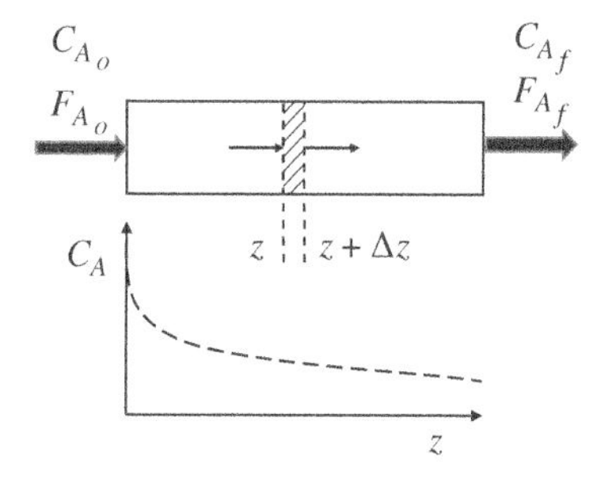

Figure 1.7 Schematic diagram of a tubular reactor.

discussed in Chapter 4. The effects of axial mixing and radial variations of temperature and concentration on the reactor performance are discussed in Chapter 8.

In the one-dimensional design approach, only axial variations of concentrations and the temperature are considered. Concentrations of the reactants decrease along the reactor in such a tubular reaction vessel. Species conservation equation for reactant A within the differential volume element between positions z and $z + \Delta z$ (shown in Figure 1.7) can be written as follows:

$$F_A\big|_z - F_A\big|_{z+\Delta z} + R_A \Delta V = 0 \tag{1.26}$$

Dividing each term by ΔV and taking the limit as ΔV approaching zero, the following differential equation can be obtained for the variation of the molar flow rate of A along the reactor.

$$\lim_{\Delta V \to 0}\left(\frac{F_A\big|_z - F_A\big|_{z+\Delta z}}{\Delta V}\right) + R_A = -\frac{dF_A}{dV} + R_A = 0 \tag{1.27}$$

Equation (1.27) is the one-dimensional design equation for a tubular flow reactor. The molar flow rate of A can be expressed in terms of the concentration of reactant A as follows:

$$F_A = QC_A = U_o A_c C_A \tag{1.28}$$

Here, A_c is the cross-sectional area of the reactor.

In writing the expression given in Eq. (1.28), any mixing or diffusion flux in the axial direction is neglected, and it is assumed that all of the species in the reactor move with the same velocity U_o. This assumption is equivalent to the statement that there is no distribution of residence times of species in the reactor. This is generally called plug-flow assumption. Hence, a tubular reactor with no mixing and diffusion in the axial direction and with no radial variations of concentrations and temperature is called a plug-flow reactor. Substitution of Eq. (1.28) into Eq. (1.27) gives the following differential equation:

$$-\frac{d(QC_A)}{dV} + R_A = 0 \tag{1.29}$$

For the constant volumetric flow rate case, Eq. (1.29) can be expressed as

$$-Q\frac{dC_A}{dV} + R_A = 0 \Rightarrow -U_o\frac{dC_A}{dz} + R_A = 0 \tag{1.30}$$

Equation (1.30) is the plug-flow design equation (with constant volumetric flow rate case), and it involves the following assumptions:

- Steady-flow tubular reactor
- No variation of concentrations and temperature in the radial direction
- No mixing and diffusional transport terms in the axial direction, hence no distribution of residence times of species in the reactor
- Constant volumetric flow rate

The reaction rate term in Eq. (1.30) depends upon concentrations of species, as well as the reaction temperature at the specific location within the reactor. In the case of an isothermal operation, integration of Eq. (1.30) can be performed by taking the reaction rate constant k as constant. If the operation of the reactor is not isothermal, the simultaneous solution of the species conservation equation with the energy balance equation is needed. Heat effects in tubular reactors and treatment of non-isothermal reactors are discussed in Chapter 7.

Instead of expressing the molar flow rate of reactant A as a product of the volumetric flow rate and its concentration, it may also be expressed in terms of fractional conversion of reactant A.

$$F_A = F_{A_0}(1 - x_A)$$

In this case, the differential species conservation equation (1.27) becomes

$$F_{A_0}\frac{dx_A}{dV} + R_A = 0 \tag{1.31}$$

This relation does not involve the assumption of the constant volumetric flow rate. Rearrangement of Eq. (1.31) gives

$$\frac{dV}{F_{A_0}} = \frac{dx_A}{-R_A} \tag{1.32}$$

Integration of both sides of Eq. (1.32) from inlet to the outlet of the tubular plug-flow reactor yields

$$\frac{V}{F_{A_0}} = \int_0^{x_{A_f}} \frac{dx_A}{-R_A} \tag{1.33}$$

The volume of the plug-flow reactor can then be found by performing the integration in Eq. (1.33). Further details of plug-flow reactors are given in Chapter 4.

The concentration of reactant A can be expressed in terms of fractional conversion as follows:

$$C_A = \frac{F_A}{Q} = \frac{F_{A_0}(1 - x_A)}{Q} \tag{1.34}$$

If the volumetric flow rate Q is constant along the reactor, the concentration of reactant A can be expressed in terms of fractional conversion as

$$C_A = C_{A_0}(1 - x_A) \tag{1.35}$$

By substituting Eq. (1.35) into Eq. (1.33), integration can be performed to obtain the following relation between the reactor volume and the fractional conversion of reactant A:

$$\frac{V}{F_{A_0}} = \frac{\tau}{C_{A_0}} = -\frac{1}{kC_{A_0}} \ln\left(1 - x_{A_f}\right) \tag{1.36}$$

Note that this relation is valid for a first-order reaction taking place in an ideal plug-flow reactor operating isothermally, with a constant volumetric flow rate along the reactor. Also, note that this equation is quite similar to Eq. (1.24), which is written for a batch reactor. The only difference is reaction time t in the batch reactor expression is replaced by space–time τ in the plug-flow reactor expression given in Eq. (1.36).

In the case of a non-isothermal reactor, we should know the temperature dependence of the reaction rate constant and the temperature distribution within the reactor. For the non-isothermal case, the simultaneous solution of the species conservation and the energy balance equations is needed. The design of non-isothermal reactors is discussed in Chapter 7 in detail.

1.7 Stoichiometric Relations Between Reacting Species

As discussed in Section 1.1, for a single-reaction system, the stoichiometric relations between the rates of reactants and the products can be written as follows:

$$\frac{R_i}{\nu_i} = \frac{R_j}{\nu_j} = \cdots = \frac{R_k}{\nu_k} = R$$

Hence, for an arbitrary reaction between reactants A and B producing C and D, the relation between the rates of different species is

$$\frac{R_A}{\nu_A} = \frac{R_B}{\nu_B} = \frac{R_C}{\nu_C} = \frac{R_D}{\nu_D} \tag{1.37}$$

Here, ν_i is the stoichiometric coefficient of species i. The sign of ν_i is negative for the reactants and positive for the products.

1.7.1 Batch Reactor Analysis

For a batch reactor, species conservation equations for different species involved in the reaction are written as given in Eq. (1.12).

$$R_A V = \frac{dn_A}{dt};\ R_B V = \frac{dn_B}{dt}; R_C V = \frac{dn_C}{dt}; R_D V = \frac{dn_D}{dt} \tag{1.38}$$

Combining these species conservation equations with the relations given in Eq. (1.37) for the rates of different species gives the following relationships:

$$\frac{1}{\nu_A}\frac{dn_A}{dt} = \frac{1}{\nu_B}\frac{dn_B}{dt} = \frac{1}{\nu_C}\frac{dn_C}{dt} = \frac{1}{\nu_D}\frac{dn_D}{dt} \tag{1.39}$$

Multiplying each term by dt and performing the integrations from initial time to an arbitrary time t gives the following relations:

$$\frac{\left(n_A - n_A^o\right)}{\nu_A} = \frac{\left(n_B - n_B^o\right)}{\nu_B} = \frac{\left(n_C - n_C^o\right)}{\nu_C} = \frac{\left(n_D - n_D^o\right)}{\nu_D} \tag{1.40}$$

Here, n_i^o and n_i are the number of moles of species i at a reaction time zero and reaction time t, respectively. In the case of the constant volume of the reaction medium, Eq. (1.40) can be expressed in terms of initial and final concentrations in the batch reactor as follows:

$$\frac{\left(C_A - C_A^o\right)}{\nu_A} = \frac{\left(C_B - C_B^o\right)}{\nu_B} = \frac{\left(C_C - C_C^o\right)}{\nu_C} = \frac{\left(C_D - C_D^o\right)}{\nu_D} \tag{1.41}$$

In a system with a single independent reaction, the solution of only one of the species conservation equations (1.38) is enough to determine the product distribution. The rest of the concentrations of the species involved in the reaction can then be obtained using the stoichiometric relations given in Eq. (1.41).

1.7.2 Steady-Flow Analysis for a CSTR

Species conservation equations for a perfectly mixed continuous-tank reactor (CSTR) for different species involved in the reaction can be written as given in Eq. (1.16).

$$R_{A_f} = \frac{(F_{A_f} - F_{A_o})}{V}; R_{B_f} = \frac{(F_{B_f} - F_{B_o})}{V}; R_{C_f} = \frac{(F_{C_f} - F_{C_o})}{V}; R_{D_f} = \frac{(F_{D_f} - F_{D_o})}{V} \quad (1.42)$$

The following expressions can be written by combining the species conservation equations given in Eq. (1.42) with the stoichiometric relations given in Eq. (1.37) for the reaction rates of different species

$$\frac{(F_{A_f} - F_{A_o})}{\nu_A} = \frac{(F_{B_f} - F_{B_o})}{\nu_B} = \frac{(F_{D_f} - F_{D_o})}{\nu_D} = \frac{(F_{C_f} - F_{C_o})}{\nu_C} \quad (1.43)$$

For the case of the constant volumetric flow rate, Eq. (1.43) reduces to

$$\frac{(C_{A_f} - C_{A_o})}{\nu_A} = \frac{(C_{B_f} - C_{B_o})}{\nu_B} = \frac{(C_{D_f} - C_{D_o})}{\nu_D} = \frac{(C_{C_f} - C_{C_o})}{\nu_C} \quad (1.44)$$

These relations are quite similar to the relationships obtained for a batch reactor. Again, for a system containing a single independent reaction, the solution of only one of the species conservation equations is enough to determine the product distribution. Note that the relationships, which are given in Eq. (1.43) or Eq. (1.44), can be used to determine the molar flow rates or concentrations of the rest of the species in the exit stream of the CSTR operating at steady state. Similar relations can also be expressed for a plug-flow reactor operating at a steady-state condition.

Let us consider an arbitrary reaction of the following form:

$$a\text{A} + b\text{B} \rightleftharpoons c\text{C} + d\text{D}$$

Molar flow rates of species in the reactor exit stream can be expressed using the relationships given in Eq. (1.43), in terms of their inlet molar flow rates and the fractional conversion of the limiting reactant A, as follows:

$$F_{A_f} = F_{A_o}(1 - x_{A_f}) \quad (1.45)$$

$$F_{B_f} = F_{B_o} - \frac{b}{a} F_{A_o} x_{A_f} \quad (1.46)$$

$$F_{C_f} = F_{C_o} + \frac{c}{a} F_{A_o} x_{A_f} \quad (1.47)$$

$$F_{D_f} = F_{D_o} + \frac{d}{a} F_{A_o} x_{A_f} \quad (1.48)$$

1.7.3 Unsteady Perfectly Mixed-Flow Reactor Analysis

Let us consider a perfectly mixed continuous stirred-tank reactor operating under unsteady-state conditions. The conservation equation for the species i involved in this system can be written as follows:

$$F_{i_o} - F_{i_f} + R_{i_f} V = \frac{dn_i}{dt} \quad \Rightarrow \quad R_{i_f} = \frac{1}{V}\left[\frac{dn_i}{dt} - (F_{i_o} - F_{i_f})\right] \quad (1.49)$$

Combining the species conservation equations (1.49) with the stoichiometric relations of different species given in Eq. (1.37), the following equation is obtained:

$$\frac{1}{(V)\nu_A}\left[\frac{dn_A}{dt} - (F_{A_o} - F_{A_f})\right] = \frac{1}{(V)\nu_B}\left[\frac{dn_B}{dt} - (F_{B_o} - F_{B_f})\right] \quad (1.50)$$

Concentrations of the species within the reactor can be taken to be the same as the concentrations of the same species in the exit stream of the reactor if the perfect mixing assumption is correct. Then, for a constant volumetric flow rate case, Eq. (1.50) reduces to

$$\frac{1}{\nu_A}\left[\frac{dC_{A_f}}{dt} - \frac{Q}{V}\left(C_{A_o} - C_{A_f}\right)\right] = \frac{1}{\nu_B}\left[\frac{dC_{B_f}}{dt} - \frac{Q}{V}\left(C_{B_o} - C_{B_f}\right)\right] \tag{1.51}$$

Rearrangement of this equation gives a linear first-order ordinary differential equation for the combined variable X.

$$\frac{dX}{dt} + \frac{Q}{V}X = \frac{Q}{V}X_o \tag{1.52}$$

$$X = \left(\frac{C_{A_f}}{\nu_A} - \frac{C_{B_f}}{\nu_B}\right);\ X_o = \left(\frac{C_{A_o}}{\nu_A} - \frac{C_{B_o}}{\nu_B}\right) \tag{1.53}$$

Rearrangement of Eq. (1.52), and integration of this equation with the initial condition of $X^o = \left(\frac{C_A^o}{\nu_A} - \frac{C_B^o}{\nu_B}\right)$ gives:

$$\int_{X^o}^{X} \frac{dX}{(X_o - X)} = \int_0^t \frac{Q}{V}dt \tag{1.54}$$

$$-\ln\left(\frac{X_o - X}{X_o - X^o}\right) = \frac{Q}{V}t \Rightarrow \left(\frac{X_o - X}{X_o - X^o}\right) = \exp\left(-\frac{Q}{V}t\right) \tag{1.55}$$

Note that X_o and X^o correspond to the inlet and the initial values of the parameter X (defined by Eq. (1.53)), respectively. Substituting X and X_o from Eqs. (1.53) and (1.55) reduces to

$$\left(\frac{C_{A_o}}{\nu_A} - \frac{C_{B_o}}{\nu_B}\right) - \left(\frac{C_{A_f}}{\nu_A} - \frac{C_{B_f}}{\nu_B}\right) = \left[\left(\frac{C_{A_o}}{\nu_A} - \frac{C_{B_o}}{\nu_B}\right) - \left(\frac{C_A^o}{\nu_A} - \frac{C_B^o}{\nu_B}\right)\right]\exp\left(-\frac{Q}{V}t\right) \tag{1.56}$$

As seen in Eq. (1.56), the stoichiometric relations, which are given in Eq. (1.44) for a CSTR operating at steady state, cannot be used at unsteady-state operation. To be able to use the stoichiometric relationships given in Eq. (1.44), the right-hand side of Eq. (1.56) should be zero. This condition can be achieved in the following two cases:

- If $\left(\frac{C_{A_o}}{\nu_A} - \frac{C_{B_o}}{\nu_B}\right) = \left(\frac{C_A^o}{\nu_A} - \frac{C_B^o}{\nu_B}\right)$
- If $t \to \infty$ (this corresponds to the new steady-state condition)

The reader may refer to Aris [4] for supplementary reading on stoichiometric relations.

Problems and Questions

1.1 It is claimed that the following reactions are taking place in a reactor:

$4NH_3 + 5O_2 \rightarrow 6H_2O + 4NO$

$4NH_3 + 7O_2 \rightarrow 6H_2O + 4NO_2$

$O_2 + 2NO \rightarrow 2NO_2$

$$2NO_2 \rightarrow N_2 + 2O_2$$

$$4NH_3 + 3O_2 \rightarrow 6H_2O + 2N_2$$

$$2NO \rightarrow N_2 + O_2$$

(a) How many of these reactions are independent? Write down an independent set of reactions.

(b) Select an appropriate set of independent rates and write down the relations for the reaction rates of the other species in terms of the selected independent rates.

1.2 Experiments were performed in a tubular chemical vapor deposition (CVD) reactor to investigate the kinetics of CVD of boron from boron trichloride (BCl_3) and hydrogen. Experimental results indicated the formation of hydrochloric acid (HCl) and dichloroborane ($BHCl_2$) in the vapor phase, as well as solid boron (B) deposited on the substrate.

(a) Find out the relations between the rates of different species in this system.
(b) How many independent reactions are possible for this system?

1.3 Consider the following set of elementary gas-phase reaction steps

$$Br_2 \rightarrow 2Br$$

$$Br + H_2 \rightarrow HBr + H$$

$$H + Br_2 \rightarrow HBr + Br$$

$$H + HBr \rightarrow H_2 + Br$$

$$2Br \rightarrow Br_2$$

(a) How many of these reactions are independent?
(b) Select an independent set of reaction rates of species.
(c) Write down the relations between the independent and dependent rates of species.

1.4 The following molecular species are observed in the product stream leaving the reactor in which dimethyl ether is produced from synthesis gas.

Dimethyl ether : $(CH_3)_2O$

Methanol : CH_3OH

Carbon monoxide : CO

Carbon dioxide : CO_2

Hydrogen : H_2

Water vapor : H_2O

Methane : CH_4

(a) How many independent reactions should be considered in the direct synthesis of dimethyl ether from synthesis gas?
(b) Write down the relations between the rates of different species and select an independent set of reaction rates.
(c) Write down a set of independent reactions that represent this system.

1.5 Using the following data for the temperature dependence of the rate constant of a second-order reaction, evaluate the activation energy:

(a) By using the Arrhenius form of temperature dependence (Eq. 1.5)
(b) By using the temperature dependence predicted by the collision theory (Eq. (1.6)).

Compare the results obtained by these two approaches.

Temperature (K)	**592**	**603**	**627**	**651**	**656**
k ((mol/cm^3)$^{-1}$ s^{-1})	520	755	1700	4020	5030

1.6 A reaction with the following stoichiometric equation has the following rate expression:

$$A + 3B \rightarrow 2C; \quad -R_A = kC_AC_B$$

What is the rate expression for reactant B if the stoichiometry of the reaction is written as follows?

$$\frac{1}{2}A + \frac{3}{2}B \rightarrow C$$

1.7 Consider a first-order reaction taking place in a constant volume batch reactor. The fractional conversion of the reactant was 0.3 at a reaction time of four minutes. Evaluate the fractional conversion values that would be obtained at reaction times of 10 and 18 min in the batch reactor. Compare these results with the fractional conversion values that would be obtained in a CSTR and a plug-flow reactor at space times of 10 and 18 min, respectively. Critically discuss the results and the reasons for differences, if any.

1.8 Production of synthesis gas from biogas involves dry reforming of methane. The main reactions involved in this process are reported as follows:

$$CH_4 + CO_2 \rightleftarrows 2CO + 2H_2$$

$$H_2 + CO_2 \rightleftarrows CO + H_2O$$

$$4H_2 + CO_2 \rightleftarrows CH_4 + 2H_2O$$

$$3H_2 + CO \rightleftarrows CH_4 + H_2O$$

$$CH_4 + H_2O \rightleftarrows CO + 3H_2$$

$$2CO \rightleftarrows C + CO_2$$

$CH_4 \rightleftarrows C + 2H_2$

$C + H_2O \rightleftarrows CO + H_2$

(a) How many of these reactions are independent?
(b) Write down a suitable set of independent reactions.
(c) Select a set of independent rates, and express the rates of other species in terms of the rates of the selected independent rates.

1.9 The extraction of chlorides by hydrogen is quite common in semiconductor processing. Let us consider the following reaction sequence for the reaction of $SiCl_4$ and H_2.

$SiCl_4 + H_2 \rightarrow SiHCl_3 + HCl$

$SiHCl_3 + H_2 \rightarrow SiH_2Cl_2 + HCl$

$SiH_2Cl_2 + H_2 \rightarrow SiH_3Cl + HCl$

$SiH_3Cl + H_2 \rightarrow SiH_4 + HCl$

$SiH_4 \rightarrow Si(c) + 2H_2$

$SiCl_4 + 2H_2 \rightarrow Si(c) + 4HCl$

$SiHCl_3 + H_2 \rightarrow Si(c) + 3HCl$

(a) How many of these reactions are independent?
(b) Write down a suitable set of independent reactions.
(c) Select a set of independent rates, and express the rates of other species in terms of the rates of the selected independent rates.

References

1 Silbey, R.J., Alberty, R.A., and Bawendi, M.G. (2004). *Physical Chemistry*. Hoboken, NJ: Wiley.
2 Boudart, M. (1968). *Kinetics of Chemical Processes*. Englewood Cliffs, NJ: Prentice-Hall.
3 Laidler, K.J. and King, M.C. (1983). *J. Phys. Chem.* **87**: 2657–2664.
4 Aris, R. (1989). *Elementary Chemical Reactor Analysis*. Boston: Butterworths.

2

Reversible Reactions and Chemical Equilibrium

Most of the chemical reactions are reversible. When the forward rate of a reaction is equal to the backward rate, the net rate of the reaction becomes zero, and the reaction is considered to be at equilibrium. In general, a closed system is said to be at equilibrium when it does not tend to change its properties with time. The fractional conversion of reactants to products reached at equilibrium is the maximum possible conversion that can be achieved at the given reaction temperature and pressure. As discussed later in this chapter, equilibrium conversion cannot be exceeded at the operating conditions of the reactor. Achievement of conversion values over the equilibrium value by the use of multifunctional reactors is discussed in Chapter 13.

2.1 Equilibrium and Reaction Rate Relations

Equilibrium calculations are critical in reactor design to determine the maximum possible extent of chemical conversion. For some reactions, the rate of reaction may be relatively high. However, if the equilibrium conversion is low, that reaction is controlled by the equilibrium. The equilibrium conversion may be very high in some other reactions, and the reaction rate controls the conversion. As illustrated in Figure 2.1, equilibrium conversion may be very high for some reactions and complete conversion of the limiting reactant may be approached. These reactions can be considered irreversible. The reaction rate controls the fractional conversion achieved in these reactions. However, for some other reactions, equilibrium conversion is relatively low, and the increase of reaction rate does not help to increase the final conversion level that can be reached.

Let us consider a reversible reaction with the following stoichiometry:

$$a\mathrm{A} + b\mathrm{B} \rightleftarrows c\mathrm{C} + d\mathrm{D}$$

Equilibrium relation in terms of concentrations of the species involved in this reaction can be written as follows:

$$K_C = \frac{C_{C_e}^c C_{D_e}^d}{C_{A_e}^a C_{B_e}^b} \text{ (at equilibrium)} \tag{2.1}$$

On the other hand, the reversible rate expression of this reaction can be expressed in power-law form as

$$-R_A = k_f C_A^n C_B^m C_C^p C_D^q - k_b C_A^{n'} C_B^{m'} C_C^{p'} C_D^{q'} \tag{2.2}$$

Fundamentals of Chemical Reactor Engineering: A Multi-Scale Approach, First Edition. Timur Doğu and Gülşen Doğu.

Companion website: www.wiley.com/go/dogu/chemreacengin

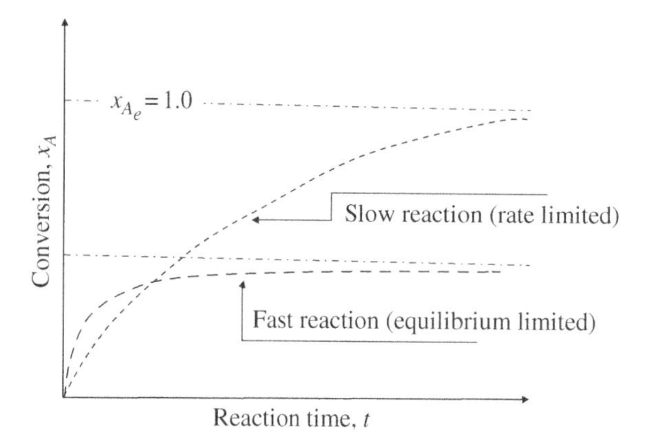

Figure 2.1 Comparison of conversion behavior for rate-limited and equilibrium-limited reactions in a batch reactor.

Here, n, m, p, and q are the reaction orders with respect to A, B, C, D in the forward direction and n', m', p', and q' are the corresponding reaction orders in the backward direction. Concentration dependence of the reaction rate is generally determined experimentally. The reaction orders are not necessarily equal to the stoichiometric coefficients. At equilibrium, the reaction rate becomes zero. Hence, we can write the equilibrium condition from the reaction rate expression as follows:

$$k_f C_{A_e}^n C_{B_e}^m C_{C_e}^p C_{D_e}^q - k_b C_{A_e}^{n'} C_{B_e}^{m'} C_{C_e}^{p'} C_{D_e}^{q'} = 0 \tag{2.3}$$

$$\frac{k_f}{k_b} = \frac{C_{A_e}^{n'} C_{B_e}^{m'} C_{C_e}^{p'} C_{D_e}^{q'}}{C_{A_e}^n C_{B_e}^m C_{C_e}^p C_{D_e}^q} \tag{2.4}$$

Comparison of Eq. (2.4) with Eq. (2.1), which is expressed for the stoichiometry of the given reaction, indicates that

$$K_C = \frac{k_f}{k_b} \tag{2.5}$$

and

$$a = (n - n'), \quad b = (m - m'), \quad c = (p' - p), \quad d = (q' - q) \tag{2.6}$$

Using these relations, we can predict the reverse rate law knowing the forward rate expression. For instance, for the following reaction between the reactants A and B

$$\mathrm{A + B \rightleftharpoons C + D}$$

suppose the rate expression in the forward direction is given as first order with respect to the concentration of A and second order with respect to the concentration of B. In that case, the backward rate expression can be predicted as follows:

$$-R_A = k_f C_A C_B^2 - k_b C_A^{n'} C_B^{m'} C_C^{p'} C_D^{q'}$$

$$\text{At equilibrium}: \frac{k_f}{k_b} = \frac{C_{A_e}^{n'} C_{B_e}^{m'} C_{C_e}^{p'} C_{D_e}^{q'}}{C_{A_e} C_{B_e}^2}$$

Since the equilibrium relation for this stoichiometric reaction is

$$K_C = \frac{C_{C_e} C_{D_e}}{C_{A_e} C_{B_e}}$$

we should have

$$\frac{k_f}{k_b} = K_C; p' = 1, q' = 1, n' = 0 \text{ and } m' = 1$$

The rate expression of this reaction should then be expressed as follows:

$$-R_A = k_f C_A C_B^2 - k_b C_B C_C C_D$$

2.2 Thermodynamics of Chemical Reactions

A chemical reaction can be represented by the following expression:

$$0 = \sum_i \nu_i A_i \tag{2.7}$$

Here, ν_i represents the stoichiometric coefficient of species A_i. The value of ν_i is positive for the products and negative for the reactants. At any extent of the chemical conversion, Gibbs free energy of the system is:

$$G = \sum_i n_i \mu_i \tag{2.8}$$

Here, n_i is the number of moles and μ_i is the chemical potential of the species i involved in the reaction (products and reactants). The reaction extent (ξ) is defined as follows:

$$\xi = \frac{n_i - n_i^o}{\nu_i} = \frac{n_j - n_j^o}{\nu_j} \tag{2.9}$$

Here, n_i^o is the number of moles of species i at the start of the reaction.

Substitution of n_i from Eq. (2.9) into Eq. (2.8) gives Gibbs free energy of the system in terms of the reaction extent as

$$G = \sum_i n_i^o \mu_i + \xi \sum_i \nu_i \mu_i \text{ (at the reaction temperature and pressure)} \tag{2.10}$$

According to the general equilibrium criterion from thermodynamics, Gibbs free energy of the system should be minimum at equilibrium [1–3]. As the reaction proceeds, the Gibbs free energy of the reaction decreases, and when equilibrium is attained, it reaches its minimum value. Thermodynamics tells us that any system is at equilibrium if the change in Gibbs free energy of the system is zero. Hence, the change in Gibbs free energy with the reaction extent should be zero when equilibrium is reached.

$$\left(\frac{\partial G}{\partial \xi}\right)_{T,P} = 0; \text{ (at equilibrium)} \tag{2.11}$$

The derivative of Gibbs free energy with respect to the reaction extent can be expressed from Eq. (2.10) as follows:

$$\left(\frac{\partial G}{\partial \xi}\right)_{T,P} = \sum_i \nu_i \mu_i = \Delta G \tag{2.12}$$

Here, ΔG is the Gibbs free energy change of the reaction. It is equal to the Gibbs free energies of the products minus the Gibbs free energies of the reactants, which are multiplied by the respective stoichiometric coefficients. For a chemical reaction to proceed spontaneously, ΔG of the reaction should be negative ($\Delta G < 0$). When equilibrium is reached ΔG becomes zero.

$$\Delta G = 0 \text{ (at equilibrium)} \tag{2.13}$$

For a pure ideal gas, chemical potential corresponds to the molar Gibbs free energy, by definition. Hence, molar Gibbs free energy $(\hat{G}_i)$ for species i at constant temperature can be expressed as

$$d\hat{G}_i = d\mu_i = \hat{V}_i dP_i \text{ (at constant } T) \tag{2.14}$$

Here, $\hat{V}_i$ is the molar volume of species i. The following expression can be written by expressing the molar volume in terms of pressure and temperature using the ideal gas law:

$$d\mu_i = \frac{R_g T}{P_i} dP_i = R_g T d \ln P_i \tag{2.15}$$

Integration of Eq. (2.15) between the pressure values of P_i^o and P_i, and choosing P_i^o as the standard pressure of one atmosphere, the following relation is obtained for the chemical potential of an ideal gas:

$$\mu_i - \mu_i^o = R_g T \ \ln P_i \tag{2.16}$$

Here, μ_i^o and μ_i are the chemical potentials of species i at P_i^o and P_i, respectively. In this equation P_i is the normalized pressure of species i with respect to one atmosphere. The substitution of Eq. (2.16) into Eq. (2.12) gives the following expression for the Gibbs free energy change of the reaction:

$$\Delta G = \sum_i \nu_i \mu_i = \sum_i \nu_i \left(\mu_i^o + R_g T \ \ln P_i\right) = \sum_i \nu_i \mu_i^o + R_g T \sum_i \nu_i \ln P_i \tag{2.17}$$

Here, P_i is the partial pressure of species i. The first term on the right side of this equation is the Gibbs free energy of reaction at the standard state of one atmosphere. Thus, Eq. (2.17) can be rearranged as

$$\Delta G = \Delta G^o + R_g T \ln \prod_i P_i^{\nu_i} \tag{2.18}$$

When the reaction reaches equilibrium, the Gibbs free energy of the reaction (ΔG) becomes zero. In this case, the partial pressures of gases are their equilibrium values at the reaction temperature. Thus, at equilibrium

$$\Delta G^o = -R_g T \ \ln \prod_i P_{i_e}^{\nu_i} \tag{2.19}$$

The term $\prod_i P_{i_e}^{\nu_i}$ is constant, which is called the equilibrium constant (K_p) of the reaction in terms of partial pressures in an ideal gas mixture.

$$\Delta G^o = -R_g T \ \ln K_p \tag{2.20}$$

This relation can be used to determine the equilibrium constant at the standard state. For a gas-phase reaction with the following stoichiometry:

$$a_1 \mathrm{A}_1 + a_2 \mathrm{A}_2 + \cdots + a_n \mathrm{A}_n \rightleftarrows b_1 \mathrm{B}_1 + b_2 \mathrm{B}_2 + \cdots + b_n \mathrm{B}_n$$

The equilibrium constant in terms of the equilibrium partial pressures is written as

$$K_p = \left[\frac{P_{B_1}^{b_1} P_{B_2}^{b_2} \ldots P_{B_n}^{b_n}}{P_{A_1}^{a_1} P_{A_2}^{a_2} \ldots P_{A_n}^{a_n}}\right]_e$$

Since ΔG^o is the Gibbs free energy of reaction evaluated at the standard state of one atmosphere, it is a function of temperature only. Therefore, K_p for an ideal gas reaction mixture is a function of temperature only. It does not depend upon the total pressure of the system. The temperature in Eq. (2.20) is the one at which ΔG^o term is evaluated. The Gibbs free energy of a reaction at standard state is calculated from the standard Gibbs free energies of formations of reactants and products as follows:

$$\Delta G^o = \sum_i \nu_i \Delta G^o_{i,f} \tag{2.21}$$

Standard Gibbs free energies of formations of species can be found in the literature [1–3].

Chemical potential expression, in terms of the partial pressure of an ideal gas, is derived above and expressed by Eq. (2.16). The general expression for the chemical potential of a species i in different phases (gas, liquid, solid) is expressed in terms of its activity as

$$\mu_i = \mu_i^o + R_g T \ln a_i \tag{2.22}$$

The activity of species i for a nonideal gas is the ratio of its fugacity in the mixture to its fugacity in the standard state. The standard state for a real gas is chosen as a state of unit fugacity, and fugacity has the same unit as pressure. In other words, for real gases, the pressure is replaced by fugacity. The ratio of fugacity to pressure is defined as fugacity coefficient $\left(\gamma = \frac{f}{P}\right)$. As pressure decreases and temperature increases, gas behaves more ideally and fugacity coefficient approaches one. The fugacity coefficient is a function of both temperature and pressure of the system. Evaluation of the fugacity coefficients using the reduced temperature ($T_{red} = T/T_c$) and reduced pressure ($P_{red} = P/P_c$) values is available in the thermodynamics books [1, 2]. Here, T_c and P_c are the critical temperature and critical pressure of species i, respectively. The equilibrium constant in the chemical equilibrium relation given by Eq. (2.20) should, in general, be written as follows:

$$\Delta G^o = -R_g T \ln K \tag{2.23}$$

2.3 Different Forms of Equilibrium Constant

For any gas-, liquid-, or solid-phase reaction, the equilibrium constant K is generally expressed in terms of activities of species involved in the reaction and denoted as K_a. K_a is pressure independent. It is a function of temperature only. If the reaction is taking place in the gas phase, K_a reduces to K_f (equilibrium constant in terms of fugacities). If the gas mixture is an ideal one, K_f reduces to K_p (equilibrium constant in terms of partial pressures). The utilization of the appropriate form of the equilibrium constant depends upon the reaction system under consideration. For a gas-phase reaction with the following stoichiometry

$$a\text{A} + b\text{B} \rightleftarrows c\text{C} + d\text{D}$$

equilibrium constant should, in general, be expressed in terms of fugacities of the reactants and the products as follows:

$$K = K_f = \left(\frac{f_C^c f_D^d}{f_A^a f_B^b}\right)_e \tag{2.24}$$

Fugacities of substances are related to the composition and pressure of the mixture using the Lewis–Randall fugacity rule. This rule states that the fugacity of a species i in the mixture is related to its mole fraction by the following relation:

$$f_i = f_{i,\text{pure}}(T, P)y_i \tag{2.25}$$

Here, $f_{i,\text{ pure}}(T, P)$ is the fugacity of pure species i, at the same temperature and total pressure of the system. Substituting $f_{i,\text{ pure}}(T, P)$ using the definition of fugacity coefficient as, $f_{i,\text{ pure}} = \gamma_i P$, fugacity of species i becomes

$$f_i = \gamma_i P y_i \tag{2.26}$$

Here, P is the total pressure of the mixture. Using the fugacity relations given by Eq. (2.26), Eq. (2.24) becomes

$$K_f = \left[\frac{(P\gamma_C y_C)^c (P\gamma_D y_D)^d}{(P\gamma_A y_A)^a (P\gamma_B y_B)^b}\right]_e \tag{2.27}$$

Here, y_i values are the mole fractions of the species at equilibrium. Equilibrium relation can then be expressed in terms of mole fractions of the reactants and the products as follows:

$$K = K_f = \left(\frac{\gamma_C^c \gamma_D^d}{\gamma_A^a \gamma_B^b}\right)\left(\frac{y_C^c y_D^d}{y_A^a y_B^b}\right)_e P^{(c+d)-(a+b)} = K_\gamma K_y P^{\sum_i \nu_i} \tag{2.28}$$

Here, K is a function of temperature only. However, a change of pressure may change the value of K_y. If the gas mixture is an ideal one, the equilibrium constant (K_p) can be written in terms of partial pressures as

$$K = K_p = \left(\frac{P_{C_e}^c P_{D_e}^d}{P_{A_e}^a P_{B_e}^b}\right) = \left(\frac{y_C^c y_D^d}{y_A^a y_B^b}\right)_e P^{(d+c)-(a+b)} = K_y P^{\sum_i \nu_i} \tag{2.29}$$

The equilibrium constant can also be written in terms of concentrations of species as

$$K_C = \left(\frac{C_{C_e}^c C_{D_e}^d}{C_{A_e}^a C_{B_e}^b}\right) \tag{2.30}$$

This approach is quite common for liquid-phase reactions. For ideal gas-phase reactions, the relation between K_C and K_p can be expressed as follows:

$$K_C = \left(\frac{C_{C_e}^c C_{D_e}^d}{C_{A_e}^a C_{B_e}^b}\right) = \left(\frac{(P_{C_e}/R_gT)^c (P_{D_e}/R_gT)^d}{(P_{A_e}/R_gT)^a (P_{B_e}/R_gT)^b}\right) = \left(\frac{P_{C_e}^c P_{D_e}^d}{P_{A_e}^a P_{B_e}^b}\right)(R_gT)^{-\sum_i \nu_i}$$
$$= K_P (R_gT)^{-\sum_i \nu_i} \tag{2.31}$$

2.4 Temperature Dependence of Equilibrium Constant and Equilibrium Calculations

The temperature dependence of equilibrium constant can be derived using the fundamental equations of thermodynamics. Taking the derivative of ln K with respect to temperature (from Eq. (2.23)), we obtain

$$-R_g\frac{d\ln K}{dT} = \frac{d}{dT}\left(\frac{\Delta G^o}{T}\right) = \frac{d(\Delta G^o)}{TdT} - \frac{\Delta G^o}{T^2} \tag{2.32}$$

Gibbs free energy is related to temperature and pressure by the following relation:

$$dG = VdP - SdT$$

Thus, the derivative of standard Gibbs free energy change of reaction with respect to temperature at constant pressure is

$$\left(\frac{d(\Delta G^o)}{dT}\right)_P = -\Delta S^o \tag{2.33}$$

Using the ($\Delta G^o = \Delta H^o - T\Delta S^o$) relation from thermodynamics, we obtain the following expression:

$$\left(\frac{d(\Delta G^o)}{dT}\right)_P = \frac{(\Delta G^o - \Delta H^o)}{T} \tag{2.34}$$

Combining this relation with Eq. (2.32), we get

$$\frac{d\ln K}{dT} = \frac{\Delta H^o}{R_g T^2} \tag{2.35}$$

This equation is known as the van't Hoff relation, where ΔH^o is the standard heat of reaction at temperature T. This equation can also be written in the following form:

$$\frac{d\ln K}{d\left(\frac{1}{T}\right)} = -\frac{\Delta H^o}{R_g} \tag{2.36}$$

Integration of Eq. (2.35) yields the relation for the temperature dependence of the equilibrium constant. If the heat of reaction is constant, the dependence of ln K on $1/T$ is linear, with a slope of ($-\Delta H^o/R_g$). However, if it is temperature dependent, integration of Eq. (2.35) should be made accordingly.

For an exothermic reaction, the heat of reaction is negative, causing a decrease of the equilibrium constant with an increase in temperature (Figure 2.2). As a result, equilibrium conversion also decreases with an increase in temperature. However, for an endothermic reaction, the reverse is true. In that case, the heat of reaction is positive and the equilibrium constant, and hence the equilibrium conversion increases with an increase in temperature.

The equilibrium constant for any reaction is calculated using Eq. (2.23). The value of the standard free energy of reaction at the reaction temperature should be known for calculating the equilibrium constant. But, Gibbs free energies of formations of substances are not available at all temperatures to calculate ΔG^o of reaction. Gibbs free energy of formations of substances are usually available at a reference temperature (T_R), which is usually chosen as 298 K. Thus, ΔG^o of reaction can easily be calculated at this reference temperature. The value of equilibrium constant at 298 K can then be evaluated using Eq. (2.23). Van't Hoff equation (Eq. (2.35)) should be used to calculate the

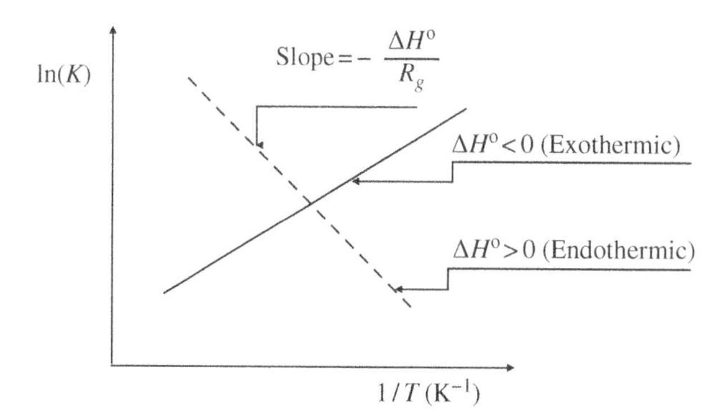

Figure 2.2 Temperature dependence of equilibrium constant.

equilibrium constant at any other reaction temperature. Integration of van't Hoff equation between the reference temperature and the reaction temperature gives a relation for the temperature dependence of equilibrium constant.

$$\ln \frac{K_T}{K_{T_R}} = \int_{T_R}^{T} \frac{\Delta H^o(T)}{T^2 R_g} dT \tag{2.37}$$

If the heat of reaction is constant, integration of this equation gives

$$\ln \frac{K_T}{K_{T_R}} = -\frac{\Delta H^o}{R_g}\left(\frac{1}{T} - \frac{1}{T_R}\right) \tag{2.38}$$

where ΔH^o may be temperature dependent. Therefore, the temperature dependence of ΔH^o should be determined. The heat of reaction at the standard state is calculated using the following relation:

$$\Delta H^o = \sum_i \nu_i \Delta H^o_{i,f} \tag{2.39}$$

Standard heats of formations of some substances are available in the literature [1–3]. To calculate the standard heat of reaction at any temperature, the path illustrated in Figure 2.3 can be considered. Since enthalpy is a state function, the enthalpy change of a process, going from the initial state (i) to the final state (f), is the same regardless of the path. Therefore,

$$\Delta H_1 + \Delta H^o_T = \Delta H^o_{T_R} + \Delta H_2 \tag{2.40}$$

Hence, the heat of reaction at a temperature of T can be expressed as follows:

$$\Delta H^o_T = \Delta H^o_{T_R} + \Delta H_2 - \Delta H_1 \tag{2.41}$$

Here, ΔH_2 and ΔH_1 are the sensible enthalpy changes of products and reactants, as a result of a change of temperature from T_R to T, respectively.

$$\Delta H^o_T = \Delta H^o_{T_R} + \int_{T_R}^{T} \sum_i \nu_i C_{p_i} dT = \Delta H^o_{T_R} + \int_{T_R}^{T} \Delta C_p dT \tag{2.42}$$

If ΔC_p is zero, then the heat of the reaction can be taken as constant. If ΔC_p is not zero, then the standard heat of reaction should be expressed as a function of temperature using Eq. (2.42), and it

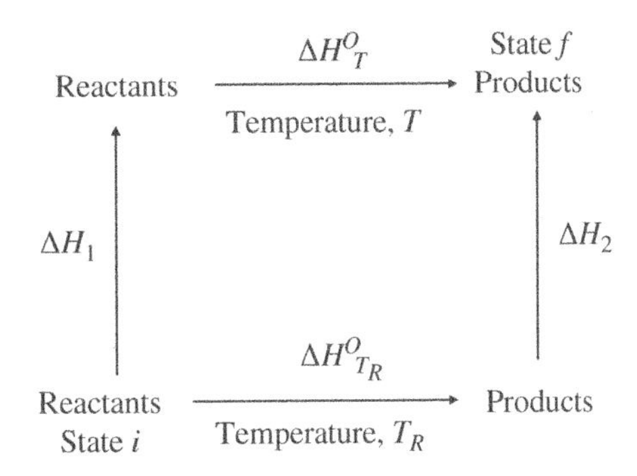

Figure 2.3 Schematic illustration for the evaluation of heat of reaction at an arbitrary temperature.

should be substituted into Eq. (2.37) to evaluate the temperature dependence of the equilibrium constant. Integration of Eq. (2.37) between the limits of the reference temperature T_R and reaction temperature T gives the equilibrium constant at the desired temperature. The temperature dependence of C_p values of different species can be estimated using the empirical relations given as

$$C_{p,i} = a + bT + cT^2 + dT^3 + \cdots \tag{2.43}$$

Example 2.1 Water–Gas Shift Reaction

Water–gas shift reaction (WGSR) is an important reversible reaction involved in steam reforming and gasification reactions in hydrogen production.

$$CO + H_2O \rightleftarrows H_2 + CO_2 \tag{R.2.1}$$

The composition of the feed stream to a flow reactor is given as 40% CO and 60% H_2O. Evaluate the equilibrium conversion of CO at 200, 500, and 700 °C (at atmospheric pressure) using the data given in Tables 2.1 and 2.2.

Solution

Standard free energy change of reaction

$$\Delta G^o = \Delta G^o_{CO_2,f} + \Delta G^o_{H_2,f} - \Delta G^o_{CO,f} - \Delta G^o_{H_2O,f} = -28\,600\ \text{J/mol}.$$

Evaluation of equilibrium constant at 298 K using Eq. (2.23):

$$\Delta G^o_{298} = -R_g T \ln K; -28\,600 = -(8.314)(298)\ln K_{298}; K_{298} = 10.3 \times 10^4$$

Table 2.1 Thermodynamic Data for Example 2.1 [1].

Species	Standard heat of formations ΔH^o_{if} at 298 K (J/mol)	Standard free energy of formations ΔG^o_{if} at 298 K (J/mol)
CO	−110 500	−137 200
H_2O	−241 800	−228 600
CO_2	−393 500	−394 400

Table 2.2 Heat Capacity Data for Example 2.1 [1].

Species	a	$b \times 10^2$	$c \times 10^5$	$d \times 10^9$
CO	28.142	0.167	0.537	−2.221
H_2O	32.218	0.192	1.055	−3.593
H_2	29.088	−0.192	0.400	−0.870
CO_2	22.243	5.977	−3.499	7.464

Standard heat of reaction at 298 K

$$\Delta H^o_{298} = \Delta H^o_{CO_2,f} + \Delta H^o_{H_2,f} - \Delta H^o_{CO,f} - \Delta H^o_{H_2O,f} = -41\,200\ \text{J/mol}.$$

This result indicates that WGSR is an exothermic reaction.

The temperature dependence of heat of reaction can be estimated from Eq. (2.42) as

$$\Delta H^o_T = \Delta H^o_{298} + \int_{298}^{T} \Delta C_p dT = \Delta H^o_{298} + \int_{298}^{T} (\Delta a + \Delta bT + \Delta cT^2 + \Delta dT^3) dT$$

$$\Delta H^o_T = -40529 - 9.029T + 2.713 \times 10^{-2} T^2 - 1.564 \times 10^{-5} T^3 + 3.10 \times 10^{-9} T^4$$

Using the van't Hoff relation (Eq. (2.37)), the equilibrium constant can be evaluated at any temperature.

$$\ln K = \ln K_{298} + \int_{298}^{T} \frac{(-40529 - 9.029T + 2.713 \times 10^{-2} T^2 - 1.564 \times 10^{-5} T^3 + 3.10 \times 10^{-9} T^4)}{R_g T^2} dT$$

Equilibrium constant values evaluated at 200, 500, and 700 °C are as follows:

Temperature (°C)	25	200	500	700
Equilibrium constant (K)	10.3×10^4	140	8.0	0.93

Note that, for this exothermic reaction, the equilibrium constant decreases with an increase in temperature. The molar flow rates of the species at equilibrium can be related to the fractional conversion of CO as follows:

$$\text{CO}: F_{A_e} = F_{A_o}(1 - x_{A_e})$$

$$H_2O: F_{B_e} = F_{B_0} - F_{A_0} x_{A_e}$$

$$CO_2 \text{ and } H_2: F_{C_e} = F_{D_e} = F_{A_0} x_{A_e}$$

Hence, with the ideal gas assumption, the equilibrium relation of this reaction can be written in terms of partial pressures of species as follows:

$$K = K_p = \left(\frac{P_{CO_2} P_{H_2}}{P_{CO} P_{H_2O}}\right)_e = \left(\frac{y_{CO_2} y_{H_2}}{y_{CO} y_{H_2O}}\right)_e$$

Expressing mole fractions of species as the ratios of their molar flow rates to the total molar flow rate, equilibrium constant can be expressed in terms of the equilibrium fractional conversion of CO.

$$K_p = \frac{(F_{A_0}x_{A_e})^2}{F_{A_0}(1-x_{A_e})(F_{B_0}-F_{A_0}x_{A_e})} = \frac{(x_{A_e})^2}{(1-x_{A_e})\left(\frac{F_{B_0}}{F_{A_0}}-x_{A_e}\right)}$$

Equilibrium fractional conversion values evaluated at different temperatures are as follows:

Temperature (°C)	25	200	500	700
Equilibrium conversion (x_{A_e})	~1.0	0.98	0.86	0.41

The temperature dependence of the equilibrium constant for exothermic and endothermic reactions can also be related to the activation energies of the forward and the backward reaction rate constants. As illustrated in Figure 2.4, in the case of exothermic reactions, the activation energy of the backward reaction rate constant is higher than the activation energy of the forward reaction rate constant. Hence, with an increase in temperature, the ratio of forward and backward rate constants decreases. The reverse is true for an endothermic reaction.

$$\frac{k_f}{k_b} = \frac{k_{of}}{k_{ob}} \exp\left(-\frac{(E_{a_f}-E_{b_b})}{R_g T}\right) \tag{2.44}$$

If $(E_{a_f}-E_{a_b}) > 0$ (endothermic reaction), $\frac{k_f}{k_b}$ increases as T increases.

If $(E_{a_f}-E_{a_b}) < 0$ (exothermic reaction), $\frac{k_f}{k_b}$ decreases as T increases.

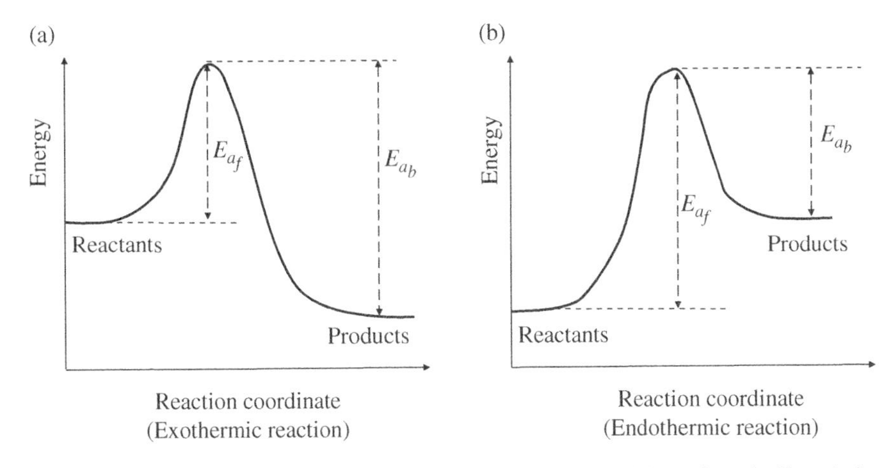

Figure 2.4 Schematic representation of activation energies for (a) exothermic (b) endothermic reactions.

Example 2.2 Methanol Synthesis

Methanol is a valuable chemical that can be produced from synthesis gas through a catalytic process. It has a high octane number, hence can be used as a motor vehicle fuel. Ethylene and other olefins can also be produced from methanol through the methanol to olefins (MTO) process. The conventional method of methanol synthesis involves the reaction of carbon monoxide with hydrogen in a fixed-bed tubular catalytic reactor, using Cu/ZnO/alumina-based solid catalysts.

$$CO + 2H_2 \rightleftharpoons CH_3OH; \ \Delta H^o_{298} = -90.2 \text{ kJ/mol} \tag{R.2.2}$$

Table 2.3 Thermodynamic data for Example 2.2, at 298 K [1]:

	ΔG_f^o (kJ/mol)	ΔH_f^o (kJ/mol)	Average molar heat capacity $C_{p,\,av}$ (J/mol K)
Methanol	−162.0	−200.7	61.2
Carbon monoxide, CO	−137.2	−110.5	30.3
Hydrogen, H_2	–	–	29.0

We expect to have some carbon dioxide, in addition to carbon monoxide and hydrogen, in the synthesis gas. Carbon dioxide in the synthesis gas may also react with hydrogen giving methanol. The occurrence of the WGSR is also expected to take place in the presence of carbon dioxide. The occurrence of these side reactions is not considered in this example.

Problem statement

It is asked to plot equilibrium conversion versus temperature curves for methanol synthesis (R.2.2) at different pressures, using the data given in Table 2.3. Do not consider the occurrence of any possible side reactions. Discuss the effect of pressure on the equilibrium conversion of CO if the feed stream to a tubular flow reactor contains only 30% CO and 70% H_2.

Solution

Gibbs free energy change of the reaction at standard state is evaluated as

$$\Delta G_{298}^o = -24.8\ \text{kJ/mol}$$

Then, the equilibrium constant at 298 K is found as

$$\Delta G^o = -R_g T \ln K \Rightarrow \quad -24\,800 = -8.314(298)\ln K \quad \Rightarrow K_{298} = 2.2\times 10^4$$

This result shows that the equilibrium constant of methanol synthesis is very high at 298 K. However, the Cu/ZnO/Alumina catalyst used in methanol synthesis shows good activity only at temperatures higher than 500 K. A typical reaction temperature is 540 K.

The temperature dependence of equilibrium constant can be estimated from the van't Hoff relation. For the integration of Eq. (2.37), we need the molar heat capacity values of each species. In general, molar heat capacity values are temperature dependent. However, in this example, average C_p values are given. The temperature dependence of the heat of reaction can then be estimated as follows:

$$\Delta H_T^o = \Delta H_{298}^o + \int_{298}^{T} \Delta C_p dT = -90\,200 + \int_{298}^{T} (61.2 - 30.3 - 2\times 29.0)dT$$

$$\Delta H_T^0 = -82\,124 - 27.1T \tag{2.45}$$

Hence, the temperature dependence of equilibrium constant is evaluated from the van't Hoff relation as follows:

$$\ln\frac{K_T}{K_{298}} = \int_{T_o}^{T} \frac{\Delta H_T^o}{R_g T^2}dT = \int_{298}^{T} \frac{(-82\,124 - 27.1T)}{8.314T^2}dT = 9878\left(\frac{1}{T} - \frac{1}{298}\right) - 3.26\ln\left(\frac{T}{298}\right) \tag{2.46}$$

The equilibrium constant for the methanol synthesis reaction can be expressed in terms of fugacities of species as

$$K = \left(\frac{f_C}{f_A f_B^2}\right)_e$$

Here, carbon monoxide, hydrogen, and methanol are denoted by A, B, and C, respectively.

$$A + 2B \rightleftarrows C$$

The equilibrium constant can then be expressed in terms of mole fractions of species as

$$K = \left(\frac{f_C}{f_A f_B^2}\right)_e = \frac{1}{P^2}\left(\frac{\gamma_C}{\gamma_A \gamma_B^2}\right)\left(\frac{y_C}{y_A y_B^2}\right)_e = \frac{1}{P^2}K_\gamma K_y \tag{2.47}$$

Here, γ_i is the fugacity coefficient of species i.

Mole fractions of species involved in this reaction can be related to the fractional conversion of the limiting reactant. The molar flow rates of species involved in this reaction in a tubular flow reactor and the total molar flow rate can be expressed as follows:

$$F_A = F_{A_o}(1 - x_A)$$

$$F_B = (F_{B_o} - 2F_{A_o}x_A)$$

$$F_C = F_{A_o}x_A$$

$$F_{\text{Total}} = (F_{A_o} + F_{B_o} - 2F_{A_o}x_A) = (F_o - 2F_{A_o}x_A)$$

Mole fractions of the species can then be expressed as

$$y_{CO} = y_A = \frac{F_A}{F_{\text{Total}}} = \frac{F_{A_o}(1 - x_A)}{(F_{A_o} + F_{B_o} - 2F_{A_o}x_A)} = \frac{y_{A_o}(1 - x_A)}{(1 - 2y_{A_o}x_A)} \tag{2.48a}$$

$$y_{H_2} = y_B = \frac{F_B}{F_{\text{Total}}} = \frac{(F_{B_o} - 2F_{A_o}x_A)}{(F_{A_o} + F_{B_o} - 2F_{A_o}x_A)} = \frac{(y_{B_o} - 2y_{A_o}x_A)}{(1 - 2y_{A_o}x_A)} \tag{2.48b}$$

$$y_{\text{MetOH}} = y_C = \frac{F_C}{F_{\text{Total}}} = \frac{F_{A_o}x_A}{(F_{A_o} + F_{B_o} - 2F_{A_o}x_A)} = \frac{y_{A_o}x_A}{(1 - 2y_{A_o}x_A)} \tag{2.48c}$$

The equilibrium relation for this reaction can then be expressed in terms of the equilibrium mole fractions, total pressure, and the fugacity coefficients as follows:

$$K = \frac{1}{P^2}\left(\frac{\gamma_C}{\gamma_A \gamma_B^2}\right)\left(\frac{y_C}{y_A y_B^2}\right)_e = \frac{1}{P^2}K_\gamma\left(\frac{x_{A_e}(1 - 2y_{A_o}x_{A_e})^2}{(1 - x_{A_e})(y_{B_o} - 2y_{A_o}x_{A_e})^2}\right) \tag{2.49}$$

Note that the equilibrium constant is a function of temperature only. Since this is an exothermic reaction, an increase in temperature causes a decrease in the value of equilibrium constant. Hence, a decrease in equilibrium conversion is expected with an increase in temperature. As can be concluded from the pressure dependence of the equilibrium relation given by Eq. (2.49), the increase in pressure increases the equilibrium conversion. To achieve high equilibrium conversion values, methanol synthesis is performed at quite high pressures ($P > 50$ bar). Since it is a high-pressure reaction, we need to use fugacities in the equilibrium calculations. Fugacity coefficients of the species should be evaluated and should be used in the equilibrium relation given by Eq. (2.49). Fugacity coefficients can be predicted from the correlations and from the charts, knowing the reduced temperature and pressure values. Reduced temperature and pressure are defined as the ratios of temperature and pressure to their critical values, respectively. The values of the critical temperature and pressure of the species involved in this reaction are given in the following table:

Critical temperature and pressure values for the species involved in methanol synthesis:

	Critical temperature (K)	Critical pressure (bar)
Methanol	512.6	81.0
CO	132.8	34.9
H_2	33.3	12.9

Fugacity coefficients of methanol and the K_γ values evaluated at some pressures and temperatures are listed in Table 2.4. Fugacity coefficients of H_2 and CO were close to one in the temperature and the pressure range listed in Table 2.4.

Equilibrium constant K is a function of temperature only. By substituting the fugacity coefficients into the equilibrium relation (Eq. (2.49)), the equilibrium conversion values are evaluated at different temperatures and pressures. Variation of equilibrium conversion of CO as a function of the reaction temperature is illustrated in Figure 2.5. As shown in this figure, high pressures are needed to achieve high equilibrium conversions at temperatures over 300 K. As for the temperature

Table 2.4 Fugacity coefficients of the species for Example 2.2 [1].

	Fugacity coefficient of methanol		K_γ	
T (K)	P = 10 atm	P = 50 atm	P = 10 atm	P = 50 atm
300	0.86	0.45	~0.86	~0.45
500	0.95	0.85	0.95	~0.85
700	0.97	0.95	0.97	0.95

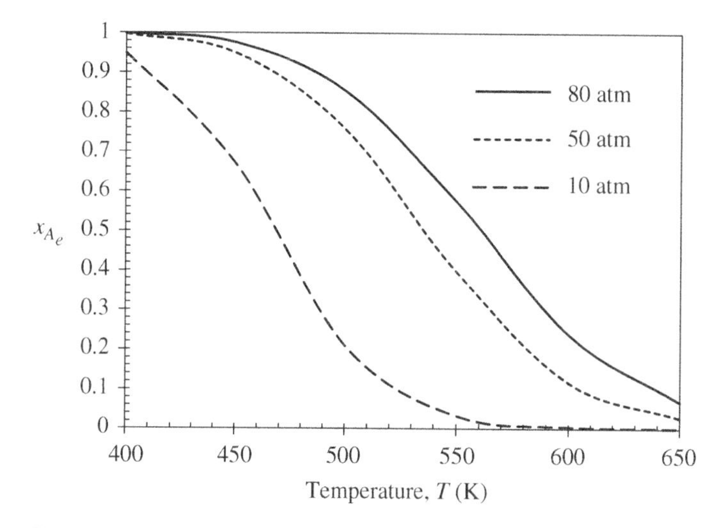

Figure 2.5 Variation of equilibrium conversion of CO to methanol as a function of temperature and pressure.

dependence of x_{A_e} is concerned, it decreases with an increase in temperature. This is due to the exothermic nature of this reaction. A typical temperature range for methanol synthesis is between 525 and 575 K. Results shown in Figure 2.5 indicated that the pressure of the methanol synthesis reactor should be selected over 50 bar. Case study example O.6.3 in the online material of this textbook may be referred to for further details.

Problems and Questions

2.1 For the water–gas shift reaction, evaluate the equilibrium constants at 298 and 1000 K and the corresponding equilibrium fractional conversion values of CO, using the thermodynamic data reported in the literature. Consider a feed stream containing 35% CO, 5% CO_2, and 55% H_2O, and 5% inerts, in these calculations

$$CO + H_2O \rightleftarrows CO_2 + H_2$$

2.2 The feed stream composition to an ammonia synthesis reactor is given as $\frac{F_{N_2,0}}{F_{H_2,0}} = \frac{1}{3}$ with no inerts. Evaluate the equilibrium conversion of nitrogen at 673 K at 50, 100, and 200 bar pressures. Compare the results and discuss them

$$N_2 + 3H_2 \leftrightarrow 2NH_3$$

2.3 Evaluate the gas-phase equilibrium conversion of ethylene to ethanol at 600 K and 30 bar. The steam to ethylene ratio in the feed stream is given as 4. Would you expect any change in the equilibrium conversion of ethylene, if this ratio is increased to 6?

$$C_2H_4 + H_2O \leftrightarrow C_2H_5OH$$

2.4 The cracking reaction of ethane produces ethylene and acetylene

$$C_2H_6 \rightleftarrows C_2H_4 + H_2$$

$$C_2H_4 \rightleftarrows C_2H_2 + H_2$$

If the reaction mixture contains 80% ethane and 20% inerts, evaluate the equilibrium conversion values of ethane to ethylene and acetylene at 1300 K (at 1 atm).

2.5 Evaluate the equilibrium conversion of $SiCl_4$ to $SiHCl_3$ at 500, 1000, and 1300 K, and discuss the effects of feed composition and temperature on the equilibrium conversion of SiC14

$$SiCl_4(g) + H_2 \rightleftarrows SiHCl_3(g) + HCl(g)$$

Take $H_2/SiCl_4$ ratio as two at the inlet of the reactor.

2.6 Oxidation of sulfur dioxide to sulfur trioxide over a vanadium pentoxide catalyst is the primary reaction in the contact process, which is developed for the production of sulfuric acid. The composition of the gases entering an adiabatic tubular reactor is given as 8% SO_2, 14% O_2, and 28% N_2. Evaluate the equilibrium conversion of SO_2 at 950 K if the reactor pressure is one atmosphere, using the following data. Note that $C_{P,\,i}$ values are given as constant average values for the temperature range of this reaction

	$\Delta H^o_{f,298}$ (kJ/mol)	$\Delta G^o_{f,298}$ (kJ/mol)	$C_{p,i}$ (J/mol K)
SO_2	−296.8	−300.2	51.4
SO_3	−395.7	−371.1	76.2
O_2			32.8
N_2			31.9

References

1 Sandler, S.I. (1999). *Chemical and Engineering Thermodynamics*. New York: Wiley.
2 Tosun, I. (2012). *Thermodynamics of Phase and Reaction Equilibria*. Amsterdam: Elsevier.
3 Smith, J.M. (1981). *Chemical Engineering Kinetics*. Singapore: McGraw Hill.

3

Chemical Kinetics and Analysis of Batch Reactors

Many laboratory-scale reactors operate in batch mode. There are no inlet and exit flow streams in batch reactors. Batch reactors are also used in industrial processes, especially in the production of specialty chemicals or chemicals at low/medium production rates. For instance, the use of batch reactors is quite common for the production of soap through the saponification of oils with sodium hydroxide. There are also high-pressure autoclave reactors to carry out gas-phase reactions in batch mode. The reaction mixture in a batch reactor is generally well mixed to achieve the uniform concentration of species and also uniform temperature. Achievement of uniform concentration and temperature in an industrial-scale batch reactor requires careful design of the mixing system of the reaction mixture. Another critical factor is the control of temperature within the batch reactor. Due to the heat of reaction of the chemical conversion, the reaction mixture should be cooled (in exothermic reactions) or should be heated (in endothermic reactions) if constant temperature operation is desired.

3.1 Kinetics and Mechanisms of Homogeneous Reactions

As discussed in Chapter 1, the reaction rate of any chemical conversion depends on the temperature and the concentrations of the reacting species in a chemical reactor. In the case of an elementary reaction, reaction orders with respect to the concentrations involved in the reaction are the same as the molecularity of the species in the reaction step, namely the stoichiometric constants in the elementary reaction step.

An elementary step is a process in which reactant molecules are converted directly to other chemical species without the formation of any intermediate species. Hence, in the case of an elementary reaction, the reaction proceeds through a single physicochemical process. However, many of the reactions do not proceed through a single physicochemical process, but involve a series of elementary steps to give the reaction products. For a reaction involving a number of elementary steps, the reaction orders for the concentrations of the species involved in this chemical conversion are not necessarily the same as the stoichiometric coefficients of the species of that overall reaction. The rate expressions of these reactions can be derived from the reaction mechanisms. Concentration and the temperature dependence of a reaction rate expression are determined by the use of experimental data, in most cases.

Experimental evidence shows that the rate laws of some reactions cannot be expressed in the form of simple power-law expressions. In some cases, the products or reactants may also appear

Fundamentals of Chemical Reactor Engineering: A Multi-Scale Approach, First Edition. Timur Doğu and Gülşen Doğu.

Companion website: www.wiley.com/go/dogu/chemreacengin

in the denominator of the rate law. In some other cases, multi-term denominators may appear in the rate expressions. Such observations can be explained by the mechanisms of the reactions, which are composed of several elementary steps. Some of such typical cases are illustrated in this section, considering the following overall reaction between reactants A and B taking place in a constant volume batch reactor.

$$\mathrm{A + B} \xrightarrow{k} \mathrm{C + D}$$

Case 1: If this reaction takes place as a result of an elementary step; that is, if the formation of C and D from A and B is through a single set of physico-chemical processes, the rate expression can be written as follows:

$$R_C = -R_A = kC_AC_B \tag{3.1}$$

Case 2: If the reaction mechanism is composed of the following two elementary steps: the first one being reversible and fast and the second one being rather slow, the rate expression can be derived as follows:

$$\mathrm{A + B} \overset{K_1}{\leftrightarrow} \mathrm{E} \quad ; \quad \text{fast, pre} - \text{equilibrium}$$

$$\mathrm{E} \xrightarrow{k_2} \mathrm{C + D} \quad ; \quad \text{slow, rate} - \text{determining step}$$

In this mechanism, E is an intermediate that is not stoichiometrically significant; that is, it does not appear in the overall stoichiometry of the reaction. The formation of intermediate E is illustrated in the energy-reaction coordinate diagram shown in Figure 3.1.

In this mechanism, the second step is the slowest, and it is considered as the rate-determining step (rds). Assuming that both forward and the reverse reactions of the first step are quite fast, this elementary step can be considered to be in equilibrium. The concentration of intermediate E can be expressed in terms of concentrations of A and B, using this pre-equilibrium assumption as follows:

$$K_1 = \frac{C_E}{C_AC_B} \Rightarrow C_E = K_1C_AC_B \tag{3.2}$$

The rate of formation of products can then be expressed from the rate-determining step as follows:

$$R_C = R_D = k_2C_E \tag{3.3}$$

Combining C_E expression given in Eq. (3.2) with Eq. (3.3), the following rate expression can be obtained for this reaction:

$$R_C = k_2K_1C_AC_B = k_{\text{obs}}C_AC_B \tag{3.4}$$

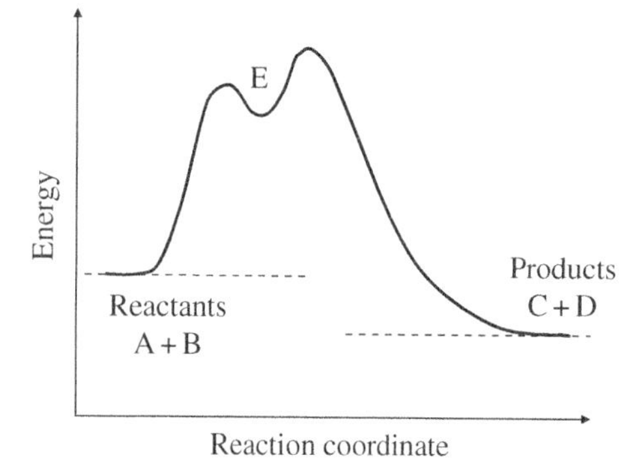

Figure 3.1 Schematic representation of the energy diagram for a reaction with an intermediate E.

The form of this rate expression is similar to the rate expression given for Case 1. However, in this case, the observed rate constant is a product of the rate constant of the rate-determining step and the equilibrium constant of the first step. Hence, the temperature dependence of the observed rate constant depends upon the activation energy of the rds, as well as the temperature dependence of the equilibrium constant of the first step.

Case 3: If one of the products of the overall reaction appears as a product of the first pre-equilibrium step of the mechanism discussed in Case 2, that product appears in the denominator of the rate law.

$$\text{A} + \text{B} \overset{K_1}{\leftrightarrow} \text{E} + \text{C} \quad ; \quad \text{fast, pre} - \text{equilibrium}$$

$$\text{E} \overset{k_2}{\rightarrow} \text{D} \quad ; \quad \text{slow, rate} - \text{determining step}$$

In this case, the concentration of intermediate E may be expressed from the pre-equilibrium relation of the first elementary step of the mechanism, as follows:

$$K_1 = \frac{C_E C_C}{C_A C_B} \Rightarrow C_E = K_1 \frac{C_A C_B}{C_C} \tag{3.5}$$

The rate of formation of products can again be expressed from the rate-determining step. The concentration of intermediate E, which is expressed in Eq. (3.5), can be inserted into this expression to obtain the rate law of the reaction.

$$R_D = k_2 C_E = k_2 K_1 \frac{C_A C_B}{C_C} \tag{3.6}$$

According to this rate expression, the formation of the product C inhibits the reaction rate.

Case 4: Another approach in obtaining the rate expression involves a pseudo-steady-state approximation for the intermediate E. If the forward and the reverse reactions of the first step of the mechanism are not fast, the pre-equilibrium assumption of this step may not be justified. Instead, the first step can be considered as a reversible reaction.

$$\text{A} + \text{B} \underset{k_{-1}}{\overset{k_1}{\rightleftarrows}} \text{E}$$

$$\text{E} \overset{k_2}{\rightarrow} \text{C} + \text{D}$$

In this case, the following relation can be written for the rate of change of concentration of intermediate E.

$$R_E = \frac{dC_E}{dt} = k_1 C_A C_B - k_{-1} C_E - k_2 C_E \cong 0 \tag{3.7}$$

According to the pseudo-steady-state approximation, it is considered that the order of magnitude of the rate of change of concentration of intermediate with time is much less than its formation and decomposition rates, which appear on the right-hand side of Eq. (3.7). This assumption is justified considering that the concentration of intermediate E is very low. Hence, an expression for the concentration of intermediate may be obtained using Eq. (3.7) as

$$C_E = \left(\frac{k_1}{k_{-1} + k_2}\right) C_A C_B \tag{3.8}$$

The rate of production of products can then be expressed as follows:

$$R_C = R_D = k_2 C_E = \left(\frac{k_2 k_1}{k_{-1} + k_2}\right) C_A C_B \tag{3.9}$$

In this case, both rate constants of the decomposition reactions of intermediate E appear in the denominator of the observed rate constant.

$$k_{obs} = \left(\frac{k_2 k_1}{k_{-1} + k_2}\right) \tag{3.10}$$

We may consider two limiting cases of this expression. If the reverse rate of the first reaction is much higher than the rate of the second elementary step ($k_{-1} \gg k_2$), the observed rate constant, which is given in Eq. (3.10), reduces to the expression obtained in Case 2, which was obtained with the pre-equilibrium assumption.

$$k_{obs} = \left(\frac{k_2 k_1}{k_{-1}}\right) = k_2 \frac{k_1}{k_{-1}} = k_2 K_1 \quad \text{if } k_{-1} \gg k_2 \tag{3.11}$$

However, if the rate of the second step of the reaction mechanism is much higher than the reverse rate of the first step, the following expression can be written for the observed rate constant of the reaction:

$$k_{obs} = k_1 \text{ if } k_2 \gg k_{-1} \tag{3.12}$$

This limiting case corresponds to the assumption that the first step of the mechanism is the slowest rate-determining step. The fast elementary steps, which appear after the slow rate-determining step, are not kinetically significant, and their rate constants do not appear in the overall rate expression.

Case 5: For some reactions, the rate expression may have a multi-term denominator containing product or reactant concentrations. A typical example is illustrated here. In this example, the first and the second steps of the mechanism are slow, and product C is produced through the first step of the mechanism.

$$A + B \underset{k_{-1}}{\overset{k_1}{\rightleftarrows}} E + C$$

$$E \xrightarrow{k_2} D$$

For this mechanism, the rate of change of concentration of intermediate E with respect to time in the batch reactor can be written using the pseudo-steady-state approximation as follows:

$$R_E = \frac{dC_E}{dt} = k_1 C_A C_B - k_{-1} C_E C_C - k_2 C_E \cong 0 \tag{3.13}$$

The concentration of intermediate E may then be expressed as

$$C_E = \left(\frac{k_1}{k_{-1} C_C + k_2}\right) C_A C_B \tag{3.14}$$

The rate of the reaction can then be expressed as follows:

$$R_D = k_2 C_E = \frac{k_1 k_2 C_A C_B}{k_{-1} C_C + k_2} \tag{3.15}$$

In this case, product C may inhibit the reaction, especially for its high concentrations.

Case 6: Another example of having multi-terms in the denominator of the rate expression is observed in enzymatic reactions. Enzymes are considered as catalysts for several biochemical reactions.

$$S \xrightarrow{e} \text{Products}$$

During such reactions, enzymes are combined with the substrate S, forming an intermediate. With the equilibrium assumption of this step, the following reaction mechanism can be written for such reactions.

$$S + e \overset{K_1}{\leftrightarrow} Se \qquad ; \text{ fast, pre}-\text{equilibrium}$$

$$Se \xrightarrow{k_2} \text{Products} \qquad ; \text{ slow, rate}-\text{determining}$$

The following expression can then be written for the concentration of the enzyme–substrate intermediate, using the equilibrium relation.

$$K_1 = \frac{C_{Se}}{C_S C_e} \Rightarrow C_{Se} = K_1 C_S C_e \tag{3.16}$$

An enzyme is a bio-catalyst and does not appear in the overall stoichiometry of the reaction, either as a reactant or as a product. Hence, the summation of the free enzyme and the combined enzyme concentrations can be taken as constant, which is equal to the initial value of the enzyme concentration in the reaction vessel.

$$C_{Se} + C_e = C_e^o \Rightarrow K_1 C_s C_e + C_e = C_e^o \tag{3.17}$$

The concentration of free enzyme in the reactor can then be expressed as follows:

$$C_e = \frac{C_e^o}{1 + K_1 C_S} \tag{3.18}$$

Then, the rate of formation of the enzymatic reaction products can be written from the second reaction (rds) of the mechanism.

$$R = k_2 C_{Se} = k_2 K_1 C_S C_e = \frac{k_2 K_1 C_e^o C_S}{1 + K_1 C_S} = \frac{k_{\text{obs}} C_S}{1 + K_1 C_S} \tag{3.19}$$

This is called the Michaelis–Menten type rate expression. For low concentrations of the substrate, this rate expression becomes first order with respect to the concentration of the substrate. However, it approaches zero order for high concentrations of S.

$$R = k_{\text{obs}} C_S \qquad \text{for} \quad K_1 C_S \ll 1 \tag{3.20}$$

$$R = \frac{k_{\text{obs}}}{K_1} = k_2 C_e^o \quad \text{for} \quad K_1 C_S \gg 1 \tag{3.21}$$

For this type of rate law, the variation of reaction rate with substrate concentration is illustrated in Figure 3.2.

Further discussions of reaction mechanisms may be found in Ref. [1].

3.2 Batch Reactor Data Analysis

A typical laboratory-scale batch reactor, which is placed into a thermostat to keep the reaction mixture temperature constant, is illustrated in Figure 3.3. Concentrations of reactants decrease as a function of reaction time in such a batch reactor. The time dependence of concentrations in a batch

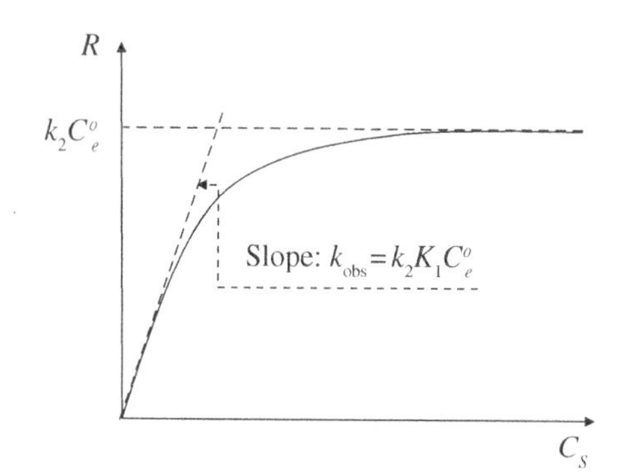

Figure 3.2 Michaelis–Menten type model for an enzyme-catalyzed reaction.

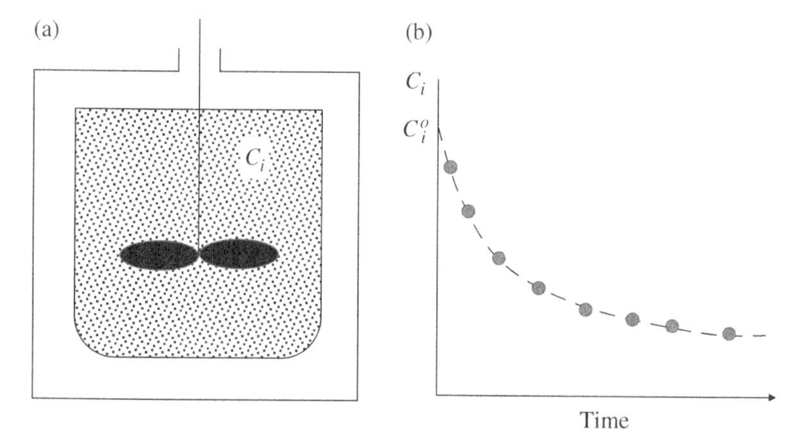

Figure 3.3 (a) Schematic representation of a batch reactor and (b) variation of concentration of a reactant as a function of reaction time.

reactor can be analyzed to obtain the reaction rate law. In the case of reactions performed in constant volume batch reactors, chemical analysis of the reaction mixture at different time intervals may be used to obtain the orders of reaction rate with respect to different species involved in the chemical conversion. In the case of gas-phase reactions with changes in the total number of moles resulting from the chemical conversion, we would expect changes in total pressure in a constant volume batch reactor. For such reactions, pressure variations within the reactor can be used to obtain reaction rate values.

As shown in Section 1.3, the conservation equation for species i in a batch reactor can be written as follows:

$$R_i = \frac{1}{V}\frac{dn_i}{dt} \tag{3.22}$$

For the constant volume case, this relation can be expressed in terms of the concentration of species i in the reactor.

$$R_i = \frac{dC_i}{dt} \tag{3.23}$$

Analysis of the data obtained for the variation of concentrations of reactants with reaction time can be performed either by an integral or a differential method. Some examples of integral and differential analysis of the rate data are illustrated in Sections 3.2.1 and 3.2.2, respectively.

3.2.1 Integral Method of Analysis

The integral method of analysis involves the integration of a postulated rate equation with the initial reactant concentrations at zero time as the initial condition. It is a convenient method, especially for the data obtained in a batch reactor. If more than one reactant species are involved in the reaction, it is sometimes decided to keep the concentrations of some of the reactants constant and simplify the problem by allowing variation of only one of the reactants as a function of time. For instance, concentrations of some of the reactants may be kept in significant excess compared to another reactant, which is called the limiting reactant. Hence, the concentrations of the reactants being in significant excess can be assumed as constant during the reaction.

3.2.1.1 First-Order Reaction

Let us consider an arbitrary irreversible reaction between reactants A and B.

$$aA + bB \rightleftharpoons cC + dD$$

If the rate expression of this reaction depends only upon the concentration of reactant A and if it is a first-order reaction, the following relation can be written for the rate of this reaction:

$$-R_A = kC_A$$

Variation of concentration of reactant A as a function of reaction time can be predicted in a constant volume batch reactor by the integration of the following expression:

$$-kC_A = \frac{dC_A}{dt} \tag{3.24}$$

$$-\int_{C_A^o}^{C_A} \frac{dC_A}{C_A} = \int_0^t kdt$$

$$-\ln\left(\frac{C_A}{C_A^o}\right) = -\ln(1-x_A) = kt \tag{3.25}$$

Analysis of the batch reactor data can then be performed using Eq. (3.25). A plot of $-\ln(1-x_A)$ as a function of reaction time should give a linear relation passing from the origin, for a first-order reaction. The slope of this linear relation gives the reaction rate constant (Figure 3.4). If the plot of $-\ln(1-x_A)$ with respect to reaction time does not give a linear relation, this implies that the first-order reaction assumption is not correct.

The half-life of the reactant A is defined as the time required to reach a fractional conversion value of 0.5 (50% conversion). For a first-order reaction, the half-life expression can then be obtained as

$$t_{1/2} = \frac{\ln 2}{k} \tag{3.26}$$

Note that, for a first-order reaction, the half-life is not dependent on the initial concentration of the reactant. However, as shown in this chapter, the half-life is a function of the initial concentration of the reactant for all other positive-order reactions.

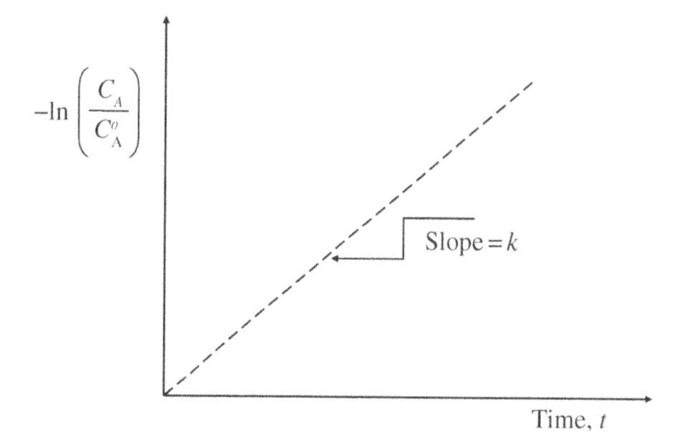

Figure 3.4 Schematic representation of the relationship between concentration and reaction time for a first-order reaction in a batch reactor.

Example 3.1 Analysis of Kinetic Data by the Integral Method
Data reported in Table 3.1 are obtained for the following reaction in a batch reactor.

$$A + B \rightarrow C + D$$

The initial concentration of reactant B is in excess in the reactor. Show that the rate of this reaction is first order with respect to the concentration of A, and evaluate the reaction rate constant.

Solution

Plot of $-\ln\left(\frac{C_A}{C_A^o}\right)$ as a function of reaction time is shown in Figure 3.5. The linear behavior of this relationship shows that the rate of this reaction is first order with respect to the concentration of A. The slope of this relation can be found by the linear least-squares regression analysis of the given data. This analysis gives the reaction rate constant as $k = 0.102\ \text{min}^{-1}$.

Details of the least-squares data analysis procedure are described in the Online Appendix O.5.1. In the case of a linear relation between the dependent variable y and independent variable x

$$y = mx + n$$

the slope and the intercept of this expression can be estimated using the experimental data as follows:

$$m = \frac{N\left(\sum_1^N (x_i y_i)\right) - \left(\sum_1^N x_i\right)\left(\sum_1^N y_i\right)}{N\left(\sum_1^N x_i^2\right) - \left(\sum_1^N x_i\right)^2}$$

$$n = \frac{\sum_1^N (y_i) - m\sum_1^N (x_i)}{N}$$

Table 3.1 Data reported for the conversion of A to C and D.

Time (min)	0	3	6	9	12	15	18
Concentration of A C_A (mol/l)	1.00	0.740	0.549	0.406	0.301	0.225	0.165

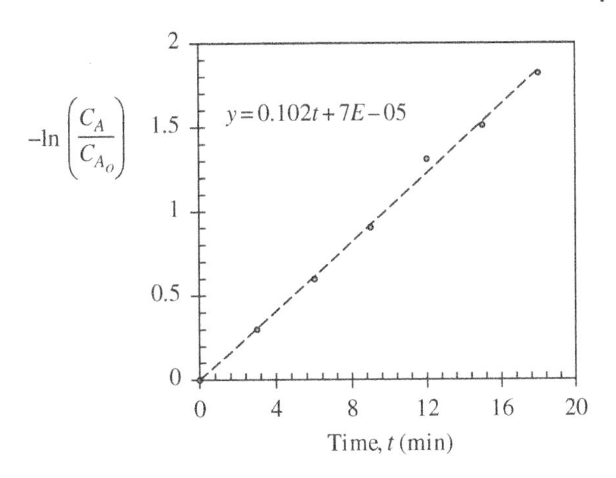

Figure 3.5 Batch reactor data analysis for Example 3.1.

Here, x_i, y_i are the experimental values of the dependent and the independent variables at the experimental point i, and N is the total number of data points.

3.2.1.2 *n*th-Order Reaction and Method of Half-Lives

Power-law type rate expressions are quite common for chemical reactions. Let us consider an nth-order reaction with respect to the concentration of reactant A.

$$-R_A = kC_A^n \tag{3.27}$$

Here, the reaction order n can be an integer or a fraction. For several catalytic reactions, fractional reaction orders are quite common. For a constant volume batch reactor, the variation of the concentration of the reactant A with time can be predicted by the integration of the species conservation expression, as follows:

$$-\frac{dC_A}{dt} = kC_A^n \tag{3.28}$$

$$-\int_{C_A^o}^{C_A} \frac{dC_A}{C_A^n} = \int_0^t kdt \tag{3.29}$$

$$\frac{1}{(C_A)^{n-1}} - \frac{1}{(C_A^o)^{n-1}} = (n-1)kt \tag{3.30}$$

Half-life expression for reactant A can then be obtained by substituting $C_A = C_A^o/2$ in Eq. (3.30).

$$t_{1/2} = \frac{(2^{(n-1)} - 1)}{(n-1)k(C_A^o)^{n-1}} \tag{3.31}$$

Note that the half-life of an nth-order reaction is inversely proportional to the $(n-1)$st power of the initial concentration of reactant A. If we take the logarithm of both sides of Eq. (3.31), we obtain the following linear relation between $\ln(t_{1/2})$ and $\ln(C_A^o)$:

$$\ln(t_{1/2}) = \ln\left(\frac{2^{(n-1)} - 1}{(n-1)k}\right) - (n-1)\ln C_A^o \tag{3.32}$$

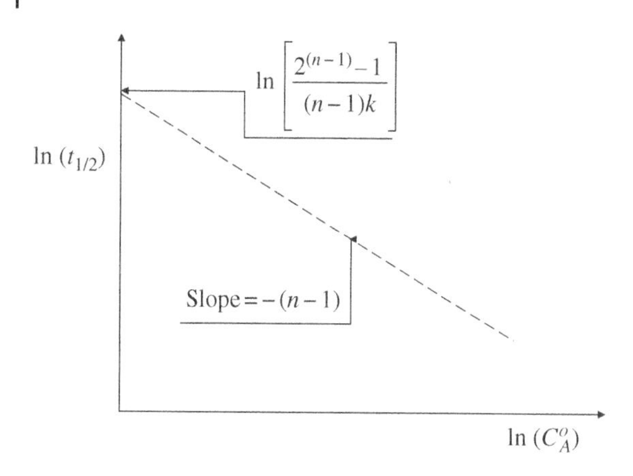

Figure 3.6 Schematic representation of half-life data analysis to determine the rate law.

If we perform a set of experiments with different initial concentrations of reactant A and determine the half-life times at each experiment, reaction order and the rate constant can be determined from the linear relation between $\ln(t_{1/2})$ and $\ln\left(C_A^o\right)$, as illustrated in Figure 3.6.

For a second-order reaction ($n = 2$), Eq. (3.30) reduces to

$$\frac{1}{(C_A)} - \frac{1}{(C_A^o)} = \frac{1}{C_A^o}\left(\frac{x_A}{1-x_A}\right) = kt \tag{3.33}$$

In this case, the plot of $\left(\frac{x_A}{1-x_A}\right)$ versus reaction time should give a linear relation with a slope of kC_A^o, as illustrated in Figure 3.7.

3.2.1.3 Overall Second-Order Reaction Between Reactants A and B

Let us consider a reaction between reactants A and B, with a rate expression being first order with respect to A and first order with respect to B. The overall order of this reaction is second order.

$$a\text{A} + b\text{B} \rightleftharpoons c\text{C} + d\text{D}$$

$$-R_A = kC_AC_B = -\frac{dC_A}{dt} \tag{3.34}$$

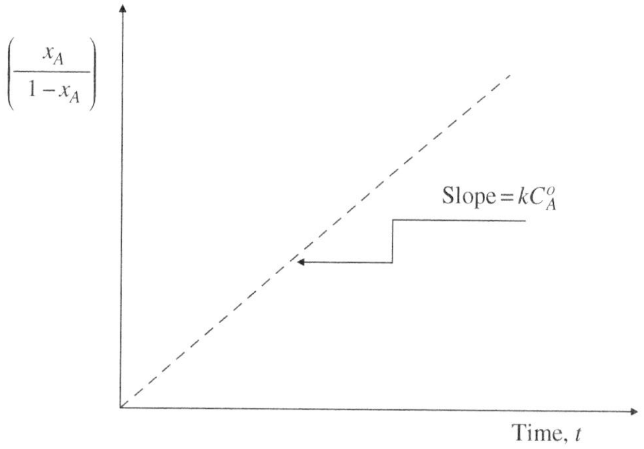

$\left(\frac{x_A}{1-x_A}\right)$

Slope $= kC_A^o$

Time, t

Figure 3.7 Schematic representation of time dependence of fractional conversion for a second-order reaction in a batch reactor.

For a constant volume problem, the concentrations of reactants A and B can be expressed in terms of fractional conversion of the limiting reactant A. Equation (3.34) can be expressed in terms of fractional conversion as

$$C_A^o \frac{dx_A}{dt} = = kC_A^o(1-x_A)\left(C_B^o - \frac{b}{a}C_A^o x_A\right) \tag{3.35}$$

$$\int_0^{x_A} \frac{dx_A}{(1-x_A)\left(C_B^o - \frac{b}{a}C_A^o x_A\right)} = \int_0^t k dt \tag{3.36}$$

The integration of Eq. (3.36) for the isothermal operation of a constant volume batch reactor can be performed using the method of partial fractions.

$$\ln\left(\frac{C_B^o - \frac{b}{a}C_A^o x_A}{C_B^o(1-x_A)}\right) = \left(C_B^o - \frac{b}{a}C_A^o\right)kt \tag{3.37}$$

Note that this expression holds with the condition of $\left(\frac{bC_A^o}{aC_B^o} \neq 1\right)$.

This analysis indicates that the relation between $\ln\left(\frac{C_B^o - \frac{b}{a}C_A^o x_A}{C_B^o(1-x_A)}\right)$ and reaction time should be linear, with a slope of $\left(C_B^o - \frac{b}{a}C_A^o\right)k$, for a reaction which is first order with respect to the concentration of A and first order with respect to the concentration of reactant B (overall order being two). The slope of this relation can then be used for the evaluation of the reaction rate constant.

Example 3.2 Data Analysis for a Second Order Reaction

Two sets of data are reported for the variation of concentration of reactant A in a constant volume batch reactor as a function of reaction time for the following reaction:

$$2A + B \rightarrow C + D$$

It is also given that the initial concentration of reactant B is 0.1 mol/l in both experiments (see Tables 3.2 and 3.3).

What is the rate law for this reaction? Evaluate the rate constant from both sets of data and compare.

Solution

It is assumed that the overall order of this reaction is 2 (first order with respect to A and first order with respect to B).

$$-R_A = -\frac{dC_A}{dt} = kC_A C_B = kC_A^o(1-x_A)\left(C_B^o - \frac{1}{2}C_A^o x_A\right)$$

Analysis for the First Set of Data

In this set of data, the initial concentration of A is given as twice the initial concentration of reactant B $\left(C_A^o = 2C_B^o\right)$. For this case, the rate expression becomes as follows:

$$-R_A = C_A^o \frac{dx_A}{dt} = kC_A^o(1-x_A)\left(\frac{1}{2}C_A^o - \frac{1}{2}C_A^o x_A\right) = k\frac{\left(C_A^o\right)^2}{2}(1-x_A)^2$$

Table 3.2 Data for Experiment 1.

Time (min)	0	1	3	5	10	20	30
C_A (mol/l)	0.200	0.178	0.146	0.126	0.090	0.061	0.043
x_A	0	0.11	0.27	0.37	0.55	0.69	0.79

For this type of second-order reaction, integration of the rate expression gives the relation between fractional conversion and reaction time as

$$\frac{1}{C_A^o}\left(\frac{x_A}{1-x_A}\right) = \frac{k}{2}t$$

Fractional conversion values of reactant A are calculated using the data given in Experiment 1 and listed in Table 3.2. The reaction rate constant is then evaluated from the slope of $\frac{2}{C_A^o}\left(\frac{x_A}{1-x_A}\right)$ vs. reaction time line (Figure 3.8) as 1.15 $(\text{mol/l})^{-1}\ \text{min}^{-1}$. The linear least square regression method is used to evaluate the slope.

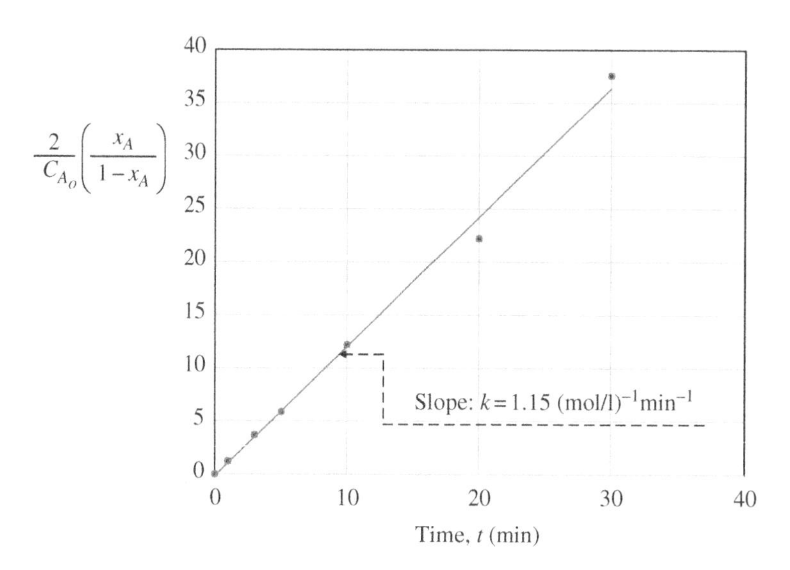

Figure 3.8 Analysis of the first set of data to evaluate the reaction rate constant.

Analysis of the Second Set of Data

In this set of data, the initial concentration of reactant A is much less than the initial concentration of reactant B. Analysis of the data can then be performed following two different approaches.

Approach 1: For this bimolecular reaction, the integral expression given in Eq. 3.37 can be used for the evaluation of the rate constant.

$$\ln\left(\frac{C_B^o - \frac{b}{a}C_A^o x_A}{C_B^o(1-x_A)}\right) = \left(C_B^o - \frac{b}{a}C_A^o\right)kt$$

Table 3.3 Data for Experiment 2.

Time (min)	0	1	3	5	10	20	30
C_A (mol/l)	0.00100	0.00088	0.00076	0.00055	0.00031	0.00010	0.00004
x_A	0	0.12	0.24	0.45	0.69	0.9	0.96

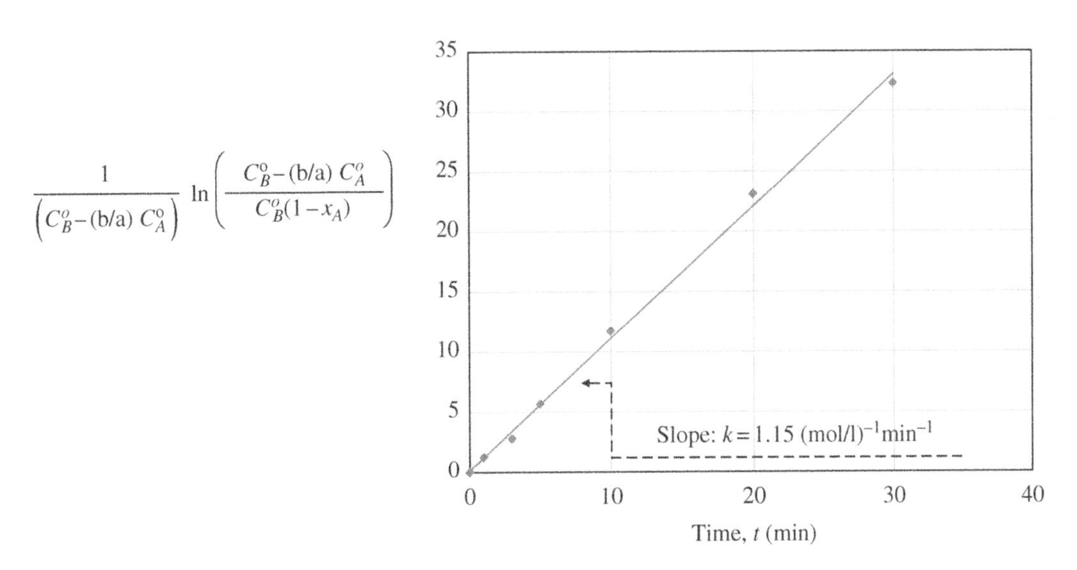

Figure 3.9 Analysis of the second set of data to evaluate the reaction rate constant.

According to this expression, the slope of the $\frac{1}{\left(C_B^o - \frac{b}{a}C_A^o\right)}\ln\left(\frac{C_B^o - \frac{b}{a}C_A^o x_A}{C_B^o(1-x_A)}\right)$ vs. reaction time line gives the rate constant as 1.15 $(\text{mol/l})^{-1}$ min^{-1} (Figure 3.9).

Approach 2: Since the initial concentration of reactant B is 100 times higher than the initial concentration of reactant A in the second set of experimental data, the concentration of B can be assumed as constant, and the reaction can be analyzed as a pseudo-first-order reaction.

$$-R_A = -\frac{dC_A}{dt} = kC_AC_B \cong \left(kC_B^o\right)C_A$$

Hence, the relation, which is given in Eq. (3.25), can be used for the evaluation of the rate constant.

$$-\ln\left(\frac{C_A}{C_A^o}\right) = -\ln(1-x_A) = \left(kC_B^o\right)t$$

According to this expression, the slope of the $-\ln(1-x_A)$ vs. reaction time relation gives the reaction rate constant as $k = 1.1\ (\text{mol/l})^{-1}\ \text{min}^{-1}$ (Figure 3.10). This value is quite close to the value evaluated using the first set of data.

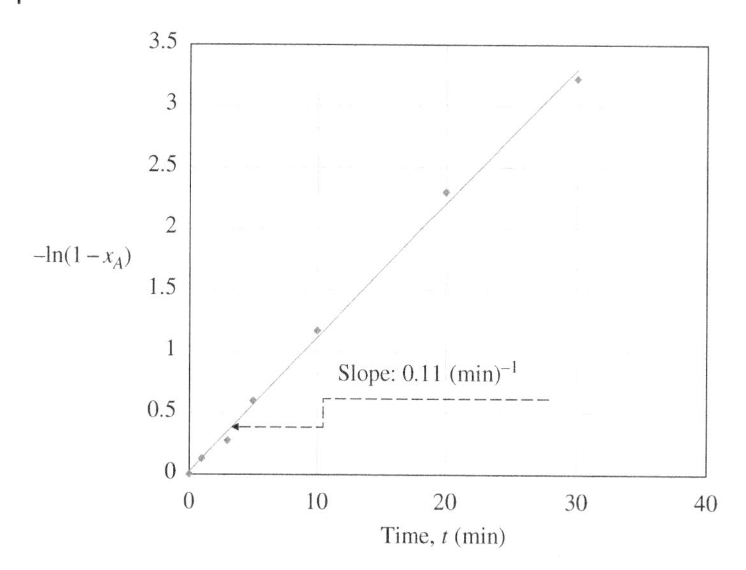

Figure 3.10 Analysis of the second set of data assuming pseudo-first-order rate law.

3.2.1.4 Second-Order Autocatalytic Reactions

In some chemical and biochemical reactions, the product may act as a catalyst in the reaction and increases the reaction rate. These reactions are called autocatalytic reactions. The overall stoichiometry of an autocatalytic reaction can be written as follows:

$$A + C \rightarrow 2C$$

In this reaction, the product C catalyzes the reaction. However, to proceed according to the given stoichiometry, product C should also be present in the reaction vessel with an initial concentration of C_C^o.

An autocatalytic reaction may also be initiated by the conversion (decomposition) of reactant A to product C as the first step of the reaction mechanism. Then product C may further react with reactant A to give the products. This reaction sequence may be expressed in terms of the following elementary reactions:

$$A \rightarrow C\,; \qquad R_{(1)} = k_1 C_A$$

$$A + C \rightarrow 2C + \cdots; \; R_{(2)} = k_2 C_A C_C$$

Here, $R_{(1)}$ and $R_{(2)}$ are the specific rates of the two reactions given above.

For this reaction system taking place in a constant volume batch reactor, the rate of decomposition of reactant A should be expressed as follows:

$$-R_A = -\frac{dC_A}{dt} = k_1 C_A + k_2 C_A C_C = k_2 C_A \left(\frac{k_1}{k_2} + C_C\right) \tag{3.38}$$

Concentrations of reactant A and the product C can be expressed in terms of fractional conversion of A as

$$C_A = C_A^o(1 - x_A) \tag{3.39}$$

$$C_C = C_C^o + C_A^o x_A \tag{3.40}$$

The following relation can also be written between the concentrations of A and C at any reaction time.

$$C_C = C_C^o + \left(C_A^o - C_A\right) \tag{3.41}$$

Then, Eq. (3.38) becomes

$$-\frac{dC_A}{dt} = k_2 C_A \left(\frac{k_1}{k_2} + \left(C_C^o + C_A^o - C_A\right)\right) \tag{3.42}$$

The integration of this equation gives a relation for the time dependence of concentrations of product C and reactant A.

$$\ln\left\{\frac{\left(\frac{k_1}{k_2} + C_C\right)C_A^o}{\left(\frac{k_1}{k_2} + C_C^o\right)C_A}\right\} = k_2\left(\frac{k_1}{k_2} + C_C^o + C_A^o\right)t \tag{3.43}$$

In many cases, the rate of the autocatalytic reaction ($R_{(2)}$) is much higher than the rate of the initiation step ($R_{(1)}$). For this case, k_1/k_2 may be taken to approach zero, and Eq. (3.43) can be simplified to the following form:

$$\ln\left(\frac{C_C}{C_A}\right) = \left(C_C^o + C_A^o\right)k_2 t + \ln\left(\frac{C_C^o}{C_A^o}\right) \tag{3.44}$$

According to this relation, a plot between the logarithm of the ratio of concentrations of C and A, and time should give a linear relation with a slope of $\left(C_C^o + C_A^o\right)k_2$ (Figure 3.11). Equation (3.44) can also be expressed in terms of fractional conversion of reactant A, as follows:

$$\ln\left(\frac{\frac{C_C^o}{C_A^o} + x_A}{1 - x_A}\right) = \left(C_C^o + C_A^o\right)k_2 t + \ln\left(\frac{C_C^o}{C_A^o}\right) \tag{3.45}$$

The rate expression for an autocatalytic reaction can also be expressed in terms of the fractional conversion of reactant A as follows:

$$\begin{aligned} -R_A &= k_1 C_A^o(1 - x_A) + k_2 C_A^o(1 - x_A)\left(C_C^o + C_A^o x_A\right) \\ &= k_1 C_A^o(1 - x_A)\left[1 + \frac{k_2}{k_1}\left(C_C^o + C_A^o x_A\right)\right] \end{aligned} \tag{3.46}$$

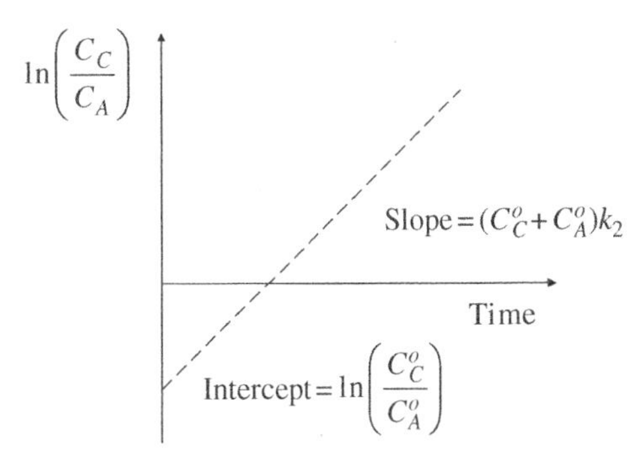

Figure 3.11 The relation between the ratio of the product and reactant concentrations and reaction time for an autocatalytic reaction.

The reaction rate expression given by Eq. (3.46) passes through a maximum for a particular value of fractional conversion of A (x_A). The maximum value of the reaction rate and the corresponding fractional conversion can be found by equating the derivative of $(-R_A)$ with respect to x_A, to zero.

$$\frac{d(-R_A)}{dx_A} = -k_1C_A^o + k_2C_A^o\left[-\left(C_C^o + C_A^o x_A\right) + C_A^o(1-x_A)\right] = 0 \tag{3.47}$$

$$x_{A_{\max}} = \frac{\left(-\frac{k_1}{k_2}\right) + \left(C_A^o - C_C^o\right)}{2C_A^o} \tag{3.48}$$

Here, $x_{A_{\max}}$ is the value of x_A when the rate is maximum.

For the case of having $\frac{k_1}{k_2} \cong 0$ and $C_C^o \ll C_A^o$, $x_{A_{max}}$ approaches 0.5.

3.2.1.5 Zeroth-Order Dependence of Reaction Rate on Concentrations

If the rate of a chemical conversion is not dependent on the concentrations of the species involved in the reaction, this reaction is called a zero-order reaction. Zero-order dependence of the reaction rate on the concentration of one of the reacting species may be observed at specific concentration ranges. In many of the reactions catalyzed by solid catalysts, adsorption of reactants on the catalyst surface and the surface reaction steps are involved, yielding Langmuir–Hinshelwood or Rideal–Eley type rate expressions having terms in the denominator of the reaction rate expression (Chapter 11). Let us consider a simple decomposition reaction of reactant A taking place on the surface of a solid catalyst, with the following rate expression:

$$-R_A = k\left(\frac{K_AC_A}{1 + K_AC_A}\right) \tag{3.49}$$

Here, K_A may correspond to the adsorption equilibrium constant of A on the catalyst surface. The limiting forms of this rate expression for very low and very high values of C_A approach first-order and zero-order reaction rate laws, respectively.

$$-R_A = kK_AC_A = k'C_A \quad \text{for} \quad K_AC_A \ll 1.0$$

$$-R_A = k \quad \text{for} \quad K_AC_A \gg 1.0$$

For a zero-order reaction, the species conservation equation for the reactant A in a constant volume batch reactor can be written as

$$-R_A = -\frac{dC_A}{dt} = k \tag{3.50}$$

Integration of this expression yields

$$C_A^o - C_A = C_A^o x_A = kt \tag{3.51}$$

Hence, we expect a linear relationship between the fractional conversion of reactant A and reaction time for a zero-order reaction.

$$x_A = \frac{k}{C_A^o}t \tag{3.52}$$

3.2.1.6 Data Analysis for a Reversible Reaction

Many of the chemical reactions are reversible. Equilibrium limitations in reversible reactions are discussed in detail in Chapter 2. Let us illustrate the analysis of a batch reactor for a first-order reversible reaction.

$$A \underset{k_b}{\overset{k_f}{\rightleftarrows}} B\,; \; -R_A = k_f C_A - k_b C_B = k_f\left(C_A - \frac{1}{K_C} C_B\right) \tag{3.53}$$

Here, k_f and k_b are the forward and the reverse reaction rate constants, and K_C is the equilibrium constant in terms of concentrations.

$$K_C = \frac{k_f}{k_b} \tag{3.54}$$

By expressing C_A and C_B in terms of fractional conversion as

$$C_A = C_A^o(1 - x_A)\,; \; C_B = C_B^o + C_A^o x_A$$

the species conservation equation for reactant A in a constant volume batch reactor can be expressed as follows:

$$-R_A = -\frac{dC_A}{dt} = C_A^o \frac{dx_A}{dt} = k_f\left(C_A^o(1 - x_A) - \frac{1}{K_C}\left(C_B^o + C_A^o x_A\right)\right) \tag{3.55}$$

As discussed in Chapter 2, forward and the reverse reaction rates become the same when equilibrium is reached. Hence, the net rate of conversion of A to B becomes zero at equilibrium.

$$-R_A = k_f C_{A_e} - k_b C_{B_e} = k_f\left(C_{A_e} - \frac{1}{K_C} C_{B_e}\right) = 0 \quad \text{(at equilibrium)}$$

Here, C_{A_e} and C_{B_e} are the concentrations of A and B at equilibrium. The equilibrium constant of a first-order reversible reaction can be expressed in terms of concentrations of species and the equilibrium conversion of reactant A to B as follows:

$$K_C = \frac{C_{B_e}}{C_{A_e}} = \frac{C_B^o + C_A^o x_{A_e}}{C_A^o(1 - x_{A_e})} \tag{3.56}$$

For the case of $C_B^o = 0$, equilibrium relations become

$$K_C = \frac{C_{B_e}}{C_{A_e}} = \frac{x_{A_e}}{(1 - x_{A_e})}\,; \; x_{A_e} = \frac{K_C}{1 + K_C} \tag{3.57}$$

Integration of Eq. (3.55) for the case of $C_B^o = 0$ yields the following expression for the time dependence of fractional conversion of A:

$$-\ln\left[1 - \frac{(1 + K_C)}{K_C} x_A\right] = \frac{(1 + K_C)}{K_C} k_f t \tag{3.58}$$

The forward reaction rate constant and the equilibrium constant can be evaluated by the regression analysis of the conversion-time data using Eq. (3.58). For very long reaction times, $(t \to \infty)$ it is expected to reach equilibrium. Hence, for a first-order reversible reaction

$$\lim_{t \to \infty} x_A = x_{A_e} = \frac{K_C}{1 + K_C} \tag{3.59}$$

3.2.2 Differential Method of Data Analysis

To perform a differential method for the analysis of the experimental data obtained in a batch reactor, point rate values should be evaluated at different concentrations. Then, these rate values can be correlated to the corresponding concentrations of the species for the evaluation of the reaction orders, using a regression procedure.

One possible procedure to evaluate the reaction rate values at different concentrations involves fitting the experimental concentration versus time data to a suitable expression (such as a polynomial expression) and evaluating the derivatives of this expression at different concentration values. Derivatives of concentration expression of the limiting reactant give us the reaction rate values at the corresponding concentrations. Another procedure to evaluate the reaction rate values involves determining the slopes of the experimental concentration versus time curve at different concentrations, using a numerical differentiation procedure, as described in Online Appendix O.5. Differential method of data analysis is illustrated in Example 3.3.

Example 3.3 Data Analysis Using Differential Method

Let us consider the data reported in Table 3.1, which are obtained in a constant volume batch reactor for the following reaction:

$$A + B \rightarrow C + D$$

The initial concentration of reactant B is given as much higher than the concentration of A in the reactor. Hence, the rate depends only on the concentration of reactant A. Find the rate law using the differential method of data analysis.

Solution

a) A polynomial fit expression of the data is obtained and illustrated in Figure 3.12.

The derivative of the polynomial expression obtained in Figure 3.12 gives an expression for the time dependence of the reaction rate of species A. Hence, the reaction rate values are evaluated

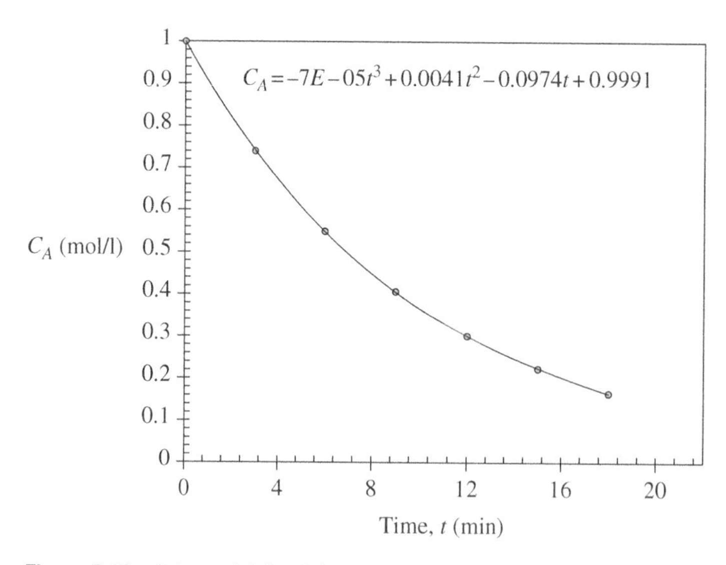

Figure 3.12 Polynomial fit of the data given in Example 3.1.

Table 3.4 Data and the rate values for the conversion of A to C and D in Example 3.3.

Time (min)	0	3	6	9	12	15	18
Concentration of A C_A (mol/l)	1.00	0.740	0.549	0.406	0.301	0.225	0.165
$-\frac{dC_A}{dt}$ (mol/l min)	0.0974	0.0750	0.0560	0.0406	0.0290	0.0210	0.0178
$\ln(C_A)$	0	−0.301	−0.599	−0.901	−1.20	−1.50	−1.8
$\ln(-R_A)$	−2.328	−2.59	−2.88	−3.20	−3.54	−3.86	−4.03

from the derivatives (slopes) of this polynomial expression at different concentrations. These values are listed in Table 3.4.

Let us assume an nth-order rate expression for this reaction.

$$-R_A = kC_A^n$$

By taking the logarithm of both sides, we obtain the following linearized relation between the logarithms of the reaction rate and the concentration of reactant A.

$$\ln(-R_A) = \ln(k) + n\ln(C_A) \tag{3.60}$$

The values of $\ln(-R_A)$ and $\ln(C_A)$ are also listed in Table 3.4. The slope of the linear relation given in Figure 3.13 gives the reaction order. The rate constant is then evaluated from the intercept of this linear relation. As shown in this figure, the order of the reaction is close to one, and the rate constant is close to 0.1. The rate expression can then be expressed as

$$-R_A = 0.099C_A$$

This result is consistent with the result obtained by the integral analysis technique in Example 3.1.

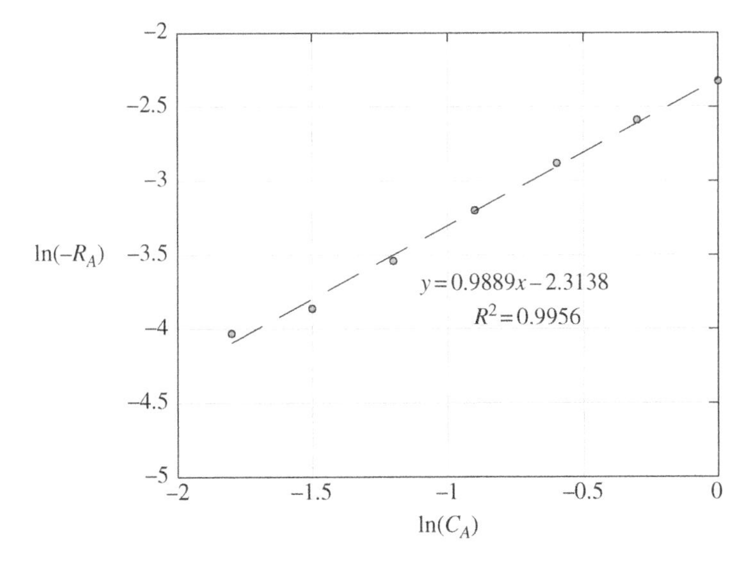

Figure 3.13 Analysis of the rate law for Example 3.3.

The second way of differential analysis involves the numerical differentiation of the given data. The reaction rate values may be evaluated at different concentrations, using the numerical differentiation procedure outlined in Online Appendix O.5.4.

3.3 Changes in Total Pressure or Volume in Gas-Phase Reactions

According to the ideal gas law, the product of total pressure and volume of the reaction mixture is proportional to the total number of moles within the reaction vessel in gas-phase reactions.

$$PV = n_T R_g T \tag{3.61}$$

For isothermal reactions with changes in the total number of moles resulting from the chemical conversion, we should observe changes in the pressure if the reaction vessel volume is kept constant. Alternatively, we should observe changes in the reaction volume of the batch system if we keep the pressure constant. Two typical gas-phase reaction examples with changes in the total number of moles are ammonia synthesis and nitrogen pentoxide decomposition.

$$N_2 + 3H_2 \rightleftarrows 2NH_3$$

$$2N_2O_5 \rightarrow 4NO_2 + O_2$$

In the case of an isothermal batch reactor with constant volume, total pressure at any fractional conversion of the limiting reactant can be expressed as follows:

$$P = P^o + (\Delta P)_{max} x_A = P^o\left(1 + \frac{(\Delta P)_{max}}{P^o} x_A\right) \tag{3.62}$$

Here, the maximum change in total pressure in the reactor is proportional to the maximum change in the total number of moles, which corresponds to the change in the total number of moles at the complete conversion of the limiting reactant at a constant temperature.

$$\varepsilon = \frac{(\Delta P)_{max}}{P^o} = \frac{(n_T)_{x_A=1} - (n_T)_{x_A=0}}{(n_T)_{x_A=0}} \tag{3.63}$$

Here, the expansion factor ε can be evaluated from the stoichiometry of the reaction, knowing the initial conditions. Hence, for an isothermal constant volume batch reactor, the following expression can be used to predict the change of pressure with fractional conversion x_A:

$$P = P^o(1 + \varepsilon x_A) \tag{3.64}$$

If the temperature of the reactor is not constant, the relation between the pressure and the fractional conversion should be expressed as follows:

$$P = P^o(1 + \varepsilon x_A)\frac{T}{T_o} \tag{3.65}$$

Alternatively, suppose the pressure of the batch reaction vessel is constant. In that case, changes in the total volume of the reaction mixture due to changes in the total number of moles in the reactor can be predicted using the following relation for an isothermal system. In the case of a variable volume batch reactor, the following expression relates total volume to fractional conversion.

$$V = V^o(1 + \varepsilon x_A) \tag{3.66}$$

To illustrate the effects of change in the total volume of the reaction mixture on the concentration of the species in the reactor, let us consider an nth-order reaction.

$$-R_A = kC_A^n = k\left(\frac{n_A}{V}\right)^n = k\left(\frac{n_A^o(1-x_A)}{V^o(1+\varepsilon x_A)}\right)^n = k\left[C_A^o\frac{(1-x_A)}{(1+\varepsilon x_A)}\right]^n \tag{3.67}$$

As discussed in Chapter 1, the species conservation equation in a well-mixed batch reactor is as follows:

$$R_A V = \frac{dn_A}{dt} \tag{3.68}$$

For an nth-order reaction, Eq. (3.68) can be expressed as

$$R_A = -kC_A^n = -k\left(\frac{n_A}{V}\right)^n = \frac{1}{V}\frac{dn_A}{dt} \tag{3.69}$$

This equation can be expressed in terms of fractional conversion of reactant as follows:

$$k\frac{(n_A^o(1-x_A))^n}{[V^o(1+\varepsilon x_A)]^{n-1}} = n_A^o\frac{dx_A}{dt} \tag{3.70}$$

The integration of this expression gives a relation for the time dependence of fractional conversion of limiting reactant A.

In the case of a simple first-order reaction, Eq. (3.69) reduces to

$$-kn_A = \frac{dn_A}{dt} \tag{3.71}$$

Integration of Eq. 3.71 gives

$$-\ln\left(\frac{n_A}{n_A^o}\right) = -\ln(1-x_A) = kt \tag{3.72}$$

Note that this expression is the same as the corresponding equation derived for the constant volume batch reactor. However, this is true only for a first-order reaction. For higher reaction orders, integration of Eq. (3.70) is needed. For instance, for a second-order reaction, ($n = 2$) the relation between the fractional conversion and reaction time can be obtained by integrating Eq. (3.70).

$$\frac{x_A(1+\varepsilon)}{(1-x_A)} + \varepsilon\ln(1-x_A) = C_A^o kt \tag{3.73}$$

In evaluating the expansion factor in gas-phase reactions, we should consider the total number of moles in the reactor, including inerts. This is illustrated in Example 3.4.

Example 3.4 Pressure Change During Reaction in a Batch Reactor

Isothermal decomposition of gaseous reactant A is investigated in a constant volume batch reactor, which initially contains 60% A and 40% inerts. Evaluate the expansion factor and the time required to increase the pressure from 1 to 1.45 bar in the reactor. Note that decomposition of A is first order with respect to its concentration, and the rate constant of the first-order reaction is given as $k = 4.7 \times 10^{-4}\ \text{s}^{-1}$.

$$2A \rightarrow 4B + C$$

Solution

Number of moles of each species in the reactor at a conversion level of x_A is

$$n_A : n_A^o(1-x_A)$$

$$n_B : 2n_A^o x_A$$

$$n_C : \frac{1}{2}n_A^o x_A$$

Inerts : n_i

The total number of moles : $n_T = \left(n_A^o + n_i\right) + \frac{3}{2}n_A^o x_A$

When the complete conversion of reactant A is reached, the total number of moles in the reactor becomes

$$n_T = \left(n_A^o + n_i\right) + \frac{3}{2}n_A^o$$

Hence, the expansion factor for this example becomes as follows:

$$\varepsilon = \frac{(n_T)_{x_A=1} - (n_T)_{x_A=0}}{(n_T)_{x_A=0}} = \frac{\left(n_A^o + n_i\right) + \frac{3}{2}n_A^o - \left(n_A^o + n_i\right)}{\left(n_A^o + n_i\right)} = \frac{3}{2}\frac{n_A^o}{\left(n_A^o + n_i\right)}$$

Since the initial mole fraction of reactant A in the reactor is given as 0.6, the expansion factor becomes 0.9. If there were no inerts in the reactor, the expansion factor would be 3/2.

Fractional conversion of gaseous reactant A, when the pressure in the reactor reached to 1.45 bar, can be evaluated from

$$\frac{P}{P^o} = (1 + \varepsilon x_A)\ ;\ 1.45 = (1 + 0.9x_A)\ \text{as}\ x_A = 0.5$$

For a first-order reaction, the relation between the fractional conversion and reaction time is given as

$$-\ln(1-x_A) = kt$$

The value of kt is then evaluated as 0.693. Knowing the rate constant ($k = 4.7 \times 10^{-4}\ s^{-1}$), the time required to increase the pressure to 1.45 bar is found as 24.6 min.

For supplementary reading, chemical reaction engineering and chemical kinetics books may be referred [1–3].

Problems and Questions

3.1 The following data were reported for the liquid phase reaction $2A \rightarrow B$. Show that the reaction is first order and evaluate the rate constant.

Time (min)	0	11	21	32	42
C_A (M)	0.62	0.45	0.32	0.23	0.17

3.2 The decomposition of reactant A to the products takes place in a liquid phase batch reactor. The initial concentration of A is 2 M. It was found out that the concentration of A was 0.8 M after a reaction period of 10 min. The reaction is given as first order with respect to A. Calculate the required reaction time for reactant concentration to decrease to 0.4 M.

3.3 The gas-phase decomposition reaction of a hydrocarbon species takes place in a constant volume batch reactor at 500 K. Initially, there is only reactant A in the reactor.

$$A \rightarrow B + C$$

Variation of the total pressure of the reactor was measured as a function of reaction time, and the following data were obtained:

Time (s)	**0**	**180**	**415**	**720**	**1095**
Total pressure (kPa)	15.9	16.4	17.0	18.4	19.1

Show that the reaction is first order and obtain the rate constant.

3.4 Treatment of a hydrocarbon (A) from an industrial operation is investigated in a liquid-phase batch reactor having a volume of 1 m^3. In this reactor, the hydrocarbon is treated with water. For safety considerations, decomposition is carried out in excess water, with dilute hydrocarbon concentrations ($C_A^o = 1\ kmol/m^3$, $C_{H_2O} = 55\ kmol/m^3$). It is known that the reaction follows elementary kinetics.

The concentration of hydrocarbon was measured with time to calculate the reaction rate constant. Under these conditions, calculate the rate constant for the reaction.

$$A + H_2O \rightarrow \text{Products}$$

Time (h)	**0**	**0.5**	**1**	**1.5**	**2**
C_A (kmol/m^3)	1.0	0.644	0.459	0.294	0.190

3.5 The following two sets of data are given for the liquid-phase reaction.

$$2A + B \rightarrow 2C$$

Set I: Initial concentrations; $C_A^o = 0.002\ M$; $C_B^o = 0.100\ M$.

Time (min)	**5**	**7**	**13**	**31**
C_A (M)	1.53×10^{-3}	1.37×10^{-3}	0.99×10^{-3}	0.38×10^{-3}

Set II: Initial concentrations; $C_A^o = 0.003\ M$; $C_B^o = 0.073\ M$.

Time (min)	**19**	**28**	**38**	**45**	**57**
C_A (M)	1.39×10^{-3}	1.01×10^{-3}	0.74×10^{-3}	0.49×10^{-3}	0.33×10^{-3}

(a) Find out the rate law and evaluate the rate constant.

(b) Calculate the time required for 30% of A to be consumed when the initial concentrations are $C_A^o = 0.050$ M; $C_B^o = 0.025$ M.

3.6 The rate law for the following reaction is given as follows:

$$A + 2B \rightarrow 2C$$

$$-R_A = k\frac{C_A C_B}{C_C}$$

Show that the batch reactor data given in the tables are consistent with the given rate law. Evaluate the rate constant by regression analysis.

Time (min)	**0**	**5**	**10**	**15**
C_A (M)	6.30×10^{-4}	3.80×10^{-4}	2.30×10^{-4}	1.36×10^{-4}

Exp. 1: Initial concentrations: $C_A^o = 6.3 \times 10^{-4}$ M; $C_B^o = 0.100\ M$; $C_C^o = 0.24$ M

Time (min)	**0**	**5**	**10**	**15**
C_B (M)	4.80×10^{-4}	3.20×10^{-4}	2.16×10^{-4}	1.45×10^{-4}

Exp. 2: Initial concentrations: $C_A^o = 0.039$ M; $C_B^o = 4.8 \times 10^{-4}\ M$; $C_C^o = 0.24$ M.
Exp. 3: Initial concentrations: $C_A^o = 3.2 \times 10^{-4}$ M; $C_B^o = 0.100$ M; $C_C^o = 0.076$ M.

$$\text{Initial rate of B} : R_B^o = -\left(\frac{dC_B}{dt}\right)_{t=0} = 2.07 \times 10^{-4}\ (\text{mol/l})/\min$$

3.7 Given the rate law:

$$-R_A = k\frac{C_A C_B}{C_C}$$

for the reaction 2A + B → 2C + D; propose two possible mechanisms.

3.8 If the rate law for the reaction

$$Fe^{++} + HNO_2 + H^+ \rightarrow Fe^{+++} + NO + H_2O$$

is given as

$$-R_{Fe^{++}} = k\frac{C_{Fe^{++}} C_{HNO_2}^2}{C_{NO}}$$

Propose a mechanism for this reaction.

3.9 From the following batch reactor initial rate data (reaction rate at the initial concentrations of reactants and products) for the liquid-phase reaction:

$$A + 2B + C \rightarrow 2D$$

Derive an expression for the rate law and calculate the rate constant.

Initial rate; $-R_A$	C_A^o (M)	C_B^o (M)	C_C^o (M)	C_D^o (M)
4.0×10^{-4}	0.010	0.010	0.010	0.010
4.0×10^{-4}	0.090	0.010	0.010	0.030
4.0×10^{-4}	0.010	0.020	0.020	0.020
8.0×10^{-4}	0.040	0.010	0.010	0.010
8.0×10^{-4}	0.040	0.020	0.010	0.020

Hint: Express the rate expression as $-R_A = kC_A^a C_B^b C_C^c C_D^d$.

3.10 Given an overall reaction

$$A + B \rightarrow C + D$$

write down possible mechanisms for the following rate laws. Indicate the assumptions involved in the derivation of these rate expressions.

(a) $-R_A = \dfrac{kC_A C_B}{1 + k' C_C}$

(b) $-R_A = \dfrac{kC_A C_B}{C_B + k' C_C}$

(c) $-R_A = \dfrac{kC_A C_B}{C_C}$

(d) $-R_A = kC_A$

(e) $-R_A = \dfrac{kC_A}{C_B}$

3.11 An isomerization reaction was carried out in a batch reactor, and the following data were obtained.

Time (min)	0	3	5	8	10	12	15
C_A (M)	4.00	2.88	2.25	1.45	0.99	0.64	0.25

(a) Evaluate the reaction order and determine the reaction rate constant using the differential method of analysis. Assume an *n*th-order reaction and perform a regression analysis to evaluate the order and the reaction rate constant.
(b) Check the result using the integral method.

3.12 Gaseous reactant A decomposes in a variable volume vessel.

$$A \rightarrow B + C$$

Due to the volume change caused by the increase of the total number of moles resulting from the conversion of reactant A to the products, the volume of the vessel increases, while the pressure remains constant at 1 atm. The reaction is given as second order in terms of reactant concentration. The temperature of the reaction zone is 400 K.

(a) Find the fractional conversion of pure A if the volume increases from 4.7 cm^3 (initial volume) to 7.2 cm^3 within a reaction period of seven minutes.

(b) Evaluate the reaction rate constant.

(c) What is the volume of the reaction zone if fractional conversion reaches 0.7?

(d) What are the reaction volume and the time required to reach the fractional conversion value of 0.7 if the initial mixture in the reaction volume contains 20% inerts and 80% A?

3.13 (a) Following rate expression is given for the gas-phase decomposition reaction of reactant A in a constant volume batch reactor.

$$\mathrm{A} \rightarrow \mathrm{B} + \frac{1}{2}\mathrm{C}\,; \quad -R_A = \frac{k_1 C_A}{1 + k_2 C_A}$$

The initial composition of the mixture in the reactor is 70% A and 30% inerts. The initial pressure is given as 1 atm. Find the fractional conversion of reactant A when the pressure in the reactor reaches 1.2 atm.

(b) Obtain an expression for the variation of fractional conversion of reactant A as a function of reaction time.

(c) If the same reaction takes place in a constant pressure reaction vessel rather than a constant volume reactor, what would be the conversion versus time relations for the following two cases (discuss your results)?

(I) $k_2 C_A \gg 1$; (II) $k_2 C_A \ll 1$

(d) If the activation energy of k_2 is given as 1/4th of the activation energy of $k_1 \left(E_{a_2} = \frac{1}{4} E_{a_1} \right)$, what would be the observed activation energy for the two limiting cases mentioned in part (c) of this problem.

3.14 Derive an equation for the half-life of an autocatalytic reaction, if the initial concentration of the product C is half of the initial concentration of the reactant A

$$\left(C_C^o = \frac{C_A^o}{2} \right).$$

$$\mathrm{A} \rightarrow \mathrm{C}\,; \quad -R_A = k C_A C_C$$

3.15 Derive an expression for the fractional conversion of reactant A in the following autocatalytic reaction in a constant volume batch reactor.

$$2\mathrm{A} \rightarrow \mathrm{C} + \mathrm{D}\,; \quad R_D = k C_A C_C$$

Initial concentrations of A and C in the reactor are C_A^o and C_C^o, respectively. You can take the initial concentration of D as zero.

Obtain an expression for the fractional conversion of A, which corresponds to the maximum value of the rate.

3.16 A gas-phase reaction of $2\mathrm{A} \rightarrow \mathrm{C} + \frac{1}{2}\mathrm{D}$ is taking place in a constant volume batch reactor. Initially pure A is charged into the reactor. A set of experiments is performed at 400 K and with different initial total pressure values, which are listed in the table. Corresponding times to reach fractional conversion values of 0.5 are also listed in the same table.

(a) Evaluate the reactor's total pressure values for the reaction times listed in the table (for $x_A = 0.5$).

(b) Find the overall order of the reaction assuming an nth-order reaction and evaluate the reaction rate constant.

Initial pressure (atm)	**0.26**	**0.32**	**0.37**	**0.42**	**0.47**
Time to achieve $x_A = 0.5$ (min)	260	185	115	105	68

References

1 Roberts, G.W. (2009). *Chemical Reactions and Chemical Reactors*. NJ: Wiley.
2 Levenspiel, O. (1999). *Chemical Reaction Engineering*. NJ, USA: Wiley.
3 Fogler, H.S. (1992). *Elements of Chemical Reaction Engineering*. NJ, USA: Prentice-Hall.

4

Ideal-Flow Reactors: CSTR and Plug-Flow Reactor Models

In many of the industrial processes, flow reactors are used for the production of chemicals. Hydrodynamics of the reaction fluid within a reaction vessel has significant effects on the performance of flow reactors. Two ideal-flow reactor models are commonly used to describe the hydrodynamics of the reaction mixture in a flow reactor [1–3]. These models are continuous perfectly stirred-tank reactor (CSTR) and ideal tubular plug-flow tubular (PFR) models. CSTRs may also be called as back-mixed flow reactors or perfectly mixed flow reactors.

4.1 CSTR Model

As illustrated in Section 1.4 of this book, stirred-tank flow reactors are quite common in chemical processes (Figure 4.1). The primary assumption of a CSTR is the perfect mixing of the reaction mixture so that there are no concentration and temperature variations within the reactor. This assumption also requires that the concentrations of species in the reactor exit stream are the same as the concentrations within the reactor. Also, the temperature of the exit stream of the reactor is assumed the same as the temperature of the fluid within the reactor. Hence, the reaction rate values can be evaluated at the reactor exit stream concentrations and temperature.

Considering an arbitrary chemical reaction between the reactants A and B

$$aA + bB \rightleftharpoons cC + dD$$

species conservation equation for the reactant A, in a CSTR operating at steady state, can be written as follows:

$$F_{A_0} - F_{A_f} + VR_{A_f} = 0 \tag{4.1}$$

Using the definition of fractional conversion of A as

$$x_{A_f} = \frac{F_{A_0} - F_{A_f}}{F_{A_0}} \tag{4.2}$$

Equation (4.1) can be rearranged as

$$\frac{V}{F_{A_0}} = \frac{x_{A_f}}{-R_{A_f}} \tag{4.3}$$

Equation (4.3) is the design equation for a CSTR operating at steady state. Evaluation of the reactor volume using Eq. (4.3) requires knowledge about the rate expression. For instance, for an irreversible nth-order reaction Eq. (4.3) reduces to,

Fundamentals of Chemical Reactor Engineering: A Multi-Scale Approach, First Edition. Timur Doğu and Gülşen Doğu.

Companion website: www.wiley.com/go/dogu/chemreacengin

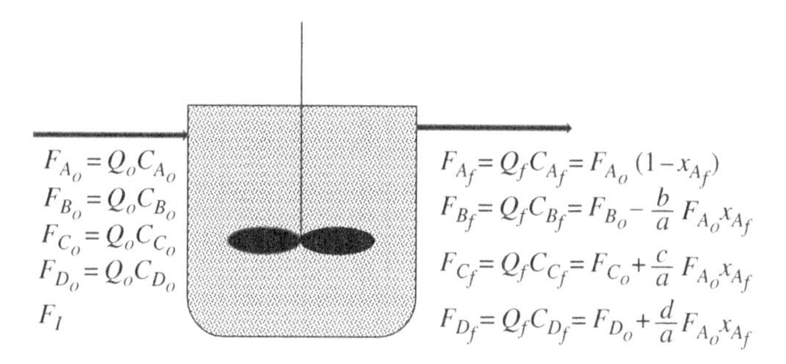

Figure 4.1 Schematic representation of a CSTR.

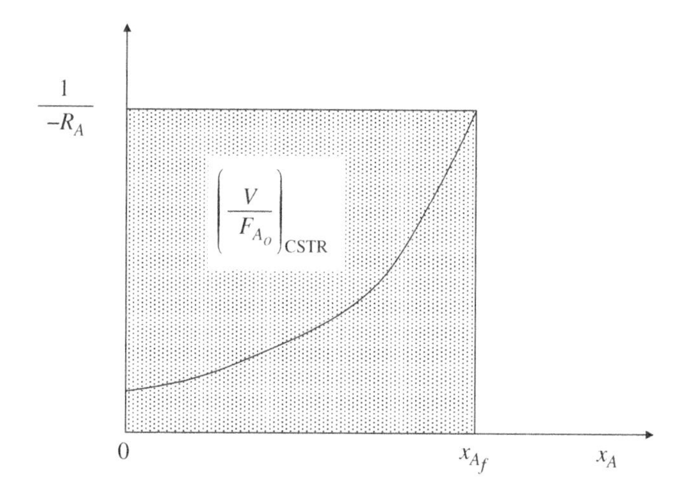

Figure 4.2 Graphical representation of CSTR volume estimation for a positive-order reaction.

$$\frac{V}{F_{A_0}} = \frac{x_{A_f}}{kC_{A_f}^n} \tag{4.4}$$

If the volumetric flow rate of the inlet and the outlet streams of the reactor is the same, the concentration of reactant A in the reactor exit stream can be expressed in terms of its fractional conversion as

$$C_{A_f} = C_{A_o}\left(1 - x_{A_f}\right)$$

For this case, the design equation becomes

$$\frac{V}{F_{A_0}} = \frac{x_{A_f}}{kC_{A_o}^n\left(1 - x_{A_f}\right)^n} \tag{4.5}$$

Graphical representation of Eq. (4.3) is shown in Figure 4.2. As illustrated in Figure 4.2, if we plot $\left(\frac{1}{-R_A}\right)$ as a function of x_A the area of the rectangle between the x_A values of zero and x_{A_f} gives $\frac{V}{F_{A_o}}$. This result indicates that, for positive-reactant-order reactions, an increase in fractional conversion causes a significant increase in reactor volume. We need infinite volume to reach the complete conversion of reactant A in a CSTR.

Determination of the reactor volume using Eq. (4.5) needs knowledge about the rate constant, which should be evaluated at the reactor temperature. Alternatively, fractional conversion of reactant A can be determined from this equation if the reactor volume and feed stream compositions are available.

Considering a first-order reaction, Eq. (4.5) reduces to

$$\frac{V}{F_{A_0}} = \frac{x_{A_f}}{kC_{A_o}(1 - x_{A_f})}. \tag{4.6}$$

Elimination of x_{A_f} from Eq. (4.6) yields the following expression for a first-order reaction:

$$x_{A_f} = \frac{k\tau}{1 + k\tau} \tag{4.7}$$

Here, τ is the space–time, which is defined as the ratio of reactor volume to inlet volumetric flow rate $\tau = \frac{V}{Q_o}$.

Instead of incorporation of the fractional conversion definition into the species conservation expression, Eq. (4.1) may also be written in terms of concentrations of species. For the constant volumetric flow rate case ($Q_f = Q_o$), the species conservation equation for the limiting reactant A can be expressed as follows:

$$Q_o(C_{A_o} - C_{A_f}) + VR_{A_f} = 0 \tag{4.8}$$

Rearrangement of this expression gives the following equation for the space–time in the reactor:

$$\tau = \frac{V}{Q_o} = \frac{(C_{A_o} - C_{A_f})}{-R_{A_f}} \tag{4.9}$$

For an nth-order reaction, this equation reduces to

$$\tau = \frac{V}{Q_o} = \frac{(C_{A_o} - C_{A_f})}{kC^n_{A_f}} \tag{4.10}$$

For gas-phase reactions, changes in the volumetric flow rate are expected due to changes in the total molar flow rate of species and changes in temperature and pressure between the inlet and outlet streams of the reactor. If the summation of the stoichiometric coefficients of the products is not the same as the summation of the stoichiometric coefficients of the reactants $[(a + b) \neq (c + d)]$, we expect a change in the total molar flow rate between the inlet and the outlet of the reactor. Similar to the approach described for a variable volume batch reactor in Chapter 3, changes in volumetric flow rate can be expressed as follows:

$$Q_f = Q_o(1 + \varepsilon x_{A_f})\frac{T_f}{T_o}\frac{P_0}{P_f} \tag{4.11}$$

Here, the volume expansion factor ε is defined as

$$\varepsilon = \frac{(F_T)_{x_A = 1} - (F_T)_{x_A = 0}}{(F_T)_{x_A = 0}} \tag{4.12}$$

If the volumetric flow rates of the inlet and outlet streams of a CSTR are not the same, the concentration of reactant A in the reactor outlet stream can be expressed in terms of its fractional conversion, as follows:

$$C_{A_f} = \frac{F_{A_f}}{Q_f} = \frac{F_{A_o}(1-x_{A_f})}{Q_o(1+\varepsilon x_{A_f})\frac{T_f}{T_o}\frac{P_o}{P_f}} = C_{A_o}\frac{T_o P_f}{T_f P_o}\frac{(1-x_{A_f})}{(1+\varepsilon x_{A_f})} \quad (4.13)$$

For the isothermal and isobaric operation cases, ($T_f = T_o$) and ($P_f = P_o$), Eq. (4.13) reduces to

$$C_{A_f} = C_{A_o}\frac{(1-x_{A_f})}{(1+\varepsilon x_{A_f})} \quad (4.14)$$

Concentrations of all the other species in the reactor exit stream can also be expressed similarly.

We need the molar flow rate values of all the species at the complete conversion of the limiting reactant A for the evaluation of the volume expansion factor. Expressing the outlet molar flow rates of the species in terms of fractional conversion of the limiting reactant (see Figure 4.1), the total molar flow rate of the reactor exit stream can be related to the fractional conversion as follows:

$$F_T = (F_{A_o} + F_{B_o} + F_{C_o} + F_{D_o} + F_i) - F_{A_o}x_{A_f}\left(1 + \frac{b}{a} - \frac{c}{a} - \frac{d}{a}\right) \quad (4.15)$$

Then, the volume expansion factor can be evaluated from

$$\varepsilon = \frac{-F_{A_o}\left(1 + \frac{b}{a} - \frac{c}{a} - \frac{d}{a}\right)}{(F_{A_o} + F_{B_o} + F_{C_o} + F_{D_o} + F_i)} = y_{A_o}\left(\frac{c}{a} + \frac{d}{a} - \frac{b}{a} - 1\right) \quad (4.16)$$

Here, y_{A_o} is the mole fraction of limiting reactant A in the feed stream of the reactor. The total molar flow rate of the inlet stream is

$$F_o = (F_{A_o} + F_{B_o} + F_{C_o} + F_{D_o} + F_i) \quad (4.17)$$

For the nth-order reaction treated above, the design equation, which is expressed by Eq. (4.5), should then be written for the variable volumetric flow rate case as follows:

$$\frac{V}{F_{A_o}} = \frac{x_{A_f}}{kC_{A_o}^n\frac{\left(1-x_{A_f}\right)^n}{\left(1+\varepsilon x_{A_f}\right)^n}} \quad (4.18)$$

For some gas-phase reactions, reaction rate expression may be available in terms of partial pressures of reactants instead of their concentrations. For instance, if the rate expression for a reaction is given as a second-order reaction in terms of the partial pressures of reactants as

$$-R_A = kP_AP_B \quad (4.19)$$

partial pressures of A and B can be expressed in terms of fractional conversion of reactant A as follows:

$$P_A = P_Ty_A = P_T\frac{F_A}{F_T} = P_T\frac{y_{A_o}(1-x_A)}{\left(1 - y_{A_o}x_A\left(1 + \frac{b}{a} - \frac{c}{a} - \frac{d}{a}\right)\right)} \quad (4.20)$$

$$P_B = P_Ty_B = P_T\frac{F_B}{F_T} = P_T\frac{\left(y_{B_o} - \frac{b}{a}y_{A_o}x_A\right)}{\left(1 - y_{A_o}x_A\left(1 + \frac{b}{a} - \frac{c}{a} - \frac{d}{a}\right)\right)} \quad (4.21)$$

Here, y_i and P_T are the mole fraction of species i and the total pressure in the reactor, respectively. In this case, the reactor design equation becomes as follows:

$$\frac{V}{F_{A_o}} = \frac{x_{A_f}}{-R_{A_f}} = \frac{x_{A_f}}{kP_{A_f}P_{B_f}} = \frac{x_{A_f}}{kP_T^2\left[\frac{y_{A_0}(1-x_A)\left(y_{B_0}-\frac{b}{a}y_{A_0}x_A\right)}{\left(1-y_{A_0}x_A\left(1+\frac{b}{a}-\frac{c}{a}-\frac{d}{a}\right)\right)^2}\right]} \tag{4.22}$$

Example 4.1 Space Time of a CSTR

Find the required space–time of a CSTR to achieve a fractional conversion of 0.75 for the limiting reactant A for an elementary reversible liquid-phase reaction. The concentrations of A and B in the reactor feed stream are given as 0.8 and 1 M, respectively.

$$A + B \rightleftarrows 2C; \quad -R_A = k_1 C_A C_B - k_{-1} C_C^2$$

$$k_1 = 20\,(\text{mol/l})^{-1}\,\text{min}^{-1};\ k_{-1} = 1\,(\text{mol/l})^{-1}\,\text{min}^{-1}.$$

Solution

Concentrations of species in the outlet stream of the reactor are

$$C_{A_f} = C_{A_0}\left(1 - x_{A_f}\right) = 0.8(1 - 0.75) = 0.2\ \text{M}$$

$$C_{B_f} = C_{B_0} - C_{A_0}x_{A_f} = 1.0 - 0.8(0.75) = 0.4\ \text{M}$$

$$C_{C_f} = 2C_{A_0}x_{A_f} = 1.2\ \text{M}$$

The rate evaluated at the reactor exit concentrations is

$$-R_{A_f} = k_1 C_{A_f} C_{B_f} - k_{-1} C_{C_f}^2 = 0.16\ \text{mol/l min}$$

Hence, the space–time in the reactor becomes

$$\tau = \frac{C_{A_0}x_{A_f}}{-R_{A_f}} = 0.8\frac{0.75}{0.16} = 3.75\ \text{minutes}$$

4.1.1 CSTR Data Analysis

Experimental fractional conversion data obtained in a CSTR may be used to evaluate the rate law of a reaction. For this purpose, a set of experiments can be performed in a CSTR at different space–times. Fractional conversion values obtained at different space–times can be used to determine the corresponding reaction rate values. Hence, using these data, the rate expression of the reaction can be predicted. This procedure is illustrated in Example 4.2.

Example 4.2 Rate Law from CSTR Data

For a gas-phase reaction $A \rightarrow B + C$, experimental data for the fractional conversion of reactant A are obtained in a CSTR at different volumetric flow rates of the feed stream. Evaluate the order of the reaction and the reaction rate constant. The concentration of A in the inlet stream of the reactor is given as $C_{A_o} = 0.0032$ mol/l, and the reactor volume is 0.1 m^3 (Table 4.1).

Table 4.1 Experimental data for Example 4.2.

Q_o (m^3/min)	2.36×10^{-1}	1.96×10^{-2}	7.40×10^{-3}	2.23×10^{-3}	5.21×10^{-4}
x_{A_f}	0.22	0.63	0.74	0.89	0.96

Solution

Using the species conservation equation for a CSTR, reaction rate values can be evaluated for each data point.

$$-R_{A_f} = \frac{x_{A_f} F_{A_o}}{V} = \frac{x_{A_f} C_{A_o} Q_o}{V} = \frac{x_{A_f} C_{A_o}}{\tau}$$

The relation between the reactor outlet concentration of reactant A and fractional conversion should be used in this evaluation.

$$C_{A_f} = C_{A_o} \frac{(1 - x_{A_f})}{(1 + \varepsilon x_{A_f})}$$

In these calculations, the volume expansion factor is taken as $\varepsilon = 1$ for this reaction. The following reaction rate values are then obtained for each conversion level in the reactor.

$-R_{A_f}$ (mol/l min)	1.66×10^{-3}	3.95×10^{-4}	1.75×10^{-4}	6.35×10^{-5}	1.60×10^{-5}
x_{A_f}	0.22	0.63	0.74	0.89	0.96
C_{A_f} (mol/l)	2.05×10^{-3}	7.26×10^{-4}	4.78×10^{-4}	1.86×10^{-4}	6.53×10^{-5}

The relation between $-\ln(-R_{A_f})$ and $-\ln(C_{A_f})$ should give a linear graph with a slope of n and an intercept of $-\ln(k)$, for an nth-order reaction (shown in Figure 4.3).

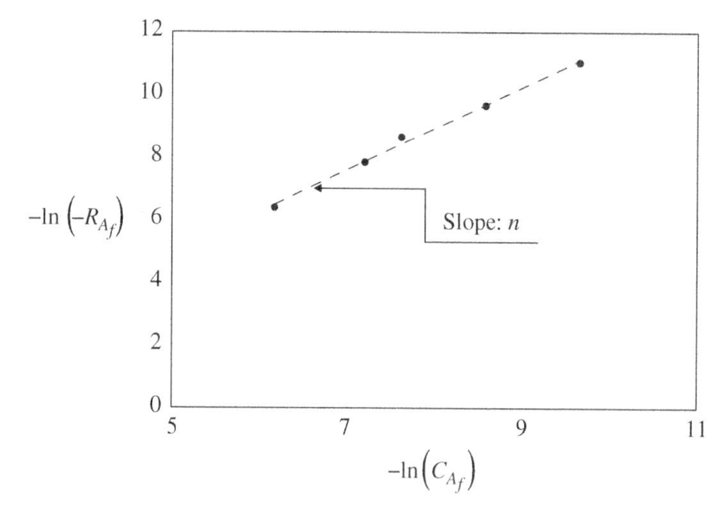

Figure 4.3 Data analysis for Example 4.2.

$$-R_{A_f} = kC^n_{A_f}; \quad \ln\left(-R_{A_f}\right) = \ln(k) + n\ln\left(C_{A_f}\right).$$

Hence, the reaction order and the rate constant were found as
$n = 1.4$ and $k = 10.7\ (\text{mol/l})^{-0.4}\ \text{min}^{-1}$, respectively.

4.2 Analysis of Ideal Plug-Flow Reactor

In many of the industrial processes, tubular reactors are used. The concentrations of the species and the temperature may change both in the axial and radial directions in tubular reactors. As discussed in Chapter 1, assuming negligible radial variations of concentrations of species and the temperature, one-dimensional species conservation equation can be written. Tubular reactors with negligible radial variations of concentrations of the species and with no mixing/diffusion in the axial direction are called ideal plug-flow reactors. According to the plug-flow reactor model, there should be no distribution of residence times of reactants in the reactor. Considering an arbitrary reaction of

$$\text{A} + \text{B} \rightarrow \text{C} + \text{D}$$

species conservation equation written for reactant A around the differential volume element between z and $z + \Delta z$ in a plug-flow reactor (Figure 4.4) is derived in Chapter 1 as

$$-\frac{dF_A}{dV} + R_A = 0 \tag{4.23}$$

Expressing the molar flow rate of A in terms of fractional conversion as $F_A = F_{A_o}(1 - x_A)$, Eq. (4.23) is rearranged to obtain the following design equation for a plug-flow reactor:

$$F_{A_o}\frac{dx_A}{dV} + R_A = 0 \qquad \frac{V}{F_{A_o}} = \int_0^{x_{A_f}} \frac{dx_A}{-R_A} \tag{4.24}$$

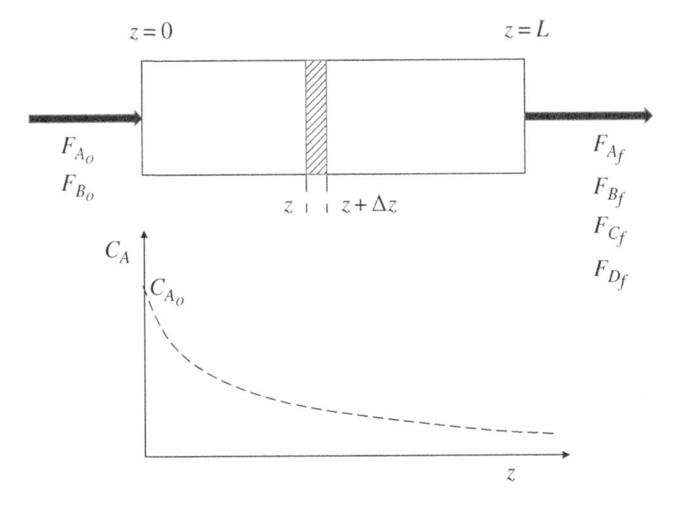

Figure 4.4 Schematic diagram of a plug-flow reactor.

Alternatively, for the constant volumetric flow rate case, the molar flow rate can be expressed as $F_A = Q_o C_A$. Substitution of this relation into Eq. (4.23) gives the species conservation equation in terms of the concentration of limiting reactant A.

$$-Q_o \frac{dC_A}{dV} + R_A = 0 \tag{4.25}$$

Rearrangement of Eq. (4.25) gives the following integral equation for designing a plug-flow reactor operating with a constant volumetric flow rate.

$$\frac{V}{Q_o} = \tau = \int_{C_{A_o}}^{C_{A_f}} \frac{dC_A}{R_A} \tag{4.26}$$

Graphically, if we plot $(1/-R_A)$ as a function of x_A, the area under the curve between the x_A values of zero and x_{A_f} gives V/F_{A_o} for a plug-flow reactor. This is illustrated in Figure 4.5 for a positive-reactant-order reaction.

Considering an nth-order reaction with respect to the concentration of reactant A, species conservation equation can be written for the constant flow rate case as follows:

$$\frac{V}{Q_o} = \tau = -\int_{C_{A_o}}^{C_{A_f}} \frac{dC_A}{kC_A^n} = \frac{1}{(n-1)k}\left(\frac{1}{(C_{A_f})^{n-1}} - \frac{1}{(C_{A_o})^{n-1}}\right) \tag{4.27}$$

Note that, during the integration of Eq. (4.26), the reaction rate constant k is taken as constant considering an isothermal operation.

In the case of gas-phase reactions with changes in the total volumetric flow rate of species flowing in the reactor, it is more convenient to use Eq. (4.24) in the design of a plug-flow reactor. For instance, for an nth-order reaction with changes in the volumetric flow rate along the reactor, Eq. (4.24) becomes

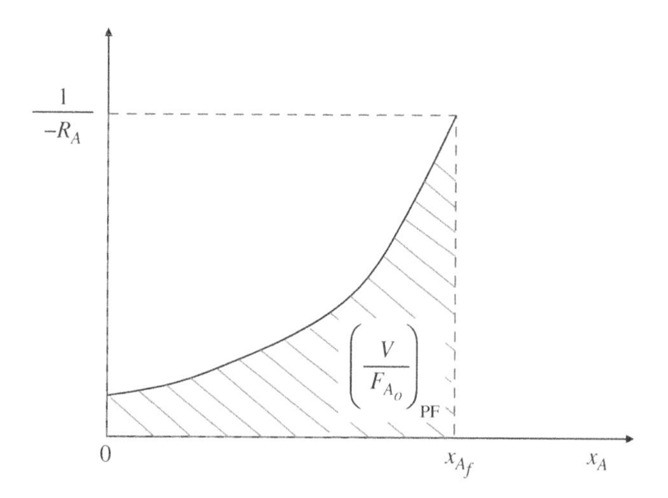

Figure 4.5 Graphical volume estimation of a plug-flow reactor for an nth-order reaction.

$$\frac{V}{F_{A_0}} = \int_0^{x_{A_f}} \frac{dx_A}{kC_{A_0}^n \frac{(1-x_A)^n}{(1+\varepsilon x_A)^n}} \tag{4.28}$$

In the case of an irreversible first-order reaction ($n = 1$), integration of Eq. (4.28) gives the following relation for the isothermal operation case:

$$\frac{V}{F_{A_0}} = \int_0^{x_{A_f}} \frac{dx_A}{kC_{A_0} \frac{(1-x_A)}{(1+\varepsilon x_A)}} = \frac{\tau}{C_{A_0}} = \left(\frac{1}{kC_{A_0}}\right)\left[-(1+\varepsilon)\ln\left(1-x_{A_f}\right) - \varepsilon x_{A_f}\right] \tag{4.29}$$

Note that, for the case of a first-order reaction with a constant volumetric flow rate, integration of Eq. (4.24) gives

$$\frac{V}{F_{A_0}} = \int_0^{x_{A_f}} \frac{dx_A}{kC_{A_0}(1-x_A)} = \left(\frac{1}{kC_{A_0}}\right)\left(-\ln\left(1-x_{A_f}\right)\right) \tag{4.30}$$

$$k\tau = -\ln\left(1-x_{A_f}\right) \tag{4.31}$$

4.3 Comparison of Performances of CSTR and Ideal Plug-Flow Reactors

The ratio of the volumes of a CSTR and an ideal plug-flow reactor for a first-order reaction can be expressed using Eqs. (4.6) and (4.31) as follows:

$$\frac{V_{\text{CSTR}}}{V_{\text{PF}}} = \frac{\frac{x_{A_f}}{\left(1-x_{A_f}\right)}}{-\ln\left(1-x_{A_f}\right)} \tag{4.32}$$

Variation of this ratio as a function of fractional conversion of A is illustrated in Figure 4.6. As shown in Figure 4.6, this ratio is higher than 1 and increases with an increase in fractional conversion [2]. For a fractional conversion value of 0.95, the ratio of the volumes of a CSTR and a plug-flow reactor is about 6.3 for a first-order reaction. This result is due to the fact that the rate

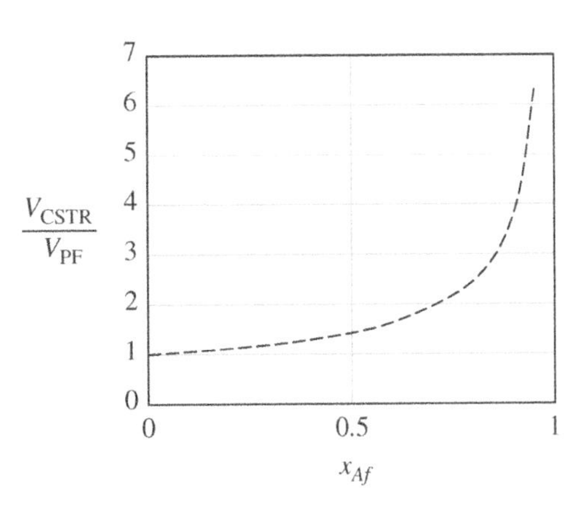

Figure 4.6 The ratio of the volumes of a CSTR and a plug-flow reactor for a first-order reaction.

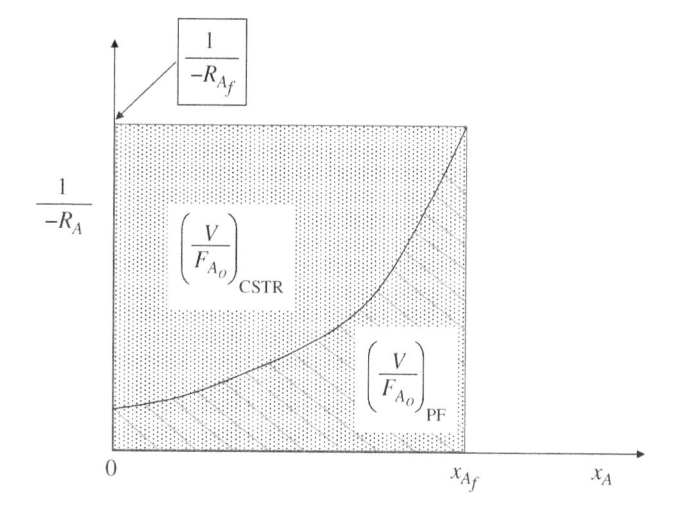

Figure 4.7 Graphical representation of volume comparison of CSTR and plug-flow reactors for a positive-reactant-order reaction.

is evaluated at the reactor exit concentrations of the reactants in a CSTR, which is much lower than the inlet concentrations.

For positive-reactant-order reactions, the reaction rate decreases as the conversion increases. In a plug-flow reactor, the concentration of reactant A and the reaction rate decreases along the reactor, reaching the reaction rate value of the CSTR at the reactor exit stream. A comparison of the volumes of CSTR and plug-flow reactors is also illustrated in Figure 4.7. This effect becomes more significant as the reaction order increases [1–3]. For instance, for a second-order reaction, the ratio of volumes of a CSTR and a plug-flow reactor is about 20 at a fractional conversion level of 0.95. This result shows that a plug-flow reactor is more efficient than a CSTR for positive-reactant-order reactions. Using the same volumes of CSTR and plug-flow reactors, a higher conversion of reactant A is expected in the plug-flow reactor.

4.4 Equilibrium and Rate Limitations in Ideal-Flow Reactors

Let us consider an elementary bi-molecular reversible second-order reaction between reactants A and B.

$$A + B \rightleftarrows C + D \quad ; \quad -R_A = k_f C_A C_B - k_b C_C C_D$$

As discussed in Chapter 2, when equilibrium is reached, forward and the reverse reaction rates become equal, and the net rate of conversion of reactants to products reaches zero. Hence, the following equilibrium relation can be written at equilibrium:

$$K_C = \frac{k_f}{k_b} = \frac{C_{C_e} C_{D_e}}{C_{A_e} C_{B_e}} \tag{4.33}$$

Expressing the concentrations of reactants and products in terms of fractional conversion of A, the equilibrium relation can be written as follows:

$$K_C = \frac{(C_{C_o} + C_{A_o}x_{A_e})(C_{D_o} + C_{A_o}x_{A_e})}{C_{A_o}(1 - x_{A_e})(C_{B_o} - C_{A_o}x_{A_e})} \tag{4.34}$$

If the reactor inlet concentrations of reactants are the same ($C_{A_o} = C_{B_o}$) and if the concentrations of C and D are zero at the inlet to the reactor, Eq. (4.34) reduces to

$$K_C = \frac{(x_{A_e})^2}{(1 - x_{A_e})^2} \tag{4.35}$$

Hence, the equilibrium conversion of A can be expressed in terms of the equilibrium constant, as follows:

$$x_{A_e} = \frac{K_C^{1/2}}{1 + K_C^{1/2}} \quad \text{(bimolecular second-order reversible reaction)} \tag{4.36}$$

A similar analysis for a reversible first-order reaction (A ⇄ B) gives the following relation for the equilibrium conversion:

$$x_{A_e} = \frac{K_C}{1 + K_C} \quad \text{(first-order reversible reaction)} \tag{4.37}$$

As discussed in Section 2.3, van't Hoff relation predicts a decrease of the equilibrium constant with an increase in temperature for exothermic reactions. On the other hand, the equilibrium constant increases with an increase in temperature for endothermic reactions. Hence, we expect a similar behavior for the temperature dependence of equilibrium conversion values. The temperature dependence of the equilibrium conversion is qualitatively illustrated in Figure 4.8a,b, for exothermic and endothermic reactions, respectively. Equilibrium conversion curves shown in Figure 4.8a,b correspond to the upper limit of achievable fractional conversion at the corresponding temperatures.

For the second-order reversible elementary reaction treated above, the rate expression can be expressed as follows:

$$-R_A = k_f C_A C_B - k_b C_C C_D = k_f \left(C_A C_B - \frac{1}{K_C} C_C C_D \right) \tag{4.38}$$

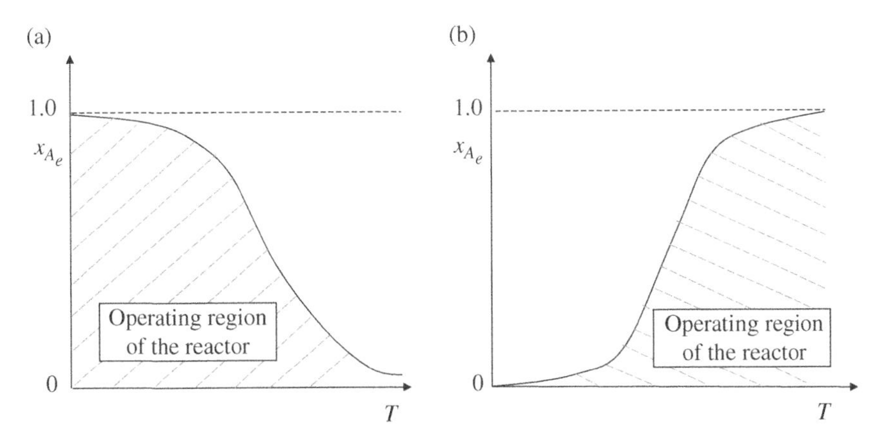

Figure 4.8 Schematic representation of temperature dependence of equilibrium conversion for (a) exothermic and (b) endothermic reactions.

This rate expression can be expressed in terms of fractional conversion of reactant A as follows:

$$-R_A = k_f C_{A_0}^2\left((1-x_A)^2 - \frac{1}{K_C}x_A^2\right) \tag{4.39}$$

In writing this equation, the inlet concentrations of reactants A and B are taken as the same. Using the K_C expression given in Eq. (4.35), Eq. (4.39) can be expressed as follows:

$$-R_A = k_f C_{A_0}^2\left((1-x_A)^2 - \frac{(1-x_{A_e})^2}{(x_{A_e})^2}x_A^2\right) = k_f C_{A_0}^2\left(\frac{(x_{A_e}-x_A)(x_{A_e}+x_A-2x_A x_{A_e})}{x_{A_e}^2}\right) \tag{4.40}$$

Species conservation equations for an isothermal ideal plug-flow reactor and a CSTR yield the following relations for this reversible second-order reaction;

Plug-flow reactor:

$$\frac{V}{F_{A_0}} = \frac{\tau}{C_{A_0}} = \int_0^{x_{A_f}} \frac{dx_A}{k_f C_{A_0}^2\left(\frac{(x_{A_e}-x_A)(x_{A_e}+x_A-2x_A x_{A_e})}{x_{A_e}^2}\right)} \tag{4.41}$$

$$2k_f C_{A_0}\left(\frac{1}{x_{A_e}}-1\right)\tau = \ln\left[\frac{(x_{A_e}-(2x_{A_e}-1)x_{A_f})}{x_{A_e}-x_{A_f}}\right] \tag{4.42}$$

$$x_{A_f} = \frac{x_{A_e}\left(\exp\left[2k_f\left(\frac{1}{x_{A_e}}-1\right)C_{A_0}\tau\right]-1\right)}{\exp\left[2k_f\left(\frac{1}{x_{A_e}}-1\right)C_{A_0}\tau\right]-(2x_{A_e}-1)} \tag{4.43}$$

CSTR:

$$\frac{V}{F_{A_0}} = \frac{\tau}{C_{A_0}} = \frac{x_{A_f}}{k_f C_{A_0}^2\left(\frac{\left(x_{A_e}-x_{A_f}\right)\left(x_{A_e}+x_{A_f}-2x_{A_f}x_{A_e}\right)}{x_{A_e}^2}\right)} \tag{4.44}$$

Corresponding equations for a first-order reversible reaction ($A \rightleftarrows B$) are

Plug-flow reactor:

$$\frac{V}{F_{A_0}} = \frac{\tau}{C_{A_0}} = \int_0^{x_{A_f}} \frac{dx_A}{k_f\left(C_{A_0}(1-x_A) - \frac{(1-x_{A_e})}{x_{A_e}}C_{A_0}x_{A_e}\right)} \tag{4.45}$$

$$\frac{k_f}{x_{A_e}}\tau = -\ln\left(1-\frac{x_{A_f}}{x_{A_e}}\right) \tag{4.46}$$

$$x_{A_f} = x_{A_e}\left[1-\exp\left(-\frac{k_f\tau}{x_{A_e}}\right)\right] \tag{4.47}$$

CSTR:

$$\frac{V}{F_{A_0}} = \frac{\tau}{C_{A_0}} = \frac{x_{A_f}}{k_f\left(C_{A_f}-\frac{1}{K_C}C_{B_f}\right)} = \frac{x_{A_f}}{k_f\left(C_{A_0}(1-x_{A_f}) - \frac{(1-x_{A_e})}{x_{A_e}}C_{A_0}x_{A_f}\right)} \tag{4.48}$$

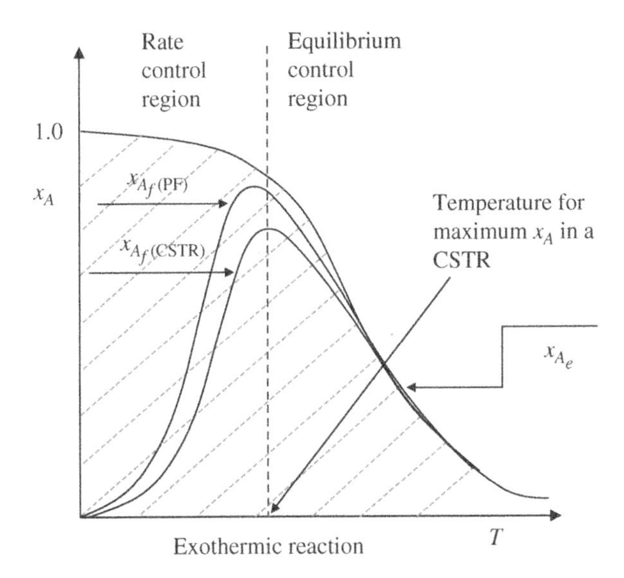

Figure 4.9 Schematic representation of equilibrium and rate-limited regions in conversion vs. reaction temperature relations for reversible exothermic reactions in reactors.

$$x_{A_f} = \frac{k_f \tau}{1 + \left(\frac{k_f \tau}{x_{A_e}}\right)} = \frac{k_f \tau}{1 + \left(\frac{1 + K_C}{K_C}\right) k_f \tau} \tag{4.49}$$

For a given value of space–time τ, Eqs. (4.43), (4.44), (4.47), and (4.49) give the relations for the temperature dependence of fractional conversion of reactant A for second- and first-order reversible reactions, respectively, in an ideal plug-flow reactor and a CSTR. For an exothermic reaction, all of these relations predict bell-shaped temperature dependencies of x_{A_f}, with a maximum at a particular value of reactor temperature. This is illustrated in Figure 4.9.

Let us explain this behavior using the x_{A_f} relation given by Eq. (4.49) for a first-order reversible reaction taking place in a CSTR. At low temperatures, equilibrium conversion approaches one. In this case, the conversion relation, which is given in Eq. (4.49), reduces to

$$x_{A_f} \approx \frac{k_f \tau}{1 + k_f \tau} = \frac{\tau k_{of} \exp\left(-\frac{E_{a_f}}{R_g T_f}\right)}{1 + \tau k_{of} \exp\left(-\frac{E_{a_f}}{R_g T_f}\right)}; \text{ (at low temperatures)} \tag{4.50}$$

This relation is valid for the case of having a negligible backward reaction. In this case, Eq. (4.50) shows an exponential increase in conversion with an increase in temperature. As the temperature increases, reverse rate also becomes important and reaction approaches equilibrium. In the case of conversion values very close to equilibrium ($x_A \rightarrow x_{A_e}$), the $\frac{k_f \tau}{x_{A_e}}$ term in the denominator of Eq. (4.49) becomes much greater than 1. Since equilibrium conversion decreases with an increase in temperature for exothermic reactions, x_{A_f} also shows a decreasing trend as equilibrium is approached. Hence, the temperature dependence of x_{A_f} passes through a maximum, as illustrated in Figure 4.9. This maximum may be evaluated by equating the derivative of x_{A_f} with respect to temperature to zero. At reaction temperatures lower than the temperature corresponding to the maximum value of x_{A_f}, the reaction is controlled by the kinetics (forward rate) of the process. However, for temperatures higher than the temperature corresponding to the maximum value of x_{A_f}, the reaction is mainly controlled by equilibrium. Similar behavior is also true for ideal plug-flow and batch reactors. In the case of endothermic reactions, both the reaction rate constant and the equilibrium constant increase with an increase in temperature. Hence, such a maximum is not observed in the x_{A_f} vs. temperature relations for the endothermic reactions (see Chapter 7).

4.5 Unsteady Operation of Reactors

4.5.1 Unsteady Operation of a Constant Volume Stirred-Tank Reactor

In this section, we will analyze a well-mixed tank reactor operating under unsteady-state conditions. At the start-up of any reactor, the unsteady operation should be considered. Also, if there is a disturbance in the feed concentration, feed temperature, or feed flow rate of a reactor operating at a steady state, we would expect a time-dependent variation of the product concentrations until a new steady state is reached [2, 3].

The species conservation equation for a perfectly stirred-tank reactor, operating at unsteady state, can be expressed as follows:

$$F_{i_o} - F_{i_f} + R_{i_f} V = \frac{dn_i}{dt} \tag{4.51}$$

This equation can be integrated to analyze the reactor performance. However, this analysis needs information about the volume of the reaction mixture. If the reactor volume is constant, Eq. (4.51) reduces to

$$F_{i_o} - F_{i_f} + R_{i_f} V = V\frac{dC_{i_f}}{dt} \tag{4.52}$$

For instance, for a first-order liquid-phase reaction taking place in such a reactor with constant volume and constant volumetric flow rate, Eq. (4.52) can be expressed as follows:

$$Q(C_{i_o} - C_{i_f}) - kC_{i_f} V = V\frac{dC_{i_f}}{dt} \tag{4.53}$$

Integration of this equation with an initial condition of, $C_{i_f} = C_{i_f}^o$ at $t = 0$, gives the following relation for C_{i_f}:

$$C_{i_f} = C_{i_f}^o \exp\left(-\frac{(1+k\tau)}{\tau}t\right) + \frac{C_{i_o}}{(1+k\tau)}\left[1 - \exp\left(-\frac{(1+k\tau)}{\tau}t\right)\right] \tag{4.54}$$

For $t = 0$, the initial concentration of the reactant in the reactor $\left(C_{i_f}^o\right)$ may be taken as zero at the start-up of such a reactor. However, if there is a disturbance in the concentration of the reactant of the feed stream at time zero, $C_{i_f}^o$ should be evaluated from the initial steady-state operation of the reactor. This is illustrated in Example 4.3.

Example 4.3 Analysis of Unsteady Operation of a CSTR

Consider a first-order liquid-phase reaction being carried out in an isothermal CSTR operating at a steady-state condition. The inlet concentration of reactant i was suddenly decreased to half of the inlet concentration of the first steady-state operation. Obtain an expression for the time dependence of reactant concentration in the reactor. Also, obtain an expression for the second steady-state concentration reached after the disturbance.

Solution

For the initial steady-state operation of the reactor, the concentration of reactant *i* in the reactor can be obtained as follows:

$$(C_{i_o} - C_{i_f}) - kC_{i_f}\tau = 0 \quad \Rightarrow \quad C_{i_f} = \frac{C_{i_o}}{(1 + k\tau)}$$

For the period after the disturbance in the feed concentration, the species conservation equation can be expressed as follows:

$$\left(\frac{C_{i_o}}{2} - C_{i_f}\right) - k\tau C_{i_f} = \tau\frac{dC_{i_f}}{dt}$$

The solution of this equation gives

$$C_{i_f} = C^o_{i_f} \exp\left(-\frac{(1 + k\tau)}{\tau}t\right) + \frac{C_{i_o}}{2(1 + k\tau)}\left[1 - \exp\left(-\frac{(1 + k\tau)}{\tau}t\right)\right]$$

Here, $C^o_{i_f}$ corresponds to the initial concentration in the reactor at the time of disturbance, which is the initial steady-state concentration. Hence, the relation becomes

$$C_{i_f} = \frac{C_{i_o}}{(1 + k\tau)}\left(\exp\left(-\frac{(1 + k\tau)}{\tau}t\right)\right) + \frac{C_{i_o}}{2(1 + k\tau)}\left[1 - \exp\left(-\frac{(1 + k\tau)}{\tau}t\right)\right]$$

$$C_{i_f} = \frac{C_{i_o}}{2(1 + k\tau)}\left[1 + \exp\left(-\frac{(1 + k\tau)}{\tau}t\right)\right]$$

The second steady-state concentration reached after the disturbance can then be estimated by substituting $t \to \infty$.

$$C_{i_{f_{ss}}} = \frac{1}{2}\frac{C_{i_o}}{(1 + k\tau)}$$

4.5.2 Semi-batch Reactors

In some cases, the outlet stream of the reactor is absent. These reactors are called as semi-batch reactors. For instance, one of the reactants (reactant A) may be continuously added to the stirred-tank reactor, which contains the second reactant with a concentration of C^o_B, while there is no outlet stream (Figure 4.10).

If we consider an elementary second-order reaction taking place in such a semi-batch reactor

$$\text{A} + \text{B} \rightarrow \text{Products}$$

species conservation equations for B and A around the reactor can be written as follows:

$$\text{Material balance for B}: R_B V = \frac{d(n_B)}{dt} \quad \Rightarrow \quad -kC_A C_B V = \frac{d(VC_B)}{dt} \tag{4.55}$$

$$\text{Material balance for A}: QC_{A_o} + R_A V = \frac{d(n_A)}{dt} \quad \Rightarrow \quad QC_{A_o} - kC_A C_B V = \frac{d(VC_A)}{dt} \tag{4.56}$$

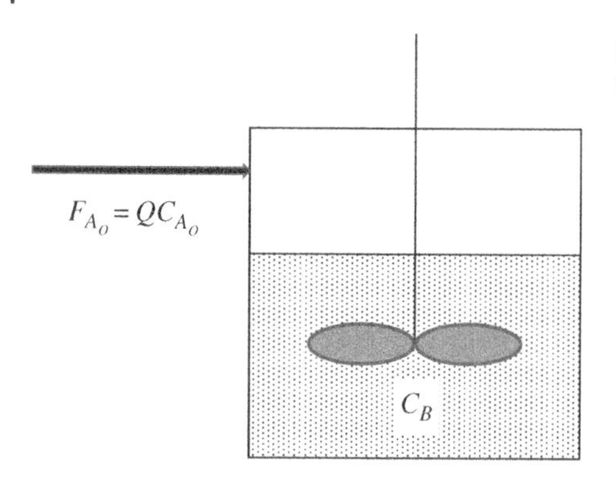

Figure 4.10 Schematic representation of a semi-batch reactor.

For this system, the volume of the reaction mixture in the reactor is not constant. Since there is no outlet stream, fluid entering into the reactor accumulates. Hence, the total mass balance around the reactor can be written as follows:

$$Q\rho = \frac{d(\rho V)}{dt} \tag{4.57}$$

For a constant density system, this equation reduces to

$$Q = \frac{d(V)}{dt}$$

The integration of Eq. (4.57) for a constant density system gives the following relation for the time variation of volume of the reaction mixture in the reactor:

$$V = V_o + Qt \tag{4.58}$$

Then, the species conservation equations for B and A become as follows:

$$-kC_AC_BV = V\frac{d(C_B)}{dt} + C_BQ \tag{4.59}$$

$$QC_{A_o} - kC_AC_BV = V\frac{d(C_A)}{dt} + C_AQ \tag{4.60}$$

Simultaneous solution of Eqs. (4.59) and (4.60), together with Eq. (4.58), would then give time dependence of concentrations of A and B.

Equations (4.55) and (4.56) may also be written as follows:

$$-k\frac{n_An_B}{V} = \frac{dn_B}{dt} \tag{4.61}$$

$$QC_{A_o} - k\frac{n_An_B}{V} = \frac{dn_A}{dt} \tag{4.62}$$

Simultaneous solution of these equations together with Eq. (4.58) will give expressions for the time dependence of n_A and n_B.

The concentration of reactant B within the reactor may be much higher than the concentration of reactant A in some cases. In such a case, the concentration of reactant B can be assumed constant, and the reaction can be assumed as pseudo-first-order with respect to the concentration of A. Simpler equations can be obtained for this special case.

4.6 Analysis of a CSTR with a Complex Rate Expression

For some non-linear rate expressions, multiple steady-state values can be obtained for the outlet concentrations of reactants and products [4]. Here, a typical example is illustrated considering a reaction with a rate expression containing an adsorption term in the denominator. Such reaction rate expressions are quite common for many of the solid-catalyzed reactions (Chapter 11).

Let us consider such a rate expression for a surface catalyzed reaction of reactant A to the products.

$$A \rightarrow \text{Products}$$

$$-R_A = \frac{kC_A}{(1 + K_A C_A)^2} \tag{4.63}$$

If this reaction is carried out in a CSTR, the species conservation equation for reactant A can be expressed for the constant volumetric flow rate case, as follows:

$$Q_o\left(C_{A_o} - C_{A_f}\right) - V\frac{kC_{A_f}}{\left(1 + K_A C_{A_f}\right)^2} = 0 \tag{4.64}$$

Rearrangement of this expression gives

$$\frac{Q_o}{V}\left(C_{A_o} - C_{A_f}\right) = \frac{kC_{A_f}}{\left(1 + K_A C_{A_f}\right)^2} \tag{4.65}$$

The right-hand side of Eq. (4.65) corresponds to the generation rate of species, while the left-hand side corresponds to the net removal rate from the reactor:

$$\text{Generation rate}: M_G = R = \frac{kC_{A_f}}{\left(1 + K_A C_{A_f}\right)^2} \tag{4.66}$$

$$\text{Removal rate}: M_R = \frac{Q_o}{V}\left(C_{A_0} - C_{A_f}\right) \tag{4.67}$$

Due to the non-linear nature of this rate expression, multiple steady-state solutions of Eq. (4.65) are possible for particular operating conditions. The following limiting forms of Eq. (4.63) can be written for the low and high concentration ranges of reactant A:

$$\text{If } K_A C_A \ll 1 \quad \Rightarrow \quad -R_A \cong kC_{A_f}$$

$$\text{If } K_A C_A \gg 1 \quad \Rightarrow \quad -R_A \cong \frac{k}{K_A^2} C_{A_f}^{-1}$$

The rate expression given in Eq. (4.63) passes through a maximum as the value of C_{A_f} increases. The maximum of M_G (maximum of the rate expression) can be found by equating its derivative to zero.

$$\frac{dM_G}{dC_{A_f}} = \frac{k\left(1 - K_A C_{A_f}\right)}{\left(1 + K_A C_{A_f}\right)^3} = 0 \left(\text{for the maximum value of} - R_{A_f}\right) \tag{4.68}$$

Hence, the maximum of the generation rate can be obtained at $\left(C_{A_f}\right)_{\max} = \frac{1}{K_A}$.

The variation of M_G and M_R as a function of C_{A_f} is qualitatively illustrated in Figure 4.11. While the M_G curve passes through a maximum, M_R gives a linear relation with a slope of $\left(-\frac{Q_o}{V}\right)$. Three possible steady-state values of C_{A_f} are illustrated as points A, B, and C in Figure 4.11.

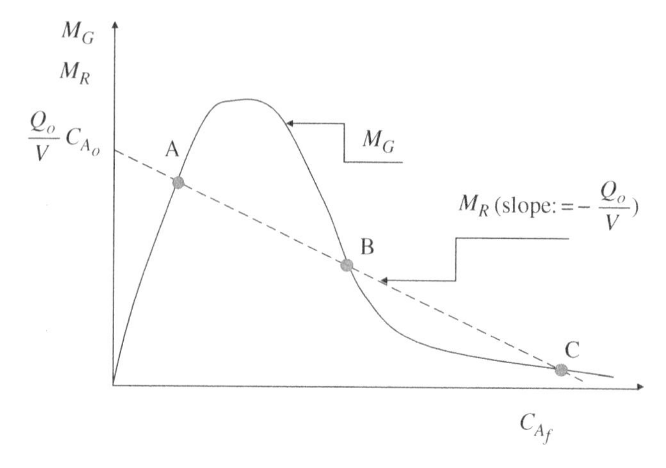

Figure 4.11 Graphical illustration of multiple steady states in a CSTR for a rate expression given in Eq. (4.63).

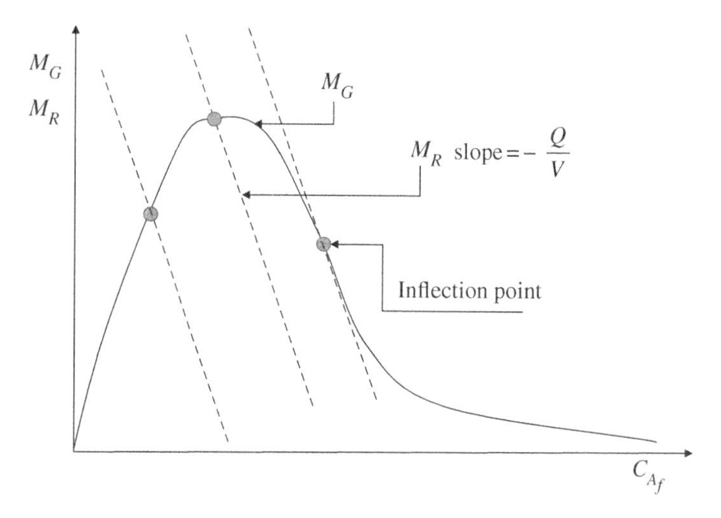

Figure 4.12 Graphical illustration of the criterion for not having a steady-state multiplicity in a CSTR.

To not have multiple steady-state solutions of Eq. (4.65) for any initial concentration of A, the absolute value of the slope of M_R expression (Eq. (4.67)) should be higher than the absolute value of the slope of M_G at its inflection point. This is illustrated in Figure 4.12. Hence, the criterion for not having multiple steady-state solutions can be expressed as follows:

$$\left| \left(\frac{dR}{dC_{A_f}} \right)_{\text{infl}} \right| \leq \left| -\frac{Q_0}{V} \right| \tag{4.69}$$

The concentration of reactant A at the inflection point of the M_G curve can be found by equating its second derivative to zero.

Problems and Questions

4.1 Experimental data obtained for the following liquid-phase reaction in a batch reactor are given in the table below. These data are obtained with the initial concentrations of A and B as 0.2 and 0.1 M, respectively.

$$2A + B \rightarrow C + D$$

x_A	0.00	0.11	0.27	0.37	0.55	0.69	0.79
$-R_A$ (mol/l s)	0.0240	0.0190	0.0128	0.0095	0.0049	0.0023	0.0011

Answer the following questions if the same reaction is carried out in a steady-flow reactor by taking the volumetric flow rate of the feed stream as 1.2 m^3/min:

(a) Find the overall order of this reaction.
(b) What is the volume of a single CSTR to achieve a fractional conversion value of 0.69?
(c) What is the volume of a single plug-flow reactor to achieve a fractional conversion value of 0.69?
(d) Compare the results obtained for CSTR and plug-flow reactors and plot the variation of the volume ratio of CSTR to plug-flow reactors as a function of fractional conversion.

4.2 Using the data reported in Example 4.2 for the gas-phase reaction $A \rightarrow B + C$ in a CSTR, answer the following questions:

(a) Evaluate the volume of a plug-flow reactor for the fractional conversion of reactant A being 0.75 for an inlet volumetric flow rate of 7.5×10^{-3} m^3/min.
(b) What is the volume of the CSTR if the feed stream contains 50% A and 50% inerts ($C_{A_0} = C_{I_0} = 0.0016$ M), and if the volumetric flow rate is 7.5×10^{-3} m^3/min.

4.3 The rate law for the following liquid-phase reaction is given as second order.

$$A + B \rightarrow C + D; \quad -R_A = 5.5 C_A C_B \text{ mol/l min}$$

The volumetric flow rate of the feed stream in a flow reactor is 1.0 l/min. The concentrations of A and B in the feed stream are given as 0.1 and 0.2 mol/l, respectively.

(a) What is the fractional conversion of reactant A that can be achieved in a plug-flow reactor having a volume of 2 l?
(b) What is the fractional conversion value of reactant A that can be achieved in a CSTR having a volume of 2 l?
(c) Plot the variation of the ratio of conversions in plug-flow and perfectly mixed reactors as a function of space–time for this system. Discuss the results.

4.4 The following elementary gas-phase reaction is taking place in a laboratory-scale spinning basket catalytic reactor, which may be modeled as a CSTR.

$$A + B \rightarrow C + D$$

The feed stream to the reactor is at 50 °C and 10 atm. It contains 5% A and 95% B. Volumetric flow rate of the feed stream and the volume of the reactor are 225 cm^3/s and 2.7 l, respectively. Experimental data obtained in this reactor at different reactor temperatures are given in the table.

Evaluate the activation energy of the reaction.

T (°C)	50	80	100	150
x_{A_f}	0.12	0.33	0.52	0.85

4.5 Derive expressions for the space–time required to reach 50% conversion of reactant A, in a CSTR, plug-flow reactor and the reaction time required in a well-mixed batch reactor if the rate expression for a liquid-phase reaction is given as follows:

$$-R_A = kC_A^{1/2}$$

4.6 An autocatalytic liquid-phase reaction is to be carried out in a flow reactor.

$$\mathrm{A} \rightarrow \mathrm{B} \quad -R_A = kC_AC_B \quad k = 4.0 \times 10^{-1} \quad (\mathrm{mol/l})^{-1}\,\mathrm{s}^{-1}$$

Feed stream to the reactor is 2.5 l/s, and feed concentration of A is 1 M. Inlet concentration of the product B is relatively low ($C_{B_o} = 0.01$ M).

(a) Plot ($-R_A$) as a function of fractional conversion of reactant A and obtain the fractional conversion value, which will maximize the reaction rate.

(b) Evaluate the volumes of CSTR and plug-flow reactors, operating at the fractional conversion value of 0.5. Compare the results and discuss them.

(c) Evaluate the volumes of CSTR and plug-flow reactors at fractional conversion values of 0.1, 0.25, 0.50, 0.70, and 0.90 and plot the ratio of CSTR and plug-flow reactor volumes as a function of fractional conversion and discuss the findings.

4.7 A second-order gaseous reactant of pure A with a concentration of 0.01 mol/l enters a perfectly mixed flow reactor (CSTR) of volume 20 l.

$$2\mathrm{A} \rightarrow \mathrm{B}; \quad -R_A = kC_A^2 \quad k = 0.1\,(\mathrm{mol/l})^{-1}\,\mathrm{s}^{-1}$$

(a) What should be the value of the inlet volumetric flow rate if the concentration of A reaches 0.005 M in the outlet stream?

(b) What is the volume of the plug-flow reactor for the conditions given in part (a) of this problem?

4.8 The following data are reported for a liquid-phase reaction of A with B taking place in a CSTR:

$$\mathrm{A} + 2\mathrm{B} \rightarrow \mathrm{C}$$

The concentration of reactant B is much higher than the concentration of reactant A. Hence, it may be considered as being constant. Inlet concentration of reactant A is given as 0.12 M, and the volume of the laboratory-scale CSTR is 2 l. Find out a rate expression for this reaction.

Q (l/min)	0.4	1.2	2.5	4.4
C_{A_f} (mol/l)	0.028	0.059	0.084	0.095

4.9 The rate expression for the liquid-phase decomposition of a reactant A to the products is given as

$$\mathrm{A} \rightarrow \mathrm{B} + \mathrm{C}; \quad -R_A = \frac{k_1 C_A^{1/2}}{1 + k_2 C_A}$$

Using the given data answer the following questions:

(a) Plot reaction rate as a function of fractional conversion of reactant A and find out the value of C_{A_f} which will give the maximum rate value.

(b) Find the space–times of a CSTR and a plug-flow reactor for the fractional conversion values of 0.2, 0.5, and 0.9 and discuss the results.

$k_1 = 0.04\ (\text{mol/l})^{1/2}\ \text{min}^{-1}$

$k_2 = 2.0\ (\text{mol/l})^{-1}$

$C_{A_o} = 1.2\ \text{M}$

4.10 Following gas-phase reaction is carried out in a plug-flow reactor. Feed stream contains 10% A, 2% C, 12% B, and 76% inerts.

$$2\mathrm{A} + \mathrm{B} \rightarrow \mathrm{C}; \quad -R_A = kC_A C_B$$

(a) Find an expression for the ratio of space–times of a CSTR and plug-flow reactors.

(b) It is desired to have a product composition of $C_{C_f} = 6C_{A_f}$. What will be the ratio of the volumes of CSTR and plug-flow reactors for this case?

4.11 For the rate expression given in Eq. (4.63), the inlet concentration of reactant A and the rate parameters k and K are given as 1.5 mol/l, 2.0 s^{-1}, and 2 l/mol, respectively. If the space–time in the reactor is three seconds, find out the steady-state value(s) of outlet concentration in a CSTR. Assume a constant volumetric flow rate in your calculations.

4.12 For the semi-batch reactor illustrated in Figure 4.10, the concentration of reactant B is much higher than the concentration of reactant A in the reactor. Hence, the rate of this reaction can be considered a pseudo-first-order with respect to reactant A and zero order with respect to B. Derive expressions for the variation of concentrations of A and B as a function of reaction time.

4.13 Consider the unsteady operation of a stirred-tank reactor illustrated in Figure 4.13. There is no input stream to the reactor, but there is an output stream with a constant flow rate. Write down the species conservation equation around the reactor for reactant A. If the reaction rate expression is first order with respect to the concentration of reactant A, derive an expression for the variation of its concentration with reaction time.

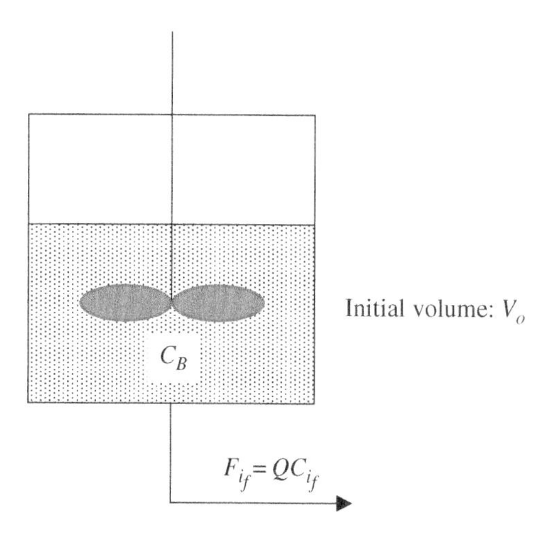

Figure 4.13 Schematic representation of the reactor in Problem 4.13.

4.14 Consider a first-order exothermic reversible reaction taking place in a CSTR.

$$A \rightleftarrows B; \quad -R_A = k\left(C_A - \frac{1}{K_C}C_B\right); \quad k = 0.31\ \exp\left(-\frac{1510}{T}\right)\ \mathrm{s}^{-1}$$

Temperature dependence of the equilibrium constant is given as

$$\ln(K_C) = 6.3 \times 10^3\left(\frac{1}{T} - \frac{1}{600}\right)$$

(a) Plot equilibrium conversion of reactant A as a function of temperature.
(b) Plot fractional conversion of reactant A achieved as a function of temperature by taking the space–time in the CSTR as 60 seconds.
(c) Evaluate the optimum conversion (highest conversion that can be achieved in this CSTR) and the corresponding reactor temperature from the graph that you plotted.
(d) What is the relative position of the constant space–time curve of a plug-flow reactor, compared to the curve obtained for a CSTR?
(e) Plot fractional conversion of reactant A as a function of temperature if a plug-flow reactor is used instead of a CSTR (space–time is again 60 seconds). Compare the maximum conversion values that can be reached in a CSTR and a plug-flow reactor.

4.15 The following data are given for a reversible decomposition reaction of reactant A ($C_{A_o} = 1$ M).

$$A \rightleftarrows B + C \quad ; \quad -R_A = k\left(C_A \frac{1}{K_C} C_B C_C\right); \quad k = 9.6\ \exp\left(-\frac{3540}{T}\right) \quad (T \text{ in K})$$

$$\ln(K_C) = 7.91 - \frac{3900}{T}$$

(a) Plot equilibrium conversion as a function of temperature.
(b) Is it an exothermic or endothermic reaction?

(c) Derive an expression relating the fractional conversion of reactant A to space–time in a CSTR.

(d) Plot fractional conversion of reactant A as a function of temperature for a space–time value of 60 seconds in a CSTR.

(e) What reactor temperature would you select if the required fractional conversion in a CSTR having a space–time of 60 seconds is 0.74?

4.16 Following reversible first-order reaction is carried out in a CSTR at 32 °C. The feed stream molar flow rate and the concentration of A in the feed stream are 80 mol/min and 1.6 mol/l, respectively. There is no product in the feed stream.

$A \rightleftarrows B; \quad k_f = 3.1\,\text{h}^{-1}; \quad K = 10.2$

(a) What is the volume of the plug-flow reactor for 75% conversion?

(b) What would be the volume for 75% conversion if the reaction was carried out in a CSTR at 32 °C?

(c) What should be the volume of the CSTR to reach equilibrium conversion?

(d) The values of the heat of reaction and the activation energy of the forward rate constant are given as −18 000 and 105 000 J/mol, respectively. Plot fractional conversion of A as a function of temperature for the reactor volume evaluated in part (b) for a CSTR, and evaluate the maximum value of the fractional conversion.

References

1 Levenspiel, O. (1999). *Chemical Reaction Engineering*. New York: Wiley.
2 Smith, J.M. (1981). *Chemical Engineering Kinetics*. Singapore: McGraw Hill.
3 Fogler, H.S. (1992). *Elements of Chemical Reaction Engineering*. NJ, USA: Prentice-Hall.
4 Perlmutter, D.D. (1972). *Stability of Chemical Reactors*. NJ, USA: Prentice Hall.

5

Multiple Reactor Systems

In many of the processes, multiple reactor combinations are used. Multiple stirred-tank reactors operating in series may be used to improve the overall efficiency of the process and to achieve higher conversions with the same total volume of the reactors. Multiple plug-flow reactors operating in series with interstage cooling or heating between the reactors are used in number of processes. In some cases, CSTR and plug-flow combinations may also be used. Modeling principles of multiple reactor combinations are illustrated in this chapter, with examples.

5.1 Multiple CSTRs Operating in Series

Let us consider a system of two perfectly stirred-tank reactors connected in series (Figure 5.1). The fractional conversions of reactant A achieved in the first and the second reactors are defined based on the inlet of the first reactor, as follows:

$$x_{A_1} = \frac{F_{A_o} - F_{A_1}}{F_{A_o}};\ x_{A_f} = \frac{F_{A_o} - F_{A_f}}{F_{A_o}} \tag{5.1}$$

The molar flow rates of reactant A at the exit of the first and the second reactors can be expressed in terms of the fractional conversion of A as

$$F_{A_1} = F_{A_o}(1 - x_{A_1});\ F_{A_f} = F_{A_o}(1 - x_{A_f}) \tag{5.2}$$

Species conservation equations for reactant A around Reactor 1 (R.1) and Reactor 2 (R.2) are

Reactor 1:

$$F_{A_o} - F_{A_1} + R_{A_1} V_1 = 0 \tag{5.3a}$$

$$F_{A_o} - F_{A_o}(1 - x_{A_1}) + R_{A_1} V_1 = 0$$

$$F_{A_o} x_{A_1} + R_{A_1} V_1 = 0 \tag{5.3b}$$

$$\frac{V_1}{F_{A_o}} = \frac{x_{A_1}}{-R_{A_1}} \tag{5.3c}$$

Fundamentals of Chemical Reactor Engineering: A Multi-Scale Approach, First Edition. Timur Doğu and Gülşen Doğu.

Companion website: www.wiley.com/go/dogu/chemreacengin

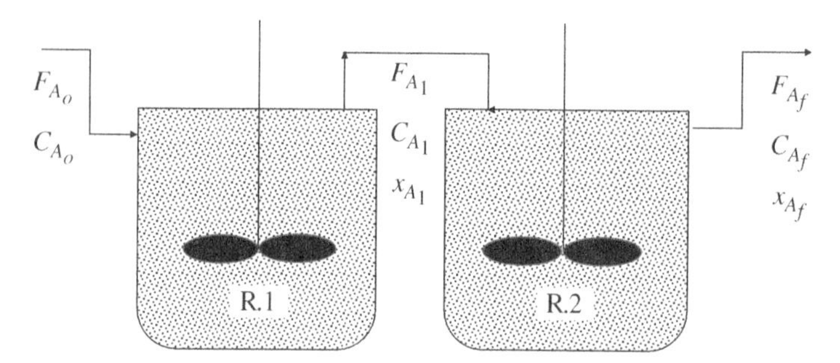

Figure 5.1 Schematic diagram of two CSTRs operating in series.

Reactor 2:

$$F_{A_1} - F_{A_f} + R_{A_f} V_2 = 0 \tag{5.4a}$$

$$F_{A_o}(1 - x_{A_1}) - F_{A_0}(1 - x_{A_f}) + R_{A_f} V_2 = 0$$

$$F_{A_o}(x_{A_f} - x_{A_1}) + R_{A_f} V_2 = 0 \tag{5.4b}$$

$$\frac{V_2}{F_{A_o}} = \frac{x_{A_f} - x_{A_1}}{-R_{A_f}} \tag{5.4c}$$

Let us illustrate the evaluation of the volumes of reactors R.1 and R.2 using the graphical approach. Consider an nth-order reaction $\left(-R_A = kC_A^n\right)$ taking place in two CSTRs operating in series. For the constant volumetric flow rate case, the design equations for reactors R.1 and R.2 can be expressed as follows:

$$\frac{V_1}{F_{A_o}} = \frac{x_{A_1}}{kC_{A_o}^n (1 - x_{A_1})^n} \tag{5.5}$$

$$\frac{V_2}{F_{A_o}} = \frac{x_{A_f} - x_{A_1}}{kC_{A_o}^n (1 - x_{A_f})^n} \tag{5.6}$$

The volumes of the reactors R.1 and R.2 can be graphically evaluated from the $(1/(-R_A))$ vs. x_A graph, as illustrated in Figure 5.2. As shown in Figure 5.2, the summation of the volumes of the two reactors operating in series is smaller than the volume of a single CSTR to obtain the same x_{A_f}.

$$\left[\frac{V_1}{F_{A_o}} + \frac{V_2}{F_{A_o}} = \left(\frac{x_{A_1}}{kC_{A_o}^n (1 - x_{A_1})^n} + \frac{x_{A_f} - x_{A_1}}{kC_{A_o}^n (1 - x_{A_f})^n}\right)\right] < \left(\frac{V_{\text{single}}}{F_{A_o}} = \frac{x_{A_f}}{kC_{A_o}^n (1 - x_{A_f})^n}\right) \tag{5.7}$$

For the constant volumetric flow rate case ($Q = Q_o$), the species conservation equations for reactors R.1 and R.2 can also be written in terms of concentrations of the reactant A in the effluent streams leaving the reactors, as follows:

$$Q(C_{A_o} - C_{A_1}) + R_{A_1} V_1 = 0; \; \tau_1 = \frac{V_1}{Q} = \frac{(C_{A_o} - C_{A_1})}{-R_{A_1}} \tag{5.8}$$

$$Q(C_{A_1} - C_{A_f}) + R_{A_f} V_2 = 0; \; \tau_2 = \frac{V_2}{Q} = \frac{(C_{A_1} - C_{A_f})}{-R_{A_f}} \tag{5.9}$$

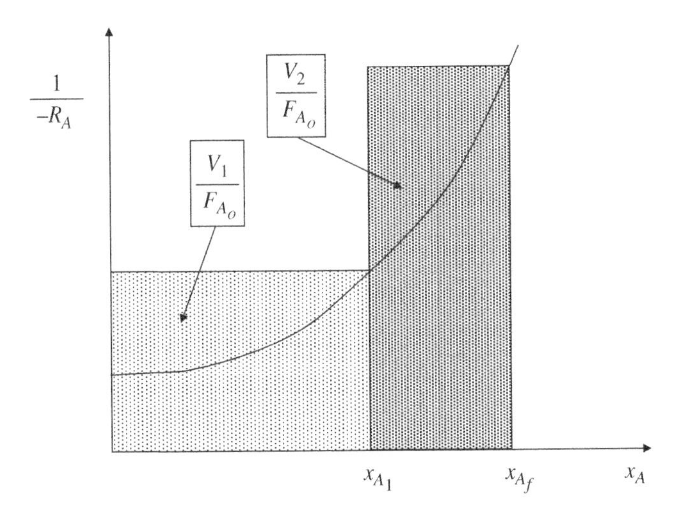

Figure 5.2 Graphical representation of reactor volumes for two CSTRs operating in series.

Considering a first-order reaction, Eqs. (5.8) and (5.9) can be expressed as

$$\tau_1 = \frac{(C_{A_0} - C_{A_1})}{kC_{A_1}} \quad \Rightarrow \quad \frac{C_{A_1}}{C_{A_0}} = \frac{1}{(1 + k\tau_1)} \tag{5.10}$$

$$\tau_2 = \frac{(C_{A_1} - C_{A_f})}{kC_{A_f}} \quad \Rightarrow \quad \frac{C_{A_f}}{C_{A_1}} = \frac{1}{(1 + k\tau_2)} \tag{5.11}$$

The following relation can then be written for the concentration of A at the exit of the second reactor by combining Eqs. (5.10) and (5.11).

$$\frac{C_{A_f}}{C_{A_0}} = \left(\frac{1}{(1 + k\tau_1)}\right)\left(\frac{1}{(1 + k\tau_2)}\right) \tag{5.12}$$

As the number of CSTRs connected in series increases, the total volume of these reactors decreases for the same overall conversion value of x_{A_f}. This is illustrated in Figure 5.3.

Species conservation equations for a four-reactor system are as follows:

$$\text{Reactor 1}: \frac{V_1}{F_{A_0}} = \frac{x_{A_1}}{-R_{A_1}} \tag{5.13}$$

$$\text{Reactor 2}: \frac{V_2}{F_{A_0}} = \frac{x_{A_2} - x_{A_1}}{-R_{A_2}} \tag{5.14}$$

$$\text{Reactor 3}: \frac{V_3}{F_{A_0}} = \frac{x_{A_3} - x_{A_2}}{-R_{A_3}} \tag{5.15}$$

$$\text{Reactor 4}: \frac{V_4}{F_{A_0}} = \frac{x_{A_f} - x_{A_3}}{-R_{A_f}} \tag{5.16}$$

As shown in Figure 5.3, for a four-CSTR system operating in series (Figure 5.4), the summation of the volumes of these CSTRs is getting close to the volume of a single plug-flow reactor having the

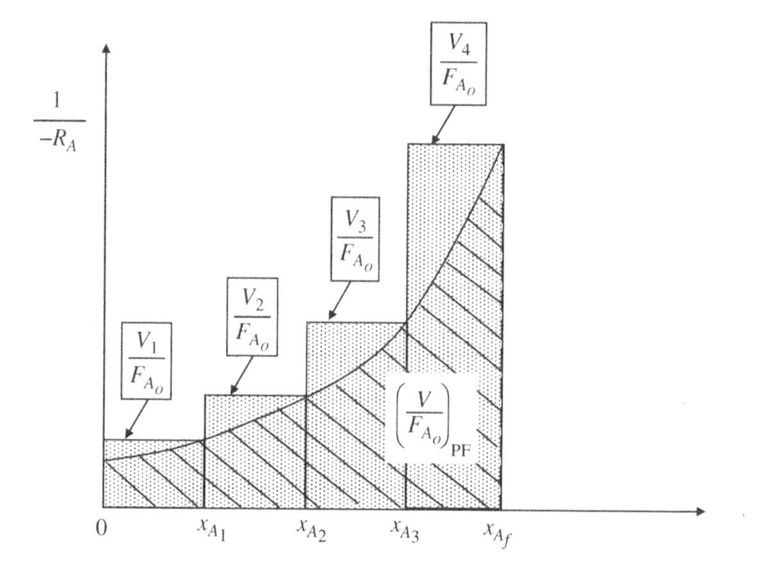

Figure 5.3 Schematic representation of the comparison of the volumes of a plug-flow reactor with the total volumes of four CSTRs operating in series.

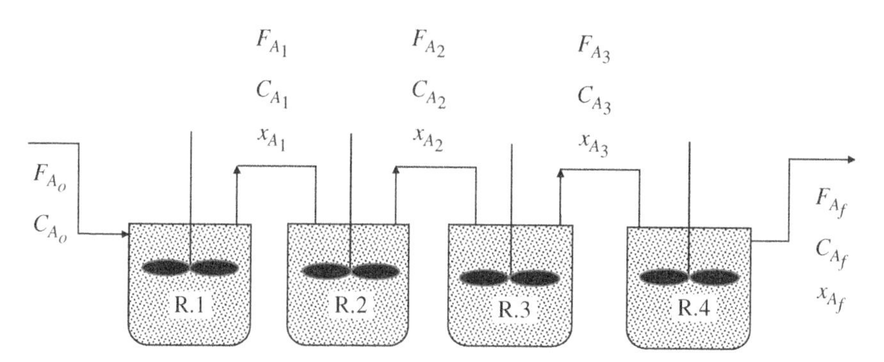

Figure 5.4 Four CSTRs operating in series.

same overall conversion as the four-reactor system. This is illustrated for a first-order reaction taking place in an N CSTR system with a constant volumetric flow rate, as follows:

$$\text{Reactor 1}: \frac{C_{A_1}}{C_{A_o}} = \frac{1}{(1 + k\tau_1)}$$

$$\text{Reactor 2}: \frac{C_{A_2}}{C_{A_1}} = \frac{1}{(1 + k\tau_2)}$$

$$\text{Reactor 3}: \frac{C_{A_3}}{C_{A_2}} = \frac{1}{(1 + k\tau_3)}$$

$$\vdots$$

$$\text{Reactor } N: \frac{C_{A_f}}{C_{A_{N-1}}} = \frac{1}{(1 + k\tau_N)}$$

If all of these expressions are multiplied side by side, we obtain the following relation:

$$\frac{C_{A_f}}{C_{A_0}} = \left(\frac{1}{(1 + k\tau_1)}\right)\left(\frac{1}{(1 + k\tau_2)}\right)\left(\frac{1}{(1 + k\tau_3)}\right)\cdots\left(\frac{1}{(1 + k\tau_N)}\right) \tag{5.17}$$

If the volumes of the CSTRs operating in series are the same, Eq. (5.17) becomes

$$\frac{C_{A_f}}{C_{A_0}} = \frac{1}{(1 + k\tau_i)^N} = \frac{1}{\left(1 + k\frac{\tau}{N}\right)^N} \tag{5.18}$$

Here, τ is the total space-time of the N reactor system.

$$\tau = \frac{V_1 + V_2 + V_3 + \cdots + V_i + \cdots + V_N}{Q} = \frac{NV_i}{Q} = \frac{V_{\text{total}}}{Q} \tag{5.19}$$

It can be shown that the limit of Eq. (5.18), as N goes to infinity, approaches the plug-flow expression for a first-order reaction.

$$\lim_{N\to\infty} \frac{C_{A_f}}{C_{A_0}} = \lim_{N\to\infty} \frac{1}{\left(1 + k\frac{\tau}{N}\right)^N} = \exp(-k\tau) \tag{5.20}$$

5.1.1 Graphical Method for Multiple CSTRs

Let us rearrange Eqs. (5.13)–(5.16) as follows:

$$-R_{A_1} = \left(\frac{F_{A_0}}{V_1}\right)x_{A_1} = \left(\frac{C_{A_0}}{\tau_1}\right)x_{A_1} \tag{5.21}$$

$$-R_{A_2} = \left(\frac{F_{A_0}}{V_2}\right)(x_{A_2} - x_{A_1}) = \left(\frac{C_{A_0}}{\tau_2}\right)(x_{A_2} - x_{A_1}) \tag{5.22}$$

$$-R_{A_3} = \left(\frac{F_{A_0}}{V_3}\right)(x_{A_3} - x_{A_2}) = \left(\frac{C_{A_0}}{\tau_3}\right)(x_{A_3} - x_{A_2}) \tag{5.23}$$

$$-R_{A_f} = \left(\frac{F_{A_0}}{V_4}\right)(x_{A_f} - x_{A_3}) = \left(\frac{C_{A_0}}{\tau_4}\right)(x_{A_f} - x_{A_3}) \tag{5.24}$$

In this graphical method, we plot the variation of $(-R_A)$ as a function of fractional conversion of the limiting reactant A. For an arbitrary rate expression, the plot of $(-R_A)$ as a function of x_A is given in Figure 5.5. Note that the slopes of the linear relations shown in Figure 5.5 can be used to evaluate the volumes of the reactors, as follows:

$$\frac{-R_{A_1}}{x_{A_1}} = \frac{F_{A_0}}{V_1};\ \frac{-R_{A_2}}{(x_{A_2} - x_{A_1})} = \frac{F_{A_0}}{V_2};\ \frac{-R_{A_3}}{(x_{A_3} - x_{A_2})} = \frac{F_{A_0}}{V_3};\ \frac{-R_{A_f}}{(x_{A_f} - x_{A_3})} = \frac{F_{A_0}}{V_4} \tag{5.25}$$

Example 5.1 Minimization of total cost for a multiple CSTR system

Following pseudo-first-order liquid-phase reaction is taking place in two identical CSTRs connected in series.

$$\text{A} + \text{B} \to \text{C};\ -R_A = kC_A;\ k = 0.02\ \text{min}^{-1}$$

Concentrations of A and B in the first reactor feed stream are 0.1 and 1.0 M, respectively. The production rate of the product C is given as 2.0×10^4 mol/day. Using the following data, evaluate

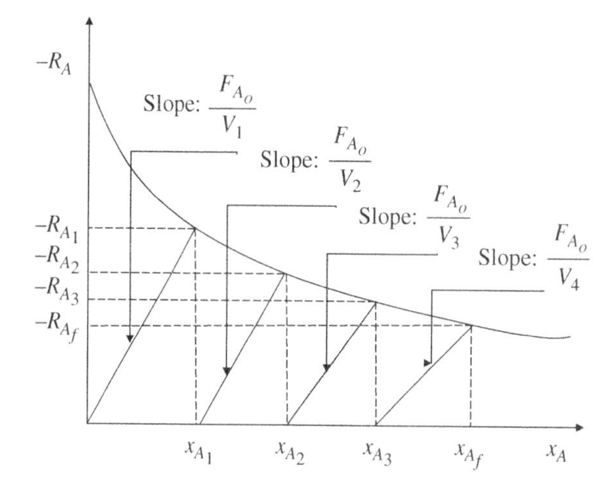

Figure 5.5 Graphical approach for multiple CSTRs operating in series.

the total volume of the reactors and the fractional conversion of A at the exit of the first and the second reactors to minimize the total cost of the process. In your calculations, consider only the cost of the reactants and the fixed cost of the reactors.

Data : Unit costs of reactants : (A) $a = 0.09$ \$/mol; (B) $b = 0.01$ \$/mol

Fixed cost of reactors : $c = 56\,000$ \$/m^3 (It is desired to depreciate the fixed cost of reactors in five years)

Solution

As shown in Eq. (5.18), for a first-order liquid-phase reaction with a constant volumetric flow rate, the ratio of concentrations at the exit of the second reactor and inlet of the first reactor is as follows:

$$\frac{C_{A_f}}{C_{A_o}} = \left(\frac{1}{(1+k\tau_i)}\right)^2; \ \tau_i = \frac{V_i}{Q} = \frac{1}{k}\left[\left(\frac{C_{A_o}}{C_{A_f}}\right)^{1/2} - 1\right]$$

$$V_{\text{total}} = 2V_i = \frac{2Q}{k}\left[\left(\frac{C_{A_o}}{C_{A_f}}\right)^{1/2} - 1\right]$$

Since $F_{C_f} = F_{A_o}x_{A_f} = QC_{A_o}x_{A_f}$, the expression for total volume can also be expressed as follows:

$$V_{\text{total}} = \frac{2F_{C_f}}{kC_{A_o}x_{A_f}}\left[\left(\frac{C_{A_o}}{C_{A_f}}\right)^{1/2} - 1\right] = \frac{2F_{C_f}}{kC_{A_o}x_{A_f}}\left[\left(\frac{1}{(1-x_{A_f})}\right)^{1/2} - 1\right]$$

Molar flow rates of A and B at the inlet of the first reactor can be expressed in terms of the production rate of C as follows:

$$F_{A_o} = \frac{F_{C_f}}{x_{A_f}}; \ F_{B_o} = F_{A_o}\frac{C_{B_o}}{C_{A_o}} = \frac{F_{C_f}}{x_{A_f}}\left(\frac{C_{B_o}}{C_{A_o}}\right)$$

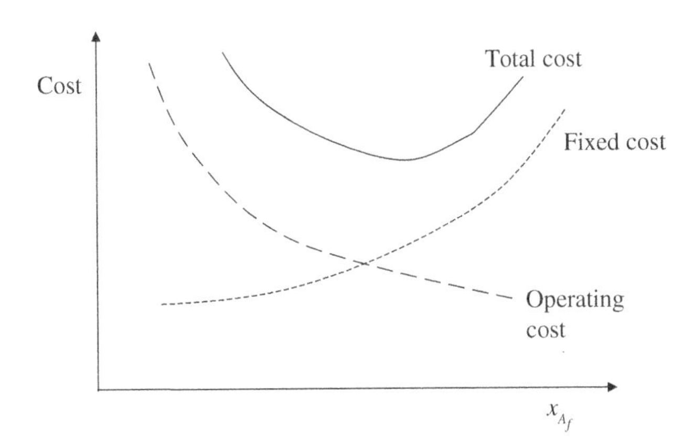

Figure 5.6 Variation of the total cost with the fractional conversion for a fixed production rate.

Using the yearly operation period of the reactors as 320 days and taking $\frac{C_{B_0}}{C_{A_0}} = 10$, total cost expression, based on one day, can be written as follows:

Total Cost = Fixed Cost of Reactors + Operation Cost (Cost of Reactants); (\$/day)

$$TC = \frac{c}{5(320)} V_{\text{total}} + aF_{A_0} + bF_{B_0}$$

$$TC = \left(\frac{c}{5(320)}\right) \frac{2F_{C_f}}{kC_{A_0}x_{A_f}} \left[\left(\frac{1}{(1-x_{A_f})}\right)^{1/2} - 1\right] + a\frac{F_{C_f}}{x_{A_f}} + b\frac{F_{C_f}}{x_{A_f}}\left(\frac{C_{B_0}}{C_{A_0}}\right)$$

For a fixed production rate, an increase in conversion causes a decrease in the feed flow rate of reactants and, hence, a decrease in the operation cost. However, an increase in conversion results in an increase in the reactor volume, which will cause an increase in fixed cost. Hence, the variation of the total cost with fractional conversion is expected to pass through a minimum (Figure 5.6). For the minimum cost, the derivative of the total cost expression with respect to x_{A_f} should be zero.

$$\frac{d(TC)}{dx_{A_f}} = \left(\frac{cF_{C_f}}{800kC_{A_0}}\right)\left[-\frac{1}{x_{A_f}^2}\left(\left(\frac{1}{(1-x_{A_f})}\right)^{1/2} - 1\right) + \frac{1}{2x_{A_f}}\left(\frac{1}{(1-x_{A_f})}\right)^{3/2}\right]$$

$$-(a + 10b)\frac{F_{C_f}}{x_{A_f}^2} = 0$$

Taking the derivative of the total cost with respect to x_{A_f} and equating to zero, fractional conversion of A, for the minimum cost, is found as about $x_{A_f} = 0.87$. The total volume of the two reactors is then evaluated as 28.3 m^3.

5.2 Multiple Plug-Flow Reactors Operating in Series

In some cases, plug-flow reactors may also be connected in series. For instance, for the catalytic oxidation of SO_2 to SO_3 in a sulfuric acid plant, plug-flow reactors may be connected in series with interstage cooling of the streams between the reactors (Figure 5.7). This kind of operation of

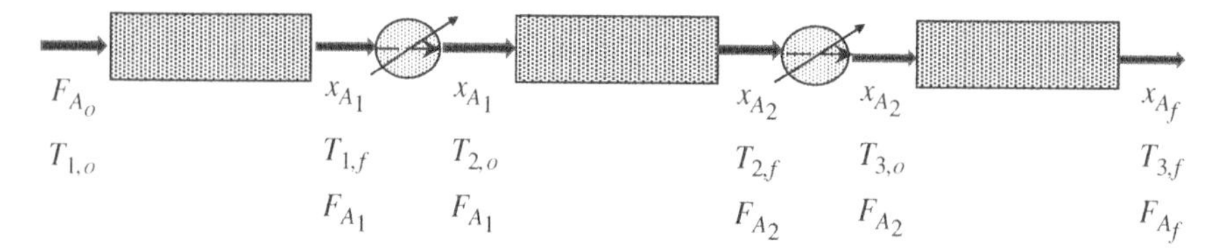

Figure 5.7 Schematic representation of three plug-flow reactors operating in series.

plug-flow reactors is further discussed in Chapter 7. Species conservation equations for the multiple plug-flow reactors shown in Figure 5.7 should be written as follows:

$$\frac{V_1}{F_{A_o}} = \int_0^{x_{A_1}} \frac{dx_A}{-R_A} \tag{5.26a}$$

$$\frac{V_2}{F_{A_o}} = \int_{x_{A_1}}^{x_{A_2}} \frac{dx_A}{-R_A} \tag{5.26b}$$

$$\frac{V_3}{F_{A_o}} = \int_{x_{A_2}}^{x_{A_f}} \frac{dx_A}{-R_A} \tag{5.26c}$$

In the case of a first-order irreversible reaction with constant volumetric flow rate, species conservation equations can be integrated to yield the following relations for the isothermal reactors R.1 and R.2:

$$\frac{V_1}{F_{A_o}} = \int_0^{x_{A_1}} \frac{dx_A}{kC_{A_o}(1-x_A)} = -\frac{1}{kC_{A_o}} \ln(1-x_{A_1}) \tag{5.27a}$$

$$\frac{V_2}{F_{A_o}} = \int_{x_{A_1}}^{x_{A_2}} \frac{dx_A}{kC_{A_o}(1-x_A)} = -\frac{1}{kC_{A_o}} \ln\frac{(1-x_{A_2})}{(1-x_{A_1})} \tag{5.27b}$$

5.3 CSTR and Plug-Flow Reactor Combinations

Reaction rate expressions containing adsorption terms in the denominator are quite common for many solid-catalyzed reactions. Let us consider such a rate expression for the catalytic decomposition of reactant A.

$$-R_A = \frac{kC_A}{(1+K_AC_A)^2} \tag{5.28}$$

Rearrangement of this expression gives

$$\frac{1}{-R_A} = \frac{(1+K_AC_A)^2}{kC_A} \tag{5.29}$$

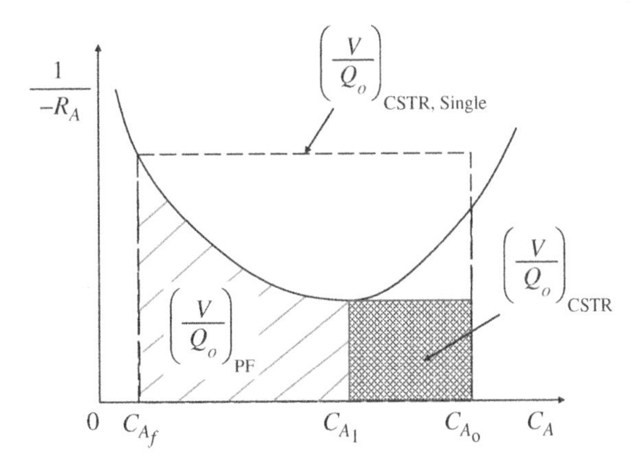

Figure 5.8 Schematic representation of the volumes of the CSTR and plug-flow reactors with the rate expression given in Eq. (5.28).

To decide whether the volume of a CSTR, a plug-flow reactor, or a combination of these reactors is the smallest for the same overall conversion of reactant A, let us use the graphical approach shown in Figure 5.8. The plot of $\left(\frac{1}{-R_A}\right)$ as a function of C_A is expected to pass through a minimum for such a rate expression. Reactor volumes for a single CSTR and a single plug-flow reactor can be evaluated from the following equations with the reaction rate expression given in Eq. (5.28).

$$\frac{V}{Q_o} = \frac{(C_{A_o} - C_{A_f})}{-R_{A_f}} = \frac{(1 + K_A C_{A_f})^2}{k C_{A_f}} (C_{A_o} - C_{A_f}) \text{ (single CSTR)} \tag{5.30}$$

$$\frac{V}{Q_o} = \int_{C_{A_o}}^{C_{A_f}} \frac{dC_A}{R_A} = \int_{C_{A_f}}^{C_{A_o}} \frac{(1 + K_A C_A)^2}{k C_A} dC_A \text{ (single plug-flow reactor)} \tag{5.31}$$

As shown in Figure 5.8, a series combination of a CSTR followed by a plug-flow reactor gives a smaller total volume than a single CSTR and also a single plug-flow reactor. Hence, it is recommended to use a CSTR between C_{A_o} and concentration C_{A_1}, which corresponds to the minimum in the $\left(\frac{1}{-R_A}\right)$ vs. concentration curve. The volume of the plug-flow reactor is smaller than the volume of a CSTR between C_{A_1} and C_{A_f}. Volumes of the reactors in the series combination can then be evaluated as follows:

$$\frac{V_1}{Q_o} = \frac{(C_{A_o} - C_{A_1})}{-R_{A_1}} = \frac{(1 + K_A C_{A_1})^2}{k C_{A_1}} (C_{A_o} - C_{A_1}) \text{ (CSTR)} \tag{5.32}$$

$$\frac{V_2}{Q_o} = \int_{C_{A_1}}^{C_{A_f}} \frac{dC_A}{R_A} = \int_{C_{A_f}}^{C_{A_1}} \frac{(1 + K_A C_A)^2}{k C_A} dC_A \text{ (plug-flow reactor)} \tag{5.33}$$

The volume of a single plug-flow reactor can be evaluated from the area under the curve between C_{A_o} and C_{A_f}, in Figure 5.8.

Further discussion of multiple reactor combinations may be found in Ref. [1,2].

Problems and Questions

5.1 The following initial-rate data are given for the vapor-phase dehydration reaction of alcohol A.

A → B + C

$(-R_A)\times 10^{-5}$ (mol/m^3 h)	4	6	8.3	8.3	5.9	4.8	4.2
P_A (atm)	2.0	3.2	7.0	9.0	13.0	20.0	30.0
$(1/(-R_A))\times 10^6)$	2.5	1.67	1.20	1.20	1.70	2.1	2.4

Answer the following questions if the feed stream is pure A and the system pressure and the temperature are given as 28 atm and 120 °C, respectively.

(a) Evaluate the space–time of a single CSTR for a fractional conversion of A as 0.6.
(b) It is proposed to use two CSTRs operating in series for this reaction. If the partial pressures of reactant A at the exit of the first and the second reactors are 13 and 7 atm, respectively, find the corresponding space-times of these reactors.
(c) Compare the total space-times of the two-reactor system described in part (b) with the space-time of the single reactor in part (a). Discuss your findings.

5.2 A first-order liquid-phase reaction is carried out in three equal-size isothermal CSTRs operating in series. The concentration of the reactant A in the feed stream is 0.2 mol/l.

$$A \rightarrow 2C\ ;\ -R_A = 2.2C_A\ \left(\text{kmol/m}^3\ \text{h}\right)$$

The production rate of C is given as 10^3 mol/min.

(a) Evaluate the volumes of the three reactors for the overall conversion of 0.75.
(b) What are the fractional conversion values at the outlet of the first and the second reactors?

5.3 Two well-mixed ideal stirred-tank reactors of volumes V_1 and $2V_1$ are available for a second-order reaction.

$$A + B \rightarrow C + D\ ;\ -R_A = kC_AC_B$$

It is also given that the inlet concentration of B is twice the inlet concentration of reactant A. Derive expressions for the overall fractional conversion of A for the following two cases in which these reactors are connected in series and discuss the results.

(a) The smaller reactor is the first stage in the series arrangement.
(b) The larger reactor is the first stage in the series arrangement.
(c) Compare the total space-time of the series arrangements with the space-time of a single CSTR to achieve the same overall conversion.

5.4 The following rate data are given for a liquid-phase decomposition reaction of A.

$-R_A$ (mol/(l min))	0.05	0.1	0.3	0.5	0.55	0.40	0.25	0.15
x_A	0	0.10	0.25	0.35	0.45	0.55	0.65	0.80

(a) What type and arrangement of ideal flow reactors would give a minimum total volume for an overall fractional conversion of 0.8? In the solution of this problem, one should consider single ideal CSTR and plug-flow reactors, as well as a series combination of ideal flow reactors.

(b) What are the fractional conversion values at the exit of each reactor in the series arrangement that is selected in part (a)? Also, calculate the volumes of single reactors and the reactors operating in series if the inlet molar flow rate of A is 10 mol/min.

5.5 The following rate expression is given for a liquid-phase decomposition of reactant A:

$$A \rightarrow B + C; \; -R_A = \frac{7C_A^{2/3}}{(1 + 10C_A)}$$

If the inlet concentration of reactant A is 1 M and the required fractional conversion is 0.95, answer the following questions:

(a) What type and arrangement of ideal flow reactors should be recommended for the minimum value of the total reactor volume?

(b) What are the space-times of the reactors in the arrangement that are proposed in part (a)?

5.6 Two equal-size CSTRs operating in series are to be used for the following liquid-phase autocatalytic reaction:

$$A \rightarrow 2C; \; -R_A = kC_AC_C; \; k = 6.0 \; (\text{mol/l})^{-1}\text{h}^{-1}$$

Inlet concentrations of A and C are given as $C_{A_0} = 1.0$ M and $C_{C_0} = 0.2$ M, respectively, and the volumetric flow rate is 10 l/min.

(a) Plot reaction rate as a function of fractional conversion of reactant A.

(b) If space-time in each CSTR is given as seven minutes, find the fractional conversion of A at the exit of the first reactor. Also, evaluate its overall conversion at the exit of the two-reactor system (you may use a graphical method for the solution).

(c) The desired overall conversion is given as 0.80. What should be the best arrangement of two flow reactors (CSTR or plug-flow reactors) to achieve minimum total reactor volume? What should be the fractional conversion at the exit of the first reactor for this case? Also, evaluate the volumes of both reactors, corresponding to the total minimum volume case.

5.7 Following gas-phase reaction takes place in two CSTRs operating in series:

$$A + B \rightarrow C \quad ; \quad -R_A = kC_A C_B;\ k = 0.3\ (\text{mol/l})^{-1}\text{s}^{-1}$$

Inlet concentrations of A and B are given as $C_{A_0} = 0.1$ M and $C_{B_0} = 0.10$ M and the volumetric flow rate is given as 20 l/h.

The volumes of the first and the second CSTRs in the series arrangement are given as 1 and 2 l, respectively. Evaluate the fractional conversion of the limiting reactant and concentrations of A, B, and C at the exit of the first and the second reactors.

5.8 An isothermal plug-flow reactor is used for the production of B from A through a first-order reaction. The outlet stream of this reactor (operating at steady state) is suddenly connected to a CSTR, having the same volume as the plug-flow reactor and operating at the same temperature. Initially, the concentration of reactant A in this CSTR is zero, but it is filled with an inert fluid of the same volume.

(a) Derive an expression for the change of the exit stream concentration of the CSTR with time.

(b) Find an expression for the overall conversion of A at the exit of the CSTR for a sufficiently long time to reach a steady state for the whole system.

References

1 Levenspiel, O. (1999). *Chemical Reaction Engineering*. Hoboken, NJ: Wiley.
2 Smith, J.M. (1981). *Chemical Engineering Kinetics*. Singapore: McGraw Hill.

6

Multiple Reaction Systems

Most of the industrially significant chemical conversions involve more than one reaction. Besides the desired product, several undesired products may also be produced through parasitic reactions. For instance, during ethylene oxide production by the oxidation of ethylene over a silver-based catalyst, total oxidation of ethylene and ethylene oxide also takes place, producing CO_2 and CO.

$$C_2H_4 + \frac{1}{2}O_2 \rightarrow C_2H_4O$$

$$C_2H_4 + 3O_2 \rightarrow 2CO_2 + 2H_2O$$

$$C_2H_4O + \frac{5}{2}O_2 \rightarrow 2CO_2 + 2H_2O$$

Another example of a multiple reaction system is the steam reforming of methane to produce hydrogen. In this case, the water–gas shift reaction (WGSR) also contributes to product distribution. Also, the occurrence of the Boudouard reaction may cause coke formation over the catalyst.

$$CH_4 + H_2O \rightleftharpoons CO + 3H_2 \quad \text{(steam reforming)}$$

$$CO + H_2O \rightleftharpoons CO_2 + H_2 \quad \text{(WGSR)}$$

$$2CO \rightleftharpoons CO_2 + C \quad \text{(Boudouard reaction)}$$

WGSR and Boudouard reaction may also contribute to the product distribution during the conversion of biogas to synthesis gas through dry reforming of methane.

$$CO_2 + CH_4 \rightleftharpoons 2CO + 2H_2 \text{ (dry reforming)}$$

Another example of a system involving multiple reactions is the synthesis of dimethyl ether (DME) from synthesis gas over a bifunctional catalyst.

$$2CO + 4H_2 \rightleftharpoons CH_3 - O - CH_3 + H_2O \text{ (DME synthesis)}$$

This overall reaction involves methanol synthesis, methanol dehydration, and WGSRs taking place at the same proximity within the reactor.

$$CO + 2H_2 \rightleftharpoons CH_3OH \quad \text{(methanol synthesis)}$$

$$2CH_3OH \rightleftharpoons (CH_3)O(CH_3) + H_2O \quad \text{(methanol dehydration)}$$

$$CO + H_2O \rightleftharpoons CO_2 + H_2 \quad \text{(WGSR)}$$

Fundamentals of Chemical Reactor Engineering: A Multi-Scale Approach, First Edition. Timur Doğu and Gülşen Doğu.

Companion website: www.wiley.com/go/dogu/chemreacengin

6.1 Selectivity and Yield Definitions

Consider a multiple reaction system containing m species and n independent reactions. The rate of formation of species i, namely R_i can be written in terms of the specific rates of reactions $R_{(j)}$ and the stoichiometric coefficients of the ith species in the jth reaction, ν_{ij}, as follows:

$$R_i = \sum_{j=1}^{n} \nu_{ij} R_{(j)} \tag{6.1}$$

Type of the reactor, hydrodynamics within the reactor, reaction temperature, and the concentrations of the reacting species may have significant effects on the maximization of the yield of the desired product. Point selectivity of the desired product B with respect to an undesired product C is defined as the ratio of reaction rates of these products (s_{BC}). Point selectivity of the desired product may also be expressed with respect to the limiting reactant A (s_{BA}). For instance, for a simple reaction system involving two parallel reactions, selectivity definitions are given as in Eqs. (6.2) and (6.3).

$$A + D \rightarrow B \text{ (B is the desired product)}$$

$$A + E \rightarrow C \text{ (undesired reaction)}$$

$$s_{BC} = \frac{R_B}{R_C} \text{ (point selectivity of product B with respect to product C)} \tag{6.2}$$

$$s_{BA} = \frac{R_B}{-R_A} \text{ (point selectivity of product B with respect to reactant A)} \tag{6.3}$$

We can also define the overall selectivity of the desired product as the ratio of the formation rate of product B to the consumption rate of the limiting reactant A in a chemical reaction vessel.

$$S_{BA} = \frac{F_{B_f}}{F_{A_o} - F_{A_f}} \tag{6.4}$$

Overall selectivity of the desired product with respect to the undesired product may also be defined as follows:

$$S_{BC} = \frac{F_{B_f}}{F_{C_f}} \tag{6.5}$$

The relations between the point and the overall selectivities depend upon the type and hydrodynamics of the reactor.

In many cases, we also make yield definition for the desired product. Yield may be expressed as the product of overall selectivity with the overall fractional conversion of the limiting reactant A.

$$Y_B = \frac{F_{B_f}}{F_{A_o}} = \left(\frac{F_{B_f}}{(F_{A_o} - F_{A_f})}\right)\left(\frac{F_{A_o} - F_{A_f}}{F_{A_o}}\right) = S_{BA} x_{A_f} \tag{6.6}$$

If we consider a simple parallel reaction system, where the desired reaction and the undesired reaction are nth and mth orders with respect to the limiting reactant A, respectively

$$A + D \rightarrow B \quad \text{(B is the desired product)}; R_B = k_1 C_A^n$$

$$A + E \rightarrow C \; ; \qquad R_C = k_2 C_A^m$$

the rate of consumption of A can be expressed as follows:

$$-R_A = k_1 C_A^n + k_2 C_A^m \tag{6.7}$$

For this system, point selectivity of the desired product B with respect to the undesired product C or with respect to the limiting reactant A can be expressed as

$$s_{BC} = \frac{R_B}{R_C} = \frac{k_1 C_A^n}{k_2 C_A^m} = \left(\frac{k_1}{k_2}\right) C_A^{(n-m)} \tag{6.8}$$

$$s_{BA} = \frac{R_B}{-R_A} = \frac{k_1 C_A^n}{k_1 C_A^n + k_2 C_A^m} = \frac{\frac{k_1}{k_2} C_A^{(n-m)}}{\frac{k_1}{k_2} C_A^{(n-m)} + 1} \tag{6.9}$$

Concentration dependence of the selectivity and the yield of the desired product strongly depends upon the reaction orders of the desired and undesired reactions. For this specific case,

if $(n - m) > 0$; high concentration of reactant A favors selectivity of the desired product
if $(n - m) < 0$; high concentration of reactant A favors selectivity of the undesired product

A similar analysis may be made for more complex reaction systems.

As seen in Eqs (6.8) and (6.9), the selectivity of the desired product also depends on the reaction temperature. The ratio of the rate constants for the desired and the undesired reactions can be expressed as

$$\frac{k_1}{k_2} = \frac{k_1^o \exp\left(-\frac{E_{a_1}}{R_g T}\right)}{k_2^o \exp\left(-\frac{E_{a_2}}{R_g T}\right)} = \frac{k_1^o}{k_2^o} \exp\left[-\frac{(E_{a_1} - E_{a_2})}{R_g T}\right] \tag{6.10}$$

If the activation energy of the desired reaction is higher than the activation energy of the undesired reaction

$$(E_{a_1} - E_{a_2}) > 0 \tag{6.11}$$

higher temperatures will favor the selectivity of the desired product. However, if the activation energy of the undesired reaction is higher than the activation energy of the desired product, lower reaction temperatures will favor the selectivity of the desired product.

6.2 Selectivity Relations for Ideal Flow Reactors

The relation between the point selectivity and the overall selectivity strongly depends upon the hydrodynamics of the reactor. For instance, for an ideal stirred-tank flow reactor (CSTR), species conservation equations for the desired product B and for the reactant A are

$$F_{B_f} = V R_{B_f} \tag{6.12}$$

$$F_{A_o} - F_{A_f} = -V R_{A_f} \tag{6.13}$$

These equations can be combined with the selectivity definition given in Eqs. (6.3) and (6.4) to obtain the following relation:

$$S_{BA} = \frac{F_{B_f}}{(F_{A_o} - F_{A_f})} = \frac{R_{B_f}}{-R_{A_f}} = s_{BA_f} \text{ (for a CSTR)} \tag{6.14}$$

Hence, for an ideal stirred-tank flow reactor (CSTR), the overall selectivity of the desired product is equal to the point selectivity evaluated at the reactor outlet conditions. Equation (6.14) can be rearranged as

$$F_{B_f} = s_{BA_f}(F_{A_o} - F_{A_f}) \tag{6.15}$$

In the case of a constant volumetric flow rate, Eq. (6.15) reduces to

$$C_{B_f} = s_{BA_f}(C_{A_o} - C_{A_f}) \text{ (for a CSTR)} \tag{6.16}$$

In the case of a tubular reactor, concentrations of the reactants and the products change along the reactor. Hence, point selectivity may also change along the reactor. In the case of an ideal plug-flow reactor (PFR), species conservation equations for the desired product B and the reactant A can be expressed as follows:

$$\frac{dF_A}{dV} = R_A \tag{6.17}$$

$$\frac{dF_B}{dV} = R_B \tag{6.18}$$

Hence, we can write the following relation for point selectivity of B with respect to A as

$$\frac{dF_B}{dF_A} = \frac{R_B}{R_A} = -s_{BA} \tag{6.19}$$

Integration of Eq. (6.19) and using the definition of overall selectivity (Eq. (6.4)) give the following relation between the overall selectivity and the point selectivity of the desired product in a PFR.

$$S_{BA} = \frac{F_{B_f}}{(F_{A_o} - F_{A_f})} = -\frac{1}{(F_{A_o} - F_{A_f})}\int_{F_{A_0}}^{F_{A_f}} s_{BA}dF_A = \frac{1}{x_{A_f}}\int_0^{x_{A_f}} s_{BA}dx_A \tag{6.20}$$

For the case of a constant volumetric flow rate within the reactor, Eq. (6.20) can be written as follows:

$$S_{BA} = \frac{QC_{B_f}}{Q(C_{A_o} - C_{A_f})} = -\frac{1}{(C_{A_o} - C_{A_f})}\int_{C_{A_0}}^{C_{A_f}} s_{BA}dC_A \tag{6.21}$$

The following expression can then be written for the reactor exit concentration of product B in a PFR, for the constant volumetric flow rate case.

$$C_{B_f} = -\int_{C_{A_0}}^{C_{A_f}} s_{BA}dC_A \tag{6.22}$$

Product distributions for the same reaction system taking place in a CSTR and a PFR indicate higher selectivity of the desired product B in the PFR if the order of the reaction rate of the desired reaction is higher than the order of the reaction rate of the undesired reaction ($n > m$). This is simply due to the lower concentration of the limiting reactant A in a CSTR than in a PFR. As discussed before, the concentrations of species within a CSTR are assumed to be equal to their concentrations in the reactor exit stream. However, in the case of a PFR, variation of concentrations is expected

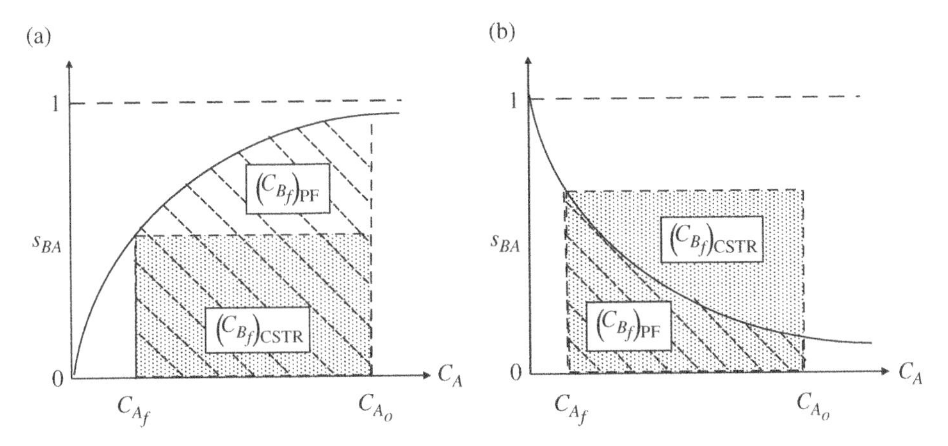

Figure 6.1 Comparison of concentrations of desired product B in a CSTR and a plug-flow reactor for a parallel reaction scheme. (a) n >m, (b) n <m.

along the reactor. Therefore, if the order of the reaction rate of the desired reaction is lower than the order of the reaction rate of the undesired reaction ($n < m$), the yield of the desired product B is expected to be higher in a CSTR than in a PFR. This fact can be graphically illustrated as described below.

If we plot point selectivity of B as a function of the concentration of reactant A, the area under the curve between the concentration values of C_{A_0} and C_{A_f} gives the outlet concentration of product B $\left(C_{B_f}\right)$ for the PFR. However, the product of s_{BA_f} with $\left(C_{A_0} - C_{A_f}\right)$ gives the value of C_{B_f} in a CSTR (Eq. (6.16)). This is illustrated in Figure 6.1. In the first illustration of Figure 6.1, point selectivity increases with an increase in the concentration of reactant A. This indicates that the reaction order of the desired reaction is higher than the reaction order of the undesired reaction for this case. As shown in Figure 6.1, $\left(C_{B_f}\right)_{PF} > \left(C_{B_f}\right)_{CSTR}$ if $n > m$. In the case of the second illustration of Figure 6.1, point selectivity of B decreases with an increase in the concentration of reactant A. This indicates that $n < m$ for this case. For the case of $n < m$, $\left(C_{B_f}\right)_{CSTR}$ is higher than $\left(C_{B_f}\right)_{PF}$.

Example 6.1 Analysis of Parallel Reactions in a CSTR

Reactants A, B, and C are fed to a CSTR with equal molar flow rates ($F_{A_0} = F_{B_0} = F_{C_0}$). The following liquid-phase parallel reactions are taking place in the reactor:

$$\text{A} + \text{C} \rightarrow \text{E}\,;\ R_E = k_1 C_A C_C$$

$$\text{B} + \text{C} \rightarrow \text{F}\,;\ R_F = k_2 C_B C_C$$

If the ratio of rate constants is $\frac{k_2}{k_1} = 0.25$ and the fractional conversion of A is given as $x_{A_f} = 0.5$, determine the fractional conversion of B and the ratio of molar flow rates of E to F in the product stream.

Solution

Point selectivity expressions for E with respect to reactant C and with respect to the other product F can be written as follows:

$$s_{EC} = \frac{R_E}{-R_C} = \frac{k_1 C_A C_C}{k_1 C_A C_C + k_2 C_B C_C} = \frac{\frac{k_1}{k_2}\frac{C_A}{C_B}}{\frac{k_1}{k_2}\frac{C_A}{C_B} + 1} = \frac{\frac{k_1}{k_2}\frac{(1-x_A)}{(1-x_B)}}{\frac{k_1}{k_2}\frac{(1-x_A)}{(1-x_B)} + 1}$$

$$s_{EF} = \frac{R_E}{R_F} = \frac{k_1 C_A C_C}{k_2 C_B C_C} = \frac{k_1}{k_2}\frac{(1-x_A)}{(1-x_B)}$$

Fractional conversions of A and B at the reactor outlet x_{A_f} and x_{B_f} are defined as follows:

$$x_{A_f} = \frac{F_{A_o} - F_{A_f}}{F_{A_o}} = \frac{F_{E_f}}{F_{A_o}};\ x_{B_f} = \frac{F_{B_o} - F_{B_f}}{F_{B_o}} = \frac{F_{F_f}}{F_{B_o}}$$

Hence, the ratio of E to F in the product stream depends upon the fractional conversion of B.

$$\frac{F_{E_f}}{F_{F_f}} = \frac{F_{A_o}}{F_{B_o}}\frac{x_{A_f}}{x_{B_f}} = \frac{F_{A_o}}{F_{B_o}}\frac{0.5}{x_{B_f}} = \frac{1}{2x_{B_f}}$$

Since the overall selectivity is equal to the point selectivity (evaluated at the reactor outlet conditions) in a CSTR, the following relations can be written:

$$S_{EF} = (s_{EF})_f; \qquad S_{EF} = \frac{F_{E_f}}{F_{F_f}} = \frac{k_1 C_{A_f} C_{C_f}}{k_2 C_{B_f} C_{C_f}}$$

$$\frac{1}{2x_{B_f}} = \frac{k_1}{k_2}\frac{(1-x_{A_f})}{(1-x_{B_f})} = \left(\frac{1}{0.25}\right)\left(\frac{0.5}{1-x_{B_f}}\right);\ x_{B_f} = 0.2$$

Hence, the ratio of E to F in the product stream becomes

$$S_{EF} = \frac{F_{E_f}}{F_{F_f}} = \frac{F_{A_o}}{F_{B_o}}\frac{x_{A_f}}{x_{B_f}} = \frac{0.5}{0.2} = 2.5$$

6.3 Design of Ideal Reactors and Product Distributions for Multiple Reaction Systems

The number of steady-state species conservation equations needed for designing a chemical reactor and evaluating the product distributions is equal to the number of independent reactions taking place in the reactor. This is illustrated for parallel and series reaction examples in a CSTR and a PFR.

6.3.1 Parallel Reactions

Let us consider two independent parallel reactions, producing desired product B and undesired product C, in a CSTR.

$$A \rightarrow B;\ R_B = k_1 C_A^2$$
$$A \rightarrow C;\ R_C = k_2 C_A$$

For this system, the fractional conversion of A to the products B and C at the outlet of the reactor can be defined as x_{1_f} and x_{2_f}, respectively.

$$x_{1_f} = \frac{F_{B_f}}{F_{A_o}};\ x_{2_f} = \frac{F_{C_f}}{F_{A_o}} \tag{6.23}$$

The overall conversion of reactant A to the products is equal to the summation of x_{1_f} and x_{2_f}

$$x_{A_f} = x_{1_f} + x_{2_f} = \frac{(F_{A_o} - F_{A_f})}{F_{A_o}} \tag{6.24}$$

Evaluation of the product distributions for this reaction system in a CSTR of known volume requires the simultaneous solution of the two species conservation equations (for instance, Eqs. (6.25) and (6.26), for A and B, respectively).

$$\mathrm{A}: F_{A_o} - F_{A_f} + VR_{A_f} = 0 \tag{6.25}$$

Note that this equation can be expressed in terms of fractional conversion of reactant A (Eq. (6.25a)). In the case of a constant volumetric flow rate, it can be expressed in terms of the concentrations of reactant A at the inlet and outlet of the reactor (Eq. (6.25b)).

$$F_{A_0}x_{A_f} - V\left(k_1C_{A_o}^2(1 - x_{A_f})^2 + k_2C_{A_o}(1 - x_{A_f})\right) = 0 \tag{6.25a}$$

$$Q(C_{A_o} - C_{A_f}) - V\left(k_1C_{A_f}^2 + k_2C_{A_f}\right) = 0 \tag{6.25b}$$

$$\mathrm{B}: -F_{B_f} + VR_{B_f} = 0 \tag{6.26}$$

Note that Eq. (6.26) can be expressed either in terms of fractional conversions (Eq. (6.26b)) or in terms of concentrations of species (Eq. (6.26a)) (if the volumetric flow rate is constant).

$$-QC_{B_f} + Vk_1C_{A_f}^2 = 0 \tag{6.26a}$$

$$-F_{A_o}x_{1_f} + Vk_1C_{A_o}^2(1 - x_{A_f})^2 = 0 \tag{6.26b}$$

Simultaneous solution of Eqs. (6.25a) and (6.26b) gives x_{A_f} and x_{1_f}. Then, product distributions at the exit of the reactor can be evaluated.

Similarly, for a PFR, a simultaneous solution of the differential species conservation equations for A and B is needed to evaluate product distributions.

$$\mathrm{A}: \quad -\frac{dF_A}{dV} + R_A = 0 \tag{6.27}$$

This equation can be expressed in terms of concentrations or fractional conversion of A, as follows:

$$-\frac{dC_A}{d\tau} - (k_1C_A^2 + k_2C_A) = 0 \tag{6.27a}$$

$$F_{A_0}\frac{dx_A}{dV} - \left(k_1C_{A_0}^2(1 - x_A)^2 + k_2C_{A_0}(1 - x_A)\right) = 0 \tag{6.27b}$$

$$\mathrm{B}: -\frac{dF_B}{dV} + R_B = 0 \tag{6.28}$$

This equation can also be expressed in terms of concentrations or fractional conversion of reactant A, as follows:

$$-\frac{dC_B}{d\tau} + k_1C_A^2 = 0 \tag{6.28a}$$

$$-F_{A_o}\frac{dx_1}{dV} + k_1C_{A_o}^2(1 - x_A)^2 = 0 \tag{6.28b}$$

For this example, point and overall selectivity definitions may also be expressed in terms of fractional conversion of reactant A.

$$s_{BA} = \frac{R_B}{-R_A} = \frac{k_1 C_A^2}{k_1 C_A^2 + k_2 C_A} = \frac{\frac{k_1}{k_2} C_{A_o}(1-x_A)}{\frac{k_1}{k_2} C_{A_o}(1-x_A) + 1} \tag{6.29}$$

In this case, a higher concentration of A favors the production of the desired product B.

In the case of another example of a two-parallel reaction system, with first-order kinetics for the production of the desired product B and second-order kinetics for the production of the undesired product C

$$A \rightarrow B;\ R_B = k_1 C_A$$

$$A \rightarrow C;\ R_C = k_2 C_A^2$$

point selectivity expression can be written as follows:

$$s_{BA} = \frac{R_B}{-R_A} = \frac{k_1 C_A}{k_1 C_A + k_2 C_A^2} = \frac{1}{1 + \frac{k_2}{k_1} C_{A_o}(1-x_A)} \tag{6.30}$$

In this case, a low concentration of reactant A favors the selectivity of the desired product B.

Variation of the point selectivity s_{BA} with respect to the fractional conversion of A is illustrated in Figure 6.2 for the two cases defined by Eqs. (6.29) and (6.30). Figure 6.2 clearly illustrates the effect of concentration of reactant A and the ratio of the rate constants on the selectivity of the desired product. Note that the ratio of the reaction rate constants is also a function of reaction temperature.

Example 6.2 Evaluation of CSTR Volume for Parallel Reactions

For the following parallel liquid-phase reactions taking place in a CSTR, the ratio of concentrations of the desired and the undesired products (C_{B_f}/C_{C_f}) is given as 2.3 in the reactor exit stream.

$$A + D \rightarrow B;\quad R_B = k_1 C_A C_D\ k_1 = 2\ (\text{l/mol})/\text{s}$$

$$A + 2D \rightarrow C;\ R_C = k_2 C_A \quad k_2 = 1\ \text{s}^{-1}$$

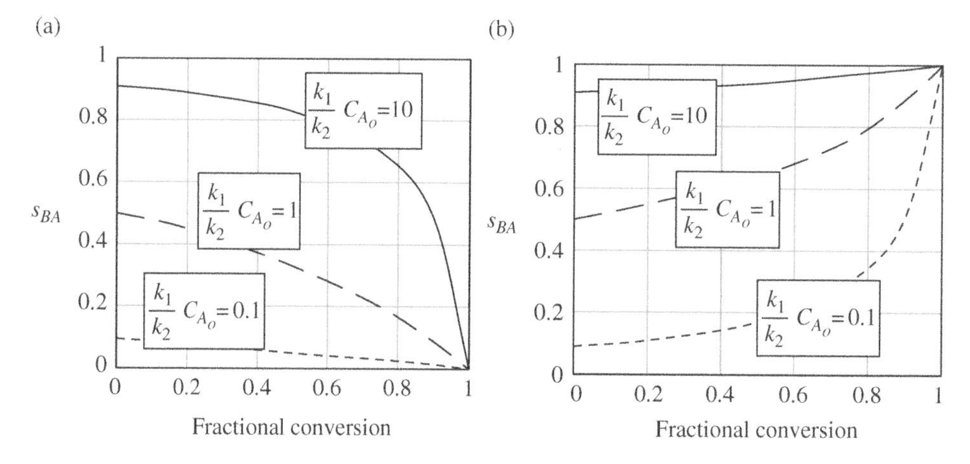

Figure 6.2 Variation of point selectivity with respect to the fractional conversion of reactant A. Case (a): $R_B = k_1 C_A^2$; $R_C = k_2 C_A$ Case (b): $R_B = k_1 C_A$, $R_C = k_2 C_A^2$

Suppose that the total volumetric flow rate of the reactor feed stream is 2 m^3/s and the inlet concentrations of A and D are 1 and 2 M, respectively. What should be the reactor volume and the overall fractional conversion of A?

Solution

Molar flow rates of reactants and products in the reactor exit stream can be expressed in terms of fractional conversions of A to B and C, as follows:

$$F_{A_f} = F_{A_o}(1 - x_{1_f} - x_{2_f}); \quad F_{D_f} = F_{D_0} - F_{A_o}x_{1_f} - 2F_{A_o}x_{2_f}$$

$$F_{B_f} = F_{A_o}x_{1_f}; \quad F_{C_f} = F_{A_o}x_{2_f}$$

$$\frac{C_{B_f}}{C_{C_f}} = \frac{F_{B_f}}{F_{C_f}} = \frac{x_{1_f}}{x_{2_f}} = 2.3$$

$$\frac{x_{1_f}}{x_{1_f} + x_{2_f}} = \frac{x_{1_f}}{x_{A_f}} = \frac{C_{B_f}}{(C_{A_0} - C_{A_f})} = \left(\frac{C_{B_f}}{C_{B_f} + C_{C_f}}\right) = \frac{2.3}{2.3 + 1.0} = 0.7$$

Hence, the overall selectivity of B with respect to A is

$$S_{BA} = \frac{F_{B_f}}{(F_{A_0} - F_{A_f})} = 0.7$$

$$\text{For a CSTR, } S_{BA} = (s_{BA})_f = \frac{R_{B_f}}{-R_{A_f}} = \frac{k_1 C_{A_f} C_{D_f}}{k_1 C_{A_f} C_{D_f} + k_2 C_{A_f}} = \frac{\frac{k_1}{k_2} C_{D_f}}{\frac{k_1}{k_2} C_{D_f} + 1} = 0.7$$

$$C_{D_f} = 1.17 \text{ M}$$

Since $C_{D_f} = (C_{D_o} - C_{A_o}x_{1_f} - 2C_{A_0}x_{2_f})$, and $\frac{x_{1_f}}{x_{2_f}} = 2.3$ fractional conversion values are evaluated as

$$x_{1_f} = 0.44, \quad x_{2_f} = 0.19, \quad x_{A_f} = 0.63$$

Reactor volume can then be determined from the species conservation equation for reactant A as follows:

$$Q(C_{A_o} - C_{A_f}) - V(k_1 C_{A_f} C_{D_f} + k_2 C_{A_f}) = 0; \; V = 1.02 \text{ m}^3$$

Example 6.3 Parallel Reactions in a Multiple CSTR System

Reactant A is converted to products B and C through two parallel reactions in a two-CSTR system operating in series (Figure 6.3). The rate expressions for these parallel reactions are given as follows:

$$\text{A} \rightarrow \text{B}; \; R_B = k_1 C_A; \; k_1 = 0.4 \text{ min}^{-1}$$

$$\text{A} \rightarrow \text{C}; \; R_C = k_2 C_A^2; \; k_2 = 0.5 \text{ (mol/l)}^{-1} \text{ min}^{-1}$$

a) Evaluate the product distribution at the exit of the second reactor, if the space–times in Reactors I and II are 2.5 and 5.0 min, respectively.

b) Find the fractional conversion values of A at the exit of the first and the second reactors.

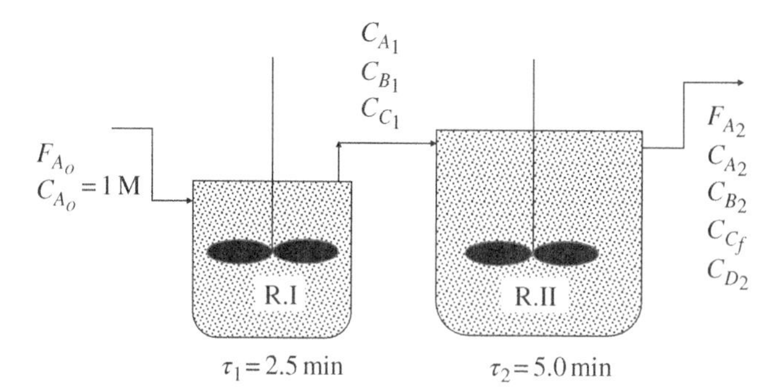

Figure 6.3 Schematic representation of the two-reactor system for Example 6.3.

Solution

a) Species conservation equations for reactant A and product B in Reactors I and II are

Reactor I

$$Q(C_{A_o} - C_{A_1}) - \left(k_1 C_{A_1} + k_2 C_{A_1}^2\right) V_1 = 0; \; (1 - C_{A_1}) - \left(0.4 C_{A_1} + 0.5 C_{A_1}^2\right) 2.5 = 0$$

$$-QC_{B_1} + k_1 C_{A_1} V_1 = 0; \qquad C_{B_1} = (0.4)(2.5) C_{A_1}$$

Reactor II

$$Q(C_{A_1} - C_{A_2}) - \left(k_1 C_{A_2} + k_2 C_{A_2}^2\right) V_2 = 0; \; (C_{A_1} - C_{A_2}) - \left(0.4 C_{A_2} + 0.5 C_{A_2}^2\right) 5.0 = 0$$

$$Q(C_{B_1} - C_{B_2}) + k_1 C_{A_2} V_2 = 0; \qquad (C_{B_1} - C_{B_2}) + (0.4)(5.0) C_{A_2} = 0$$

Concentrations of A and B at the exit of Reactors I and II can then be found as

$$C_{A_1} = 0.4\,\text{M};\, C_{B_1} = 0.4\,\text{M};\, C_{A_2} = 0.12\,\text{M};\, C_{B_2} = 0.64\,\text{M}$$

b) Fractional conversion values are

$$x_{A_1} = \frac{(F_{A_o} - F_{A_1})}{F_{A_o}} = \frac{Q(C_{A_o} - C_{A_1})}{QC_{A_o}} = \frac{1.0 - 0.4}{1.0} = 0.6$$

$$x_{A_2} = \frac{(F_{A_o} - F_{A_2})}{F_{A_o}} = \frac{C_{A_o} - C_{A_2}}{C_{A_o}} = 0.88$$

Example 6.4 Catalytic Oxidation of Ethylene in a Plug-Flow Reactor

Ethylene oxide is produced by the catalytic oxidation of ethylene over a silver-based catalyst [1, 2]. However, some of the ethylene and produced ethylene oxide may also be oxidized to carbon dioxide and water through total oxidation reactions.

$$C_2H_4 + \frac{1}{2}O_2 \rightarrow C_2H_4O \text{ (desired reaction)}$$

$$C_2H_4 + 3O_2 \rightarrow 2CO_2 + 2H_2O \text{ (undesired combustion reaction)}$$

$$C_2H_4O + \frac{5}{2}O_2 \rightarrow 2CO_2 + 2H_2O \quad \text{(undesired combustion reaction)}$$

If only ethylene, ethylene oxide, oxygen, carbon dioxide, and water are present in the reaction vessel, the following atomic conservation expressions can be written:

$$\text{C balance}: 2R_{C_2H_4} + 2R_{C_2H_4O} + R_{CO_2} = 0$$

$$\text{H balance}: 4R_{C_2H_4} + 4R_{C_2H_4O} + 2R_{H_2O} = 0$$

$$\text{O balance}: R_{C_2H_4O} + 2R_{O_2} + 2R_{CO_2} + R_{H_2O} = 0$$

Since we have three equations for the five unknown rate values, we should have two independent rates, indicating two independent reactions. The following reactions can be arbitrarily selected as independent reactions:

$$C_2H_4 + \frac{1}{2}O_2 \rightarrow C_2H_4O$$

$$C_2H_4 + 3O_2 \rightarrow 2CO_2 + 2H_2O$$

The rate expressions for ethylene oxide and carbon dioxide production are available in the literature [1]. Neglecting the further oxidation of produced ethylene oxide to carbon dioxide, the rate expressions given in the literature reduce to

$$R_{C_2H_4O} = k_1 P_{C_2H_4} P_{O_2}^{1/2}$$

$$R_{CO_2} = k_2 P_{C_2H_4} P_{O_2}$$

It is asked to set up the PFR design equations for this reactor.

Solution

To design this reactor, we need two species-conservation equations. For a PFR, the differential species conservation equations for ethylene oxide and carbon dioxide can then be expressed as follows:

$$-\frac{dF_{C_2H_4O}}{dV} + k_1 P_{C_2H_4} P_{O_2}^{1/2} = 0$$

$$-\frac{dF_{CO_2}}{dV} + k_2 P_{C_2H_4} P_{O_2} = 0$$

If we define the fractional conversion values of ethylene to ethylene oxide and carbon dioxide as x_1 and x_2, respectively, molar flow rates of species at an arbitrary location z in the reactor can be expressed as follows:

$$F_{C_2H_4} = F_{(C_2H_4)_o}(1 - x_1 - x_2) = F_{(C_2H_4)_o}\left(1 - x_{(C_2H_4)_f}\right)$$

$$F_{C_2H_4O} = F_{(C_2H_4)_o} x_1$$

$$F_{CO_2} = 2F_{(C_2H_4)_o} x_2$$

$$F_{H_2O} = 2F_{(C_2H_4)_o} x_2$$

$$F_{O_2} = (F_{O_2})_o - \frac{1}{2}F_{(C_2H_4)_o} x_1 - 3F_{(C_2H_4)_o} x_2$$

The total molar flow rate can then be expressed as follows:

$$F_T = F_{(C_2H_4)_o} + (F_{O_2})_o - \frac{1}{2}F_{(C_2H_4)_o} x_1 = F_o - \frac{1}{2}F_{(C_2H_4)_o} x_1$$

Partial pressure of ethylene, which appear in the rate expressions, can then be expressed in terms of the two unknown parameters x_1 and x_2 as

$$P_{C_2H_4} = P\frac{F_{C_2H_4}}{F_T} = P\frac{F_{(C_2H_4)_o}(1-x_1-x_2)}{F_o - \frac{1}{2}F_{(C_2H_4)_o}x_1} = P\frac{y_{(C_2H_4)_o}(1-x_1-x_2)}{1-\frac{1}{2}y_{(C_2H_4)_o}x_1}$$

Here, P and $y_{(C_2H_4)_o}$ are the total pressure and the mole fraction of ethylene at the inlet of the tubular reactor, respectively. Substitution of these partial pressure expressions into the species conservation equations yields the following differential equations:

$$-F_{(C_2H_4)_o}\frac{dx_1}{dV} + k_1P^{3/2}\left[\frac{y_{(C_2H_4)_o}(1-x_1-x_2)}{1-\frac{1}{2}y_{(C_2H_4)_o}x_1}\right]\left[\frac{(y_{O_2})_o - \frac{1}{2}y_{(C_2H_4)_o}x_1 - 3y_{(C_2H_4)_o}x_2}{1-\frac{1}{2}y_{(C_2H_4)_o}x_1}\right]^{1/2} = 0$$

$$-2F_{(C_2H_4)_o}\frac{dx_2}{dV} + k_2P^2\left[\frac{y_{(C_2H_4)_o}(1-x_1-x_2)}{1-\frac{1}{2}y_{(C_2H_4)_o}x_1}\right]\left[\frac{(y_{O_2})_o - \frac{1}{2}y_{(C_2H_4)_o}x_1 - 3y_{(C_2H_4)_o}x_2}{1-\frac{1}{2}y_{(C_2H_4)_o}x_1}\right] = 0$$

For the isothermal operation of the reactor, the rate constants can be taken as constant. These two differential equations should then be solved simultaneously to evaluate fractional conversion values and then the product distribution at the reactor outlet.

6.3.2 Consecutive Reactions

Many of the reaction systems are quite complex, involving consecutive reactions, as well as parallel reactions. A typical example of a consecutive reaction system is the dehydrogenation of ethane to ethylene and acetylene.

$$C_2H_6 \rightarrow C_2H_4 + H_2$$

$$C_2H_4 \rightarrow C_2H_2 + H_2$$

Let us consider a reaction network involving two consecutive elementary reactions

$$A \xrightarrow{k_1} B \xrightarrow{k_2} C$$

The reaction rates of different species involved in this system can be expressed as follows:

$$-R_A = k_1C_A$$

$$R_B = k_1C_A - k_2C_B$$

$$R_C = k_2C_B$$

Suppose B is the desired product. Its point selectivity with respect to the reactant A can be expressed as follows:

$$s_{BA} = \frac{R_B}{-R_A} = \frac{k_1C_A - k_2C_B}{k_1C_A}$$

The concentration of the desired product B passes through a maximum, with respect to space–time in the flow reactor for this system. Here, we have two independent reactions, which require two species-conservation equations to determine the product distributions.

Let us define the overall fractional conversion of A and the conversion of A to B in a flow reactor as

$$x_{A_f} = \frac{F_{A_o} - F_{A_f}}{F_{A_o}} \tag{6.31}$$

$$x_{1_f} = \frac{F_{B_f}}{F_{A_o}} \tag{6.32}$$

Considering a CSTR, the species conservation equations for reactant A and the desired product B can be expressed for the constant volumetric flow rate case, as follows:

$$(F_{A_o} - F_{A_f}) - k_1 C_{A_f} V = 0; \qquad F_{A_0} x_{A_f} - k_1 C_{A_o}(1 - x_{A_f})V = 0 \tag{6.33}$$

$$-F_{B_f} + (k_1 C_{A_f} - k_2 C_{B_f})V = 0; \ -F_{A_0} x_{1_f} + k_1 C_{A_o}(1 - x_{A_f})V - k_2 V C_{A_0} x_{1_f} = 0 \tag{6.34}$$

Simultaneous solution of Eqs. (6.31) and (6.32) yields the following expressions:

$$x_{A_f} = \frac{k_1 \tau}{1 + k_1 \tau} \tag{6.35}$$

$$x_{1_f} = \left(\frac{1}{1 + k_1 \tau}\right)\left(\frac{k_1 \tau}{1 + k_2 \tau}\right); \ \frac{C_{B_f}}{C_{A_o}} = x_{1_f} = \left(\frac{1}{1 + k_1 \tau}\right)\left(\frac{k_1 \tau}{1 + k_2 \tau}\right) = \frac{x_{A_f}(1 - x_{A_f})}{1 + x_{A_f}\left(\frac{k_2}{k_1} - 1\right)} \tag{6.36}$$

The variation of concentrations of A, B, and C as a function of space–time τ is illustrated qualitatively in Figure 6.4.

To find the required space–time to reach the maximum value of the concentration of the desired product B, the derivative of C_{B_f} (Eq. (6.36)) with respect to τ should be equated to zero. Alternatively, the derivative of C_{B_f} with respect to the fractional conversion of A can be evaluated from Eq. (6.36) and equated to zero to find the value of x_{A_f}, which will maximize C_{B_f}.

$$\frac{d\left(\frac{C_{B_f}}{C_{A_0}}\right)}{dx_{A_f}} = \frac{(1 - 2x_{A_f})\left(1 + x_{A_f}\left(\frac{k_2}{k_1} - 1\right)\right) - x_{A_f}(1 - x_{A_f})\left(\frac{k_2}{k_1} - 1\right)}{\left(1 + x_{A_f}\left(\frac{k_2}{k_1} - 1\right)\right)^2} = 0 \tag{6.37}$$

x_{A_f} and τ expressions corresponding to the maximum value of C_{B_f} are then found as

$$(x_{A_f})_{\max} = \frac{1}{1 + \left(\frac{k_2}{k_1}\right)^{1/2}} \tag{6.38}$$

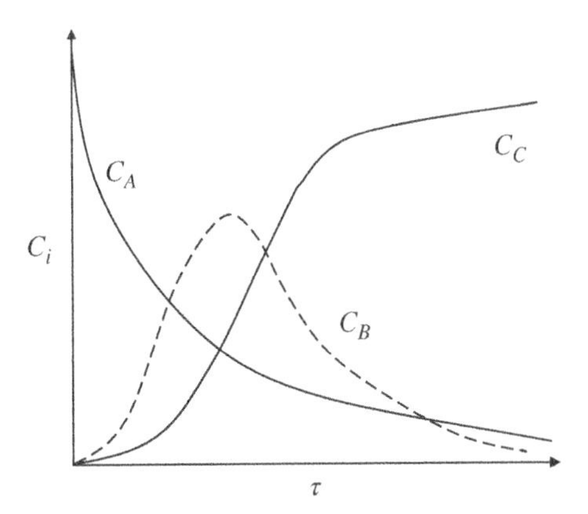

Figure 6.4 Schematic diagram of concentration variations in a consecutive reaction system.

$$\tau_{max} = (k_1 k_2)^{-1/2} \tag{6.39}$$

The maximum value of x_{1_f} can then be evaluated from the following expression:

$$\left(x_{1_f}\right)_{max} = \left(1 + \left(\frac{k_2}{k_1}\right)^{1/2}\right)^{-2} \tag{6.40}$$

As seen in Eq. (6.40), the maximum value of C_{B_f} depends upon $\frac{k_2}{k_1}$. This ratio is a function of reaction temperature.

$$\frac{k_2}{k_1} = \frac{k_{2,o}}{k_{1,o}} \exp\left[-\frac{(E_{a_2} - E_{a_1})}{R_g T}\right] \tag{6.41}$$

Suppose the difference of the activation energies of the second and the first reactions is positive $(E_{a_2} - E_{a_1}) \geq 0$. In this case, an increase in temperature causes an increase in the ratio of $\frac{k_2}{k_1}$ and hence a decrease in the concentration of the desired product B. However, if the reverse is true, an increase in temperature favors the production of the desired product B.

The concentration of the undesired product C can also be evaluated considering that

$$F_{A_o} = F_{A_f} + F_{B_f} + F_{C_f} \tag{6.42}$$

For the constant flow rate case, C_{C_f} expression becomes as follows:

$$C_{C_f} = C_{A_o} - C_{A_f} - C_{B_f} = \left(\frac{k_2}{k_1}\right) \frac{x_{A_f}^2}{1 + \left(\frac{k_2}{k_1} - 1\right) x_{A_f}} \tag{6.43}$$

Variation of C_{A_f}, C_{B_f}, and C_{C_f} as a function of fractional conversion of reactant A is illustrated in Figure 6.5.

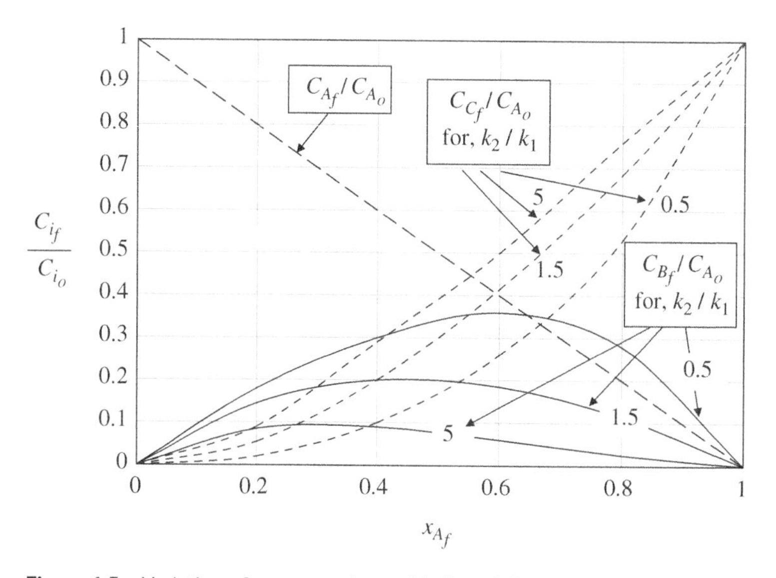

Figure 6.5 Variation of concentrations of A, B, and C as a function of fractional conversion of reactant A in a consecutive reaction system (A → B → C) in a CSTR.

Example 6.5 Analysis of a Consecutive-Parallel Reaction System in a CSTR
The following reactions are taking place in an isothermal CSTR.

$$A + B \rightarrow C + D \quad ; R_1 = -R_A = k_1 C_A C_B$$

$$C + B \rightarrow E + D \quad ; R_2 = R_E = k_2 C_C C_B$$

Suppose the ratio of reaction rate constants and the desired ratio of concentrations of C and E in the product stream are given as, $\frac{k_1}{k_2} = 4$ and 1.0, respectively. What is the total fractional conversion of A?

Solution
For this reaction system involving two independent reactions, we should write two independent species conservation equations.

$$\text{Material balance for A}: Q\left(C_{A_o} - C_{A_f}\right) - Vk_1 C_{A_f} C_{B_f} = 0 \tag{6.44}$$

$$\text{Material balance for C}: -QC_{C_f} + V\left(k_1 C_{A_f} C_{B_f} - k_2 C_{C_f} C_{B_f}\right) = 0 \tag{6.45}$$

We can also write the following stoichiometric relation:

$$\text{Concentration of E}: C_{E_f} = C_{A_o} - C_{A_f} - C_{C_f} \tag{6.46}$$

$$\text{It is also given that}: \frac{C_{C_f}}{C_{E_f}} = 1 \tag{6.47}$$

Here, we have four equations for the four unknowns, namely C_{A_f}, C_{C_f}, C_{E_f}, and $\tau k_1 C_{B_f}$, where $\tau = \frac{V}{Q}$. Note that the ratio of the rate constants is given as $\frac{k_1}{k_2} = 4$. Instead of Eq. (6.46), we can also use the species conservation equation for E as

$$-QC_{E_f} + V\left(k_2 C_{C_f} C_{B_f}\right) = 0 \tag{6.48}$$

The solution of these equations yields

$$\frac{C_{A_f}}{C_{A_o}} = 0.20; \quad \frac{C_{C_f}}{C_{A_o}} = \frac{C_{E_f}}{C_{A_o}} = 0.40; \ \tau k_1 C_{B_f} = 4.0$$

Fractional conversion of A at the exit of the reactor can then be evaluated as

$$x_{A_f} = \frac{\left(C_{A_o} - C_{A_f}\right)}{C_{A_o}} = 0.80$$

To evaluate the space–time, we need additional information for the rate constant k_1, as well as the concentration of B in the reactor.

Further discussion on multiple reactions may be found in Refs. [3–6].

Problems and Questions

6.1 For the two parallel reactions given below, plot point selectivity of desired product B (with respect to reactant A) as a function of fractional conversion of A.

$$A \rightarrow B \quad ; R_B = k_1 C_A; \ k_1 = 0.4 \text{ min}^{-1}$$

A → C ; $R_C = k_2 C_A^2$; $k_2 = 1.5\ (\text{mol/l})^{-1}\ \text{min}^{-1}$

Inlet concentration of A : $C_{A_o} = 1$ M

(a) For a fractional conversion of A being 0.75, what type of ideal reactor (CSTR or plug-flow reactor (PFR)) would give a higher concentration of desired product B (discuss)?

(b) Evaluate the concentrations of A, B, and C at the outlet of the CSTR and PFRs for the fractional conversion value of 0.75.

(c) What are the volumes of the CSTR and PFRs for $x_{A_f} = 0.75$, if the volumetric flow rate at the reactor inlet is given as 6×10^{-3} m^3/min?

6.2 The liquid-phase reaction of A with E yields three different products through three parallel reactions. The concentration of A is given as 1 M at the reactor inlet. Since the concentration of E is much higher than the concentration of A, it may be taken as constant in the reactor.

A + E → B (desired reaction); $R_B = (k_1 C_E) C_A^2 = k_1^{'} C_A^2$

A + E → C ; $R_C = (k_2 C_E) C_A = k_2^{'} C_A$

A + E → D ; $R_D = (k_3 C_E) C_A^3 = k_3^{'} C_A^3$

The ratios of apparent rate constants are given as

$$\frac{k_1^{'}}{k_2^{'}} = 40\ (\text{mol/l})^{-1}; \quad \frac{k_3^{'}}{k_2^{'}} = 100\ (\text{mol/l})^{-2}$$

(a) Plot point selectivity of B (with respect to reactant A) as a function of the concentration of A, in the concentration range between 2.0 and 0.01 M.

(b) What kind of reactor arrangement can be recommended for this process to achieve maximum concentration of desired product B at a fractional conversion of 0.99 for the limiting reactant A?

(c) Evaluate the volumes of the reactors in the arrangement proposed in part (b), if the molar feed rate of A to the system is 10^3 mol/min and $k_1^{'} = 10\ (\text{mol/l})^{-1}\ \text{min}^{-1}$.

6.3 Data reported in the table are obtained for the following parallel reaction system in a CSTR, by performing experiments at different space–times with a reactor inlet concentration of $C_{A_0} = 10$ mol/l.

A → B; desired reaction

A → C; undesired reaction

C_{A_f} (mol/l)	9.0	8.0	7.0	6.0	5.0	4.0	3.0	2.0	1.0
C_{B_f} (mol/l)	0.8	1.2	1.5	1.9	2.2	2.5	2.8	3.0	3.1

(a) Find the value of desired product B in a PFR if C_{A_0} and C_{A_f} are 8 and 2.0 M, respectively.

(b) Find the value of desired product B in a CSTR if C_{A_0} and C_{A_f} are 8 and 2.0 M, respectively.

6.4 The following rate expressions are given for the liquid-phase decomposition reactions of reactant A:

$$A \rightleftarrows B + C \quad R_B = k_1\left(C_A - \frac{C_B C_C}{K_1}\right)$$

$$A \rightarrow D + E \quad R_D = k_2 C_A$$

The ratio of the rate constants and the equilibrium constant of the first reaction are given as, $\frac{k_2}{k_1} = 0.5$ and $K_1 = 0.4$, respectively. The concentration of reactant A at the inlet of the reactor is also given as $C_{A_o} = 1$ M.

Evaluate the product distribution at the exit of a CSTR if the overall fractional conversion value of reactant A is 0.5. What is the corresponding value of fractional conversion of A to product B, $x_{1_f} = \frac{F_{B_f}}{F_{A_o}}$?

6.5 The following liquid-phase elementary reactions are taking place in a series of two equal-size CSTRs operating in series:

$$A \xrightarrow{k_1} B \xrightarrow{k_2} C\ ;\ B \xrightarrow{k_3} E + F$$

Derive an expression for the concentration of product B at the exit of the second CSTR.

6.6 Following gas-phase reactions take place in an ideal flow reactor.

$$A + B \rightarrow C + D \ ; R_C = 2x10^5 \exp\left(-\frac{7200}{T}\right) P_A P_B \text{ mol/l h (desired reaction)}$$

$$A + B \rightarrow E \ ; R_E = 12 \exp\left(-\frac{1850}{T}\right) P_A P_B \quad \text{mol/l h (T is in Kelvin)}$$

Feed stream contains equimolar quantities of A and B at 2 atm pressure.

(a) Plot point selectivity of C with respect to reactant A, as a function of its fractional conversion and discuss which type of ideal reactor (CSTR or PFR) you would prefer to maximize the desired product C.

(b) Discuss the effect of temperature on the overall selectivity of C in a CSTR and PFR. Evaluate its overall selectivity values at 600 and 700 K and discuss whether the higher or lower temperature should be preferred.

(c) If the desired overall conversion of A is 0.7 at 700 K, evaluate the ratio of production rates of C to E in a CSTR.

(d) Evaluate the space–time of reactants in a CSTR for the overall conversion of A being 0.7 at 700 K.

6.7 The following liquid-phase, consecutive elementary reactions are taking place in a flow reactor:

$$A \xrightarrow{k_1} B \rightleftarrows C;\ k_1 = 1.0\,h^{-1};\ k_2 = 0.1\,h^{-1};\ k_{-2} = 1.0\,h^{-1}$$

(a) Plot concentrations of A, B, and C as a function of space–time in a CSTR by taking $C_{A_o} = 1$ M.

(b) You are asked to set-up the necessary differential equations, which are needed to evaluate the variation of concentrations of species as a function of space–time in a PFR.

6.8 Reactant A is converted to the desired product B and undesired product C through parallel elementary reactions in two CSTRs operating in series at the same temperature. Volumes of the first and the second reactors are given as 1 and 0.8 m^3, respectively. The volumetric flow rate of the feed stream of the first reactor is given as 0.4 m^3/min. Concentrations of A, B, and C at the inlet to the first reactor are

$$C_{A_o} = 1\ \mathrm{M}, C_{B_o} = 0.1\ \mathrm{M}, C_{C_o} = 0$$

$$\mathrm{A} \rightarrow \mathrm{B}$$

$$\mathrm{A} \rightarrow \mathrm{C}$$

The concentrations of A and B at the outlet of the first reactor are given as 0.3 and 0.7 M, respectively. What should be the concentrations of A, B, and C at the outlet of the second reactor?

6.9 The following specific rate expressions are reported by Veronica et al. (*Chem. Eng. J.* 138, 602–607, 2008) for steam reforming of ethanol to produce synthesis gas:

$$C_2H_5OH + 3H_2O \rightarrow 2CO_2 + 6H_2; R_{(1)} = 5.74 \times 10^{-4} \exp\left(-\frac{1.44 \times 10^5}{R_g}\left(\frac{1}{T} - \frac{1}{873}\right)\right) P_{\mathrm{eth}}^{0.80}$$

$$C_2H_5OH + H_2O \rightarrow 2CO + 4H_2; R_{(2)} = 1.88 \times 10^{-4} \exp\left(-\frac{2.07 \times 10^5}{R_g}\left(\frac{1}{T} - \frac{1}{873}\right)\right) P_{\mathrm{eth}}^{0.75}$$

Units of rate constants of $R_{(1)}$ and $R_{(2)}$ are: mol min^{-1} $\mathrm{atm}^{-0.8}$ (mg catalyst)$^{-1}$ and mol min^{-1} $\mathrm{atm}^{-0.75}$ (mg catalyst)$^{-1}$, respectively. The bulk density of the catalyst can be taken as 1 g/cm^3. The ratio of H_2O to ethanol in the feed stream is given as 3. These rate expressions can be used at conditions away from equilibrium, at which the backward rates are negligible.

(a) Do you expect an increase or decrease in the ratio of reaction rates of CO to CO_2 with an increase in temperature? Discuss.

(b) Write down the necessary species conservation equations for the design of a PFR. Take the pressure and the isothermal temperature of the reactor as 101 kPA and 600 °C, respectively.

References

1 Ayame, A., Kano, H., Kanazuka, T. et al. (1973). *Bull. Jap. Pet. Inst.* **15**: 150–155.
2 Dogu, G. and Sozen, Z.Z. (1985). *Chem. Eng. J.* **3**: 145–150.
3 Levenspiel, O. (1999). *Chemical Reaction Engineering. NJ.* Wiley.
4 Roberts, G.W. (2009). *Chemical Reactions and Chemical Reactors. NJ.* Wiley.
5 Smith, J.M. (1981). *Chemical Engineering Kinetics*. Singapore: McGraw Hill Co.
6 Fogler, H.S. (1992). *Elements of Chemical Reaction Engineering*. New Jersey: Prentice-Hall.

7

Heat Effects and Non-isothermal Reactor Design

Due to the heat generation/consumption resulting from chemical conversions, significant temperature variations may develop within a chemical reactor. In the case of exothermic reactions, the heat generated as a result of chemical conversion increases the temperature of the reaction mixture unless all of the heat generated is transferred to the surroundings. On the other hand, heat consumed as a result of chemical conversion may decrease the temperature of the reaction mixture in the case of endothermic reactions.

Hydrodynamics of the reaction mixture within a reactor may also significantly influence the temperature gradients in the reactor. Temperature gradients are expected to develop both in the axial and radial directions in reaction vessels. Heat exchange to the surroundings from the external surface of a tubular reactor is the main reason for radial temperature gradients. Besides the effects of heat exchange through the reactor wall, the velocity profile within a tubular reactor is also expected to cause temperature variations in the radial direction. For instance, in a homogeneous laminar flow reactor, parabolic velocity distribution causes residence time distribution of the reactants within the reactor. Reactant molecules near the reactor wall will spend more time within a laminar flow reactor than the molecules close to the center of the reactor. This may cause higher conversions and higher heat generation near the reactor wall. Temperature gradients may also develop within the batch and stirred-tank flow reactors due to poor mixing.

Nonuniform temperature distributions within a reactor may cause significant changes in the rate and equilibrium parameters. For typical chemical reactions having activation energy values between 80 and 200 kJ/mol, a significant change of reaction rate constant is expected with a small change in the reactor temperature. A typical example is ammonia synthesis, for which the Arrhenius parameter γ $\left(\gamma = \frac{E_a}{R_g T_s}\right)$ is about 29.4 [1]. In this case, only a 10 °C increase in temperature from the reactor wall temperature of 450 °C causes about a 50% increase in the reaction rate constant. The reaction rate constant increases about 2.7 times if the increase of temperature in the radial direction is about 25 °C. The formation of hot spots having temperatures of about 300 °C higher than the wall temperature of 197 °C was also reported in a sulfur dioxide oxidation reactor [2, 3].

Temperature nonuniformities are quite common in tubular reactors. Due to the large heat of reactions, the approach to isothermal operation is not easy in most commercial reactors. In the case of the adiabatic operation of the reactors, significant temperature variations are expected, especially in the axial direction in tubular reactors and as a function of reaction time in well-mixed batch

Fundamentals of Chemical Reactor Engineering: A Multi-Scale Approach, First Edition. Timur Doğu and Gülşen Doğu.

Companion website: www.wiley.com/go/dogu/chemreacengin

reactors. Hence, the design of non-isothermal reactors requires the simultaneous solution of species conservation and energy balance equations. The coupling term of species conservation and energy balance equations is the reaction rate, which is a function of both temperature and concentrations of reacting species.

7.1 Heat Effects in a Stirred-Tank Reactor

The primary assumption of a continuous stirred-tank reactor (CSTR) is the perfect mixing of the reaction mixture. As a result of perfect mixing, temperature and species concentrations are assumed to be uniform within such well-mixed tank reactors. Thus, the reactor exit temperature and species concentrations are expected to be the same as the corresponding values within the reactor.

The macroscopic species conservation equation around a CSTR is written as

$$F_{A_o} - F_{A_f} + VR_{A_f} = 0 \tag{7.1}$$

$$F_{A_o}x_{A_f} = -VR_{A_f} \tag{7.2}$$

In this expression, the rate constant should be evaluated at the temperature within the reactor, which is equal to the reactor exit temperature. Hence, an energy balance is needed to determine the reactor exit temperature.

Heat effects may be quite significant within a CSTR and should be considered in the design of these reactors. Even if the uniform temperature assumption can be approached within a stirred-tank reactor, significant temperature differences are expected between the inlet stream temperature and the temperature within the reactor. The heat generated/consumed due to chemical conversions causes such differences. In some cases, cooling or heating of the well-mixed stirred-tank reactors may be required to keep the reaction temperature at the desired level (Figure 7.1). Macroscopic energy balance around a CSTR can be written by applying the first law of thermodynamics around this vessel.

The first law of thermodynamics for an open system can be expressed as follows:

$$q + W_n = \Delta H \tag{7.3}$$

Here, the heat transfer term across the system boundaries (q) is zero for an adiabatic reactor. This term is negative if the reaction mixture is cooled, and it is positive if the reaction mixture is heated during the reaction. The work term in Eq. (7.3) may be taken as zero in most of the reaction systems. Then, Eq. (7.3) reduces to $\Delta H = 0$ for an adiabatic reactor. Here, ΔH is the difference between enthalpies of the outlet and the inlet streams of the reactor.

Since enthalpy is a state function, the enthalpy change between the outlet and the inlet of the reactor may be evaluated following any path between these states. A typical path used for the evaluation of the enthalpy change is illustrated in Figure 7.2. Following this path, enthalpy balance around an adiabatic reactor becomes

$$\Delta H = \Delta H_1 + \Delta H_2 + \Delta H_3 = \sum_i \int_{T_o}^{T_R} F_{i_o} C_{p_i}\, dT + \Delta H^o_{T_R} F_{A_o} x_{A_f} + \sum_i \int_{T_R}^{T_f} F_{i_f} C_{p_i}\, dT = 0 \tag{7.4}$$

Here, T_R is the reference temperature at which enthalpy of formations of the reacting species is available. The first summation term of this expression includes sensible enthalpy change of all of the species present in the inlet stream of the reactor from inlet temperature to a reference

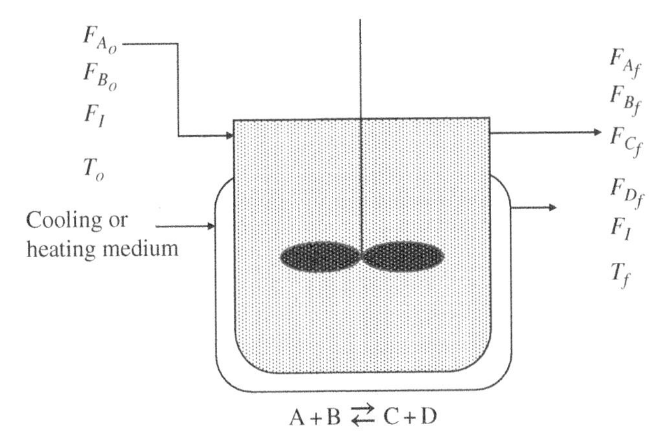

Figure 7.1 Schematic representation of a CSTR.

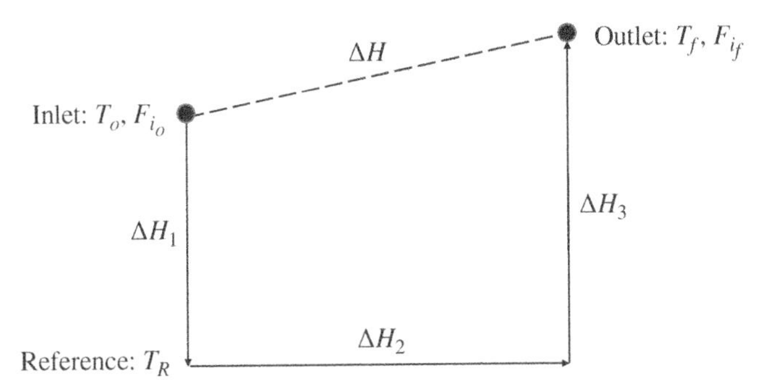

Figure 7.2 Schematic representation of a path selected to estimate enthalpy change around the reactor.

temperature. However, the last summation term includes sensible enthalpies of all of the species present in the reactor exit stream with respect to the reference. In the case of constant molar heat capacity values, enthalpy balance reduces to

$$\Delta H = \sum_i F_{i_o} C_{p_i} (T_R - T_o) + \Delta H^o_{T_R} F_{A_o} x_{A_f} + \sum_i F_{i_f} C_{p_i} \left(T_f - T_R\right) = 0 \tag{7.5}$$

Each term in this energy balance expression has the unit of J/s. Here, $\Delta H^o_{T_R}$ is the heat of the reaction at the reference temperature T_R, which is usually selected as 298 K.

Enthalpy balance around the reactor may also be written following another path, as illustrated in Figure 7.3.

$$\Delta H = \Delta H_1 + \Delta H_2 = \sum_i \int_{T_o}^{T_f} F_{i_o} C_{p_i} \, dT + \Delta H^o_{T_f} F_{A_o} x_{A_f} = 0 \tag{7.6}$$

For constant C_{p_i} values, Eq. (7.6) reduces to

$$\Delta H = \sum_i F_{i_o} C_{p_i} \left(T_f - T_o\right) + \Delta H^o_{T_f} F_{A_o} x_{A_f} = 0 \tag{7.7}$$

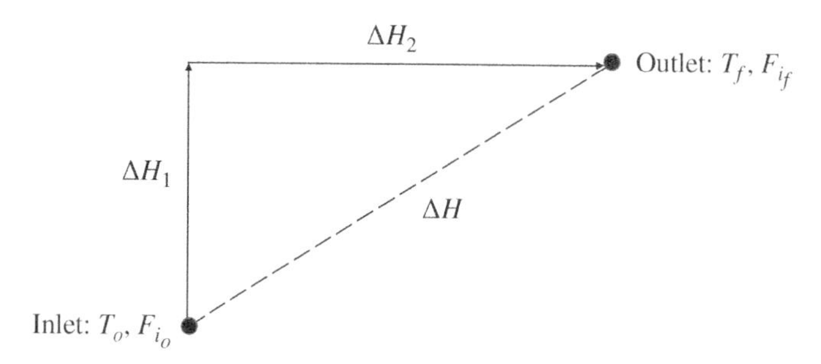

Figure 7.3 Schematic representation of a second path selected to estimate enthalpy change around the reactor.

By combining the species conservation equation (Eq. (7.2)) with the energy balance equation given by Eq. (7.7), the reaction rate term can be inserted into the energy balance expression.

$$\Delta H = \sum_i F_{i_o} C_{p_i} (T_f - T_o) + \Delta H^o_{T_f}(-R_{A_f} V) = 0 \tag{7.8}$$

Either Eq. (7.7) or (7.8) can be used to predict the reactor temperature for a given value of fractional conversion in a CSTR.

In the case of a system with constant specific heat for the reaction mixture, Eq. (7.7) reduces to

$$\Delta H = Q\rho\bar{c}(T_f - T_o) + \Delta H^o_{T_f} F_{A_o} x_{A_f} = 0 \tag{7.9}$$

In writing this equation, constant specific heat ($\bar{c}$, J/kg K) and constant density (ρ) assumptions are made for the reaction mixture of the reactor inlet stream. In this second path (Figure 7.3), the heat of reaction should be evaluated at the reactor outlet temperature T_f, which should be determined using the heat of reaction at the reference temperature and the molar heat capacities of the products and the reactants, as follows:

$$\Delta H^o_{T_f} = \Delta H^o_{T_R} + \int_{T_R}^{T_f} \Delta C_{p_i}\, dT \tag{7.10}$$

Equation (7.7) or (7.9) can be used to evaluate reaction temperatures at different fractional conversion values. Rearrangement of these equations gives the following relations:

$$x_{A_f} = \left(\frac{Q\rho\bar{c}}{-\Delta H^o_{T_f} F_{A_o}}\right)(T_f - T_0) = \left(\frac{\sum_i F_{i_o} C_{p_i}}{-\Delta H^o_{T_f} F_{A_0}}\right)(T_f - T_o) \tag{7.11}$$

These expressions indicate a linear relationship between the reaction temperature and the fractional conversion of the limiting reactant.

In the case of a non-adiabatic reactor, the heat exchange term with the surroundings can be expressed in terms of an overall heat transfer coefficient and temperature driving force between the reactor temperature and the temperature of the environment. For instance, if there is heat loss to the surroundings in an exothermic reaction system, the energy balance equation, given in Eq. (7.7), reduces to

$$\sum_i F_{i_o} C_{p_i} (T_f - T_0) + \Delta H^o_{T_f} F_{A_o} x_{A_f} = -UA_h(T - T_a) \tag{7.12}$$

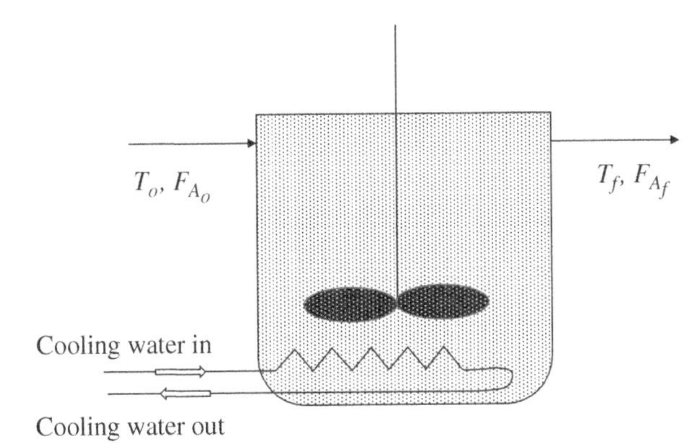

Figure 7.4 A CSTR cooled by a cooling coil.

Here, A_h is the area through which heat exchange takes place between the reactor and the environment and T_a is the temperature of the environment. If the temperature of the cooling/heating medium is not constant (as in the case of having a cooling coil in the reactor, Figure 7.4), log-mean temperature difference should be used as the driving force.

$$\sum_i F_{i_o} C_{p_i} (T_f - T_o) + \Delta H^o_{T_f} F_{A_o} x_{A_f} = -UA_h \Delta T_{LM} \tag{7.13}$$

Here, the log-mean temperature difference is defined as follows:

$$\Delta T_{LM} = \frac{(T_f - T_{c,in}) - (T_f - T_{c,out})}{\ln\left(\frac{(T_f - T_{c,in})}{(T_f - T_{c,out})}\right)} \tag{7.14}$$

Here, $T_{c,in}$ and $T_{c,out}$ are the inlet and the outlet temperatures of the fluid flowing in the cooling/heating coil.

7.2 Steady-State Multiplicity in a CSTR

Multiple steady-state conversion values and the corresponding multiple steady-state reactor temperatures are possible in reactors at certain reaction conditions [2–4]. This is illustrated in this section for a simple first-order reaction taking place in a CSTR. The species conservation equation for a CSTR can be written as follows:

$$F_{A_0} - F_{A_f} + R_{A_f} V = 0 \tag{7.15}$$

Considering a simple first-order reaction ($-R_A = kC_A$), and expressing F_{A_f} in terms of fractional conversion of reactant A, Eq. (7.15) reduces to Eq. (7.16) for a system with a constant volumetric flow rate.

$$F_{A_0} - F_{A_0}(1 - x_{A_f}) - kC_{A_0}(1 - x_{A_f})V = 0 \tag{7.16}$$

Elimination of x_{A_f} from Eq. (7.16) yields

$$x_{A_f} = \frac{k\tau}{1 + k\tau} \tag{7.17}$$

Here τ is the space-time, defined as $\tau = \frac{V}{Q}$.

The rate constant is temperature-dependent and should be evaluated at the reactor exit temperature T_f, which is equal to the reaction temperature in the vessel. Even if the value of space-time is known, evaluation of fractional conversion of reactant A from Eq. (7.17) requires the value of the rate constant, which should be calculated at the reactor exit temperature. The species conservation equation should be solved together with the energy balance around the reactor to evaluate the two unknowns, namely the reactor temperature and the fractional conversion of A. For an adiabatic CSTR, energy balance around the reactor can be written as follows:

$$\sum_i F_{i_o} C_{p_i} \left(T_f - T_o\right) = -\Delta H^o_{T_f} F_{A_o} x_{A_f} \tag{7.18}$$

In the case of having a constant volumetric flow rate, constant density, and constant specific heat of the reaction mixture, Eq. (7.18) can be simplified as follows:

$$Q\rho\bar{c}\left(T_f - T_o\right) = -\Delta H^o_{T_f} F_{A_o} x_{A_f}; \quad q_R = q_G \tag{7.19}$$

The right-hand side of this equation corresponds to the heat generation rate (J/s), and the left-hand side corresponds to the heat removal rate. Equations (7.17) and (7.19) contain the two unknown variables x_{A_f} and T_f. Substitution of x_{A_f} from Eq. (7.17) into Eq. (7.19) yields a heat generation rate term expressed in terms of temperature only, as the unknown variable.

$$q_G = -\Delta H^o_{T_f} F_{A_o} \frac{k\tau}{1 + k\tau} = -\Delta H^o_{T_f} F_{A_o} \left(\frac{\tau k_o e^{-\frac{E_a}{R_g T_f}}}{1 + \tau k_o e^{-\frac{E_a}{R_g T_f}}} \right) \tag{7.20}$$

In the case of a non-adiabatic system, heat loss to the surrounding should also be added to the heat removal rate expression.

$$q_R = Q\rho\bar{c}\left(T_f - T_0\right) + UA_h\left(T_f - T_a\right) \tag{7.21}$$

Steady-state temperature is reached within the reactor when heat generation and heat removal rate terms become equal.

$$-\Delta H^o_{T_f} F_{A_o} \left(\frac{\tau k_o e^{-\frac{E_a}{R_g T_f}}}{1 + \tau k_o e^{-\frac{E_a}{R_g T_f}}} \right) = Q\rho\bar{c}\left(T_f - T_o\right) + UA_h\left(T_f - T_a\right) \tag{7.22}$$

The plot of the heat generation rate term q_G as a function of the reactor temperature T_f gives an S-shaped curve (Figure 7.5). On the other hand, the relation between q_R and T_f is linear, with the slope and intersection values of $(Q\rho\bar{c} + UA_h)$ and $(-Q\rho\bar{c}T_o - UA_h T_a)$, respectively.

$$q_R = (Q\rho\bar{c} + UA_h)T_f - (Q\rho\bar{c}T_o + UA_h T_a) \tag{7.23}$$

Intersections of q_G and q_R correspond to the steady-state temperatures in the reactor. As illustrated in Figure 7.5, it is possible to have three intersection points under certain conditions, which correspond to three different steady-state temperatures in the reactor. The heat generation rate curve

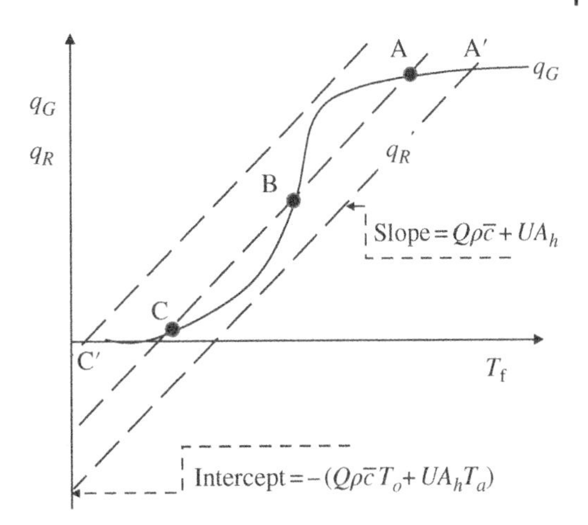

Figure 7.5 Schematic representation of heat generation and heat removal rate curves and the multiple steady-state temperatures.

asymptotically approaches to $\left(-\Delta H^o_{T_f} F_{A_o}\right)$ at high T_f values. This asymptotic limit corresponds to the complete conversion of reactant A for an irreversible reaction.

Multiple steady-state temperatures are not desired and should be avoided to not have operational problems for the reactors. Elimination of multiple steady-state temperatures can be achieved either by changing the intersection of the heat removal line or by changing its slope. An increase in the reactor inlet temperature shifts the q_R line to the right with the same slope and may help to obtain a single steady-state temperature (Figure 7.5). On the other hand, an increase in the volumetric flow rate Q increases the slope of the q_R line and also changes its intersection with the q_G curve. To not have multiple steady-state temperatures for any value of reactor inlet temperature, the slope of the q_R line should be higher than the slope of the q_G curve for all the T_f values. Multiple steady states can be avoided if the slope of q_R is higher than the slope of the S-shaped q_G curve at its inflection point.

$$\frac{dq_R}{dT} > \left.\frac{dq_G}{dT}\right|_{\text{inf}}; \quad (Q\rho\bar{c} + UA_h) > \left.\frac{dq_G}{dT}\right|_{\text{inf}} \tag{7.24}$$

The possibility of occurrence of multiple steady-state temperatures in an adiabatic CSTR is illustrated in Example 7.1.

Example 7.1 Multiple Steady-State Temperatures

Consider a simple first-order decomposition reaction of reactant A, which is taking place in an adiabatic CSTR.

$$A \rightarrow \text{products}; \quad -R_A = kC_A; \quad k = 3 \times 10^6 \exp\left(-\frac{8370}{T}\right) \text{s}^{-1}$$

The following data are given for this reaction system:

Reactor volume: 1.6 m^3
The heat of reaction: $-\Delta H^o = 1 \times 10^4$ J/mol (given as constant)
Molar heat capacity of reactant A: 90 J/mol K (it is the same for the reaction mixture)
Space-time: $\tau = 400$ seconds
Inlet molar flow rate of reactant A: $F_{A_o} = 8$ mol/s

Evaluate the fractional conversion of reactant A and the reaction temperature for the three inlet temperature values of $T_o = 335,\ 345,\ 360$ K.

Solution

Substitution of the fractional conversion of reactant A expressed by Eq. (7.17) into the energy balance yields

$$-\Delta H^o_{T_f} F_{A_0} \frac{k\tau}{1+k\tau} = F_{A_o} C_{p_A} (T_f - T_o)$$

A relation between the inlet and outlet temperatures of the reactor is then expressed by substituting the given data into this equation.

$$(10\,000)(8)\left(\frac{1.2\times 10^9 \exp\left(-\frac{8370}{T_f}\right)}{1 + 1.2\times 10^9 \exp\left(-\frac{8370}{T_f}\right)}\right) = (8)(90)(T_f - T_o)$$

Variation of the heat generation and the heat removal rate curves for the three inlet temperature values given in this example are shown in Figure 7.6. T_f values evaluated from this equation for different T_o values, and the corresponding fractional conversion values of reactant A (from Eq. (7.17)) are listed in Table 7.1.

As shown in Table 7.1, for the inlet temperature of 335 K, the fractional conversion of A is very low. This steady-state conversion corresponds to the intersection of q_R and q_G at the lower end of the S-shaped heat generation curve in Figure 7.6. The increase of the inlet temperature by 25 K caused an increase in fractional conversion to a value of about 0.95. This result illustrates the importance of the selection of the reactor inlet temperature to achieve high conversion values. A more interesting result is obtained for the inlet temperature of 345 K. In this case, three possible steady-state temperatures are obtained from the non-linear energy balance expression.

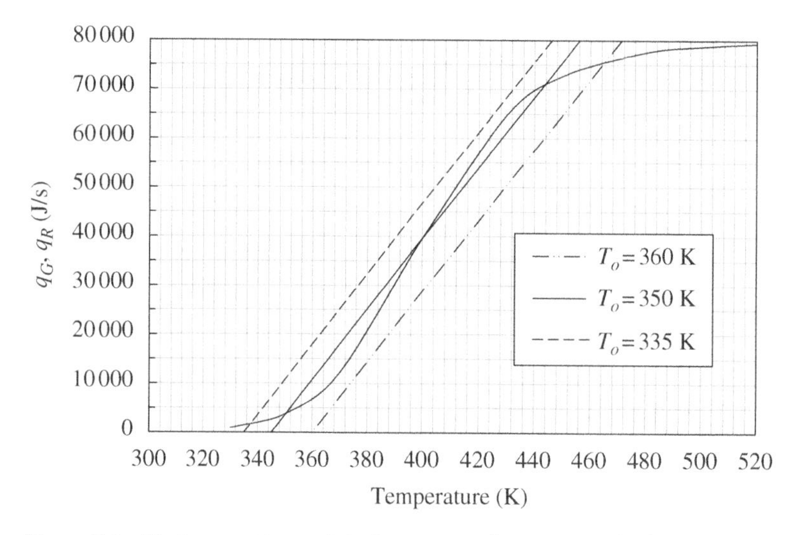

Figure 7.6 Heat generation and the heat removal rate curves for Example 7.1.

Table 7.1 Steady-state temperatures for Example 7.1.

Reactor inlet temperature, T_o (K)	Reactor temperature, T_f (K)	Fractional conversion of A, x_{A_f}
335	337.16	0.019
345	350.34	0.048
	399.96	0.495
	443.00	0.916
360	465.39	0.949

Figure 7.7 Schematic representation of heat generation and heat removal rate curves for an exothermic-reversible reaction.

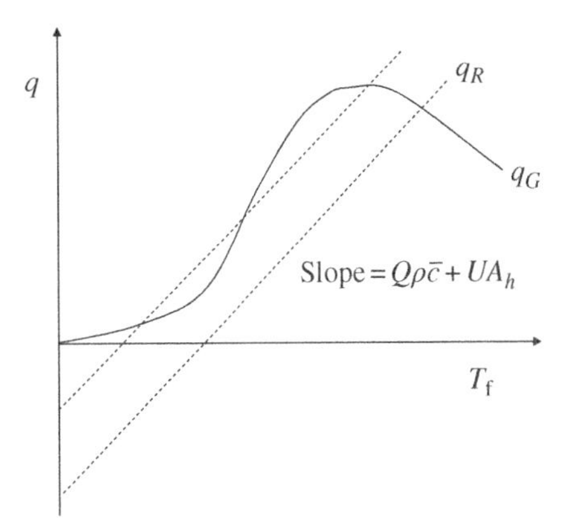

Three steady-state reactor temperatures shown in Figure 7.5 correspond to three different steady-state conversion values. This is not a desired case and should be avoided. Among the three steady-state reactor temperatures, the intersections at points A and C are stable (Figure 7.5). However, the intersection at point B is unstable since any fluctuation in the operational conditions of the reactor may cause a shift from points B to A or from B to C. At temperatures slightly higher than the reactor temperature at point B, the heat generation rate is higher than the heat removal rate. This causes an increase in temperature until point A is reached. On the contrary, at temperatures slightly lower than the reactor temperature at point B, the heat removal rate is higher than the heat generation rate. Hence, temperature and the corresponding conversion will decrease until point C is reached.

For an exothermic reversible reaction, the temperature dependence of heat generation and heat removal is qualitatively illustrated in Figure 7.7. As illustrated in Figure 7.7, the heat generation curve passes through a maximum for an exothermic-reversible reaction. At temperatures higher than the temperatures corresponding to the maximum value of q_G, reverse reaction gains importance, and equilibrium limitations control the fractional conversion reached in the reactor. Hence, a decrease in heat generation rate is expected with an increase in temperature after reaching a maximum.

7.3 One-Dimensional Energy Balance for a Tubular Reactor

For a tubular reactor with negligible radial variations of concentration and temperature, a simultaneous solution of one-dimensional energy and species conservation equations should be made to obtain temperature and concentration profiles within the reactor and to estimate the reactor volume. A decrease in the concentration of the reactant is expected along a tubular reactor. If the reactor is operating adiabatically with no heat exchange with the surroundings, a continuous increase of temperature is expected in the axial direction (Figure 7.8) for an exothermic reaction.

The rate of increase of temperature is expected to be higher in the region close to the inlet of the reactor than the exit of the reactor. This prediction is due to higher concentrations of reactants and hence higher reaction rates close to the inlet of the reactor for positive-order reactions. On the other hand, if the operation of the reactor is not adiabatic, the temperature may pass through a maximum along the reactor, where heat generation rate due to chemical conversion becomes equal to the heat removal rate from the walls of the reactor. Temperature is expected to decrease along the reactor after the achievement of this maximum. This is due to the lower heat generation rate than the heat removal rate toward the exit of the tubular reactor.

Species conservation equation for a differential volume element in a plug-flow reactor was derived in Section 4.2 as

$$-\frac{dF_i}{dV} + R_i = 0$$

The molar flow rate of species i can be expressed as a product of its flux in the axial direction $\left(N_{i_z}\right)$ with the cross-sectional area A_c of the reactor.

$$F_i = N_{i_z} A_c = N_{i_z} \pi r_o^2 \tag{7.25}$$

By substituting this relation into Eq. (4.23), the following species conservation equation is obtained:

$$-\frac{dN_{i_z}}{dz} + R_i = 0 \tag{7.26}$$

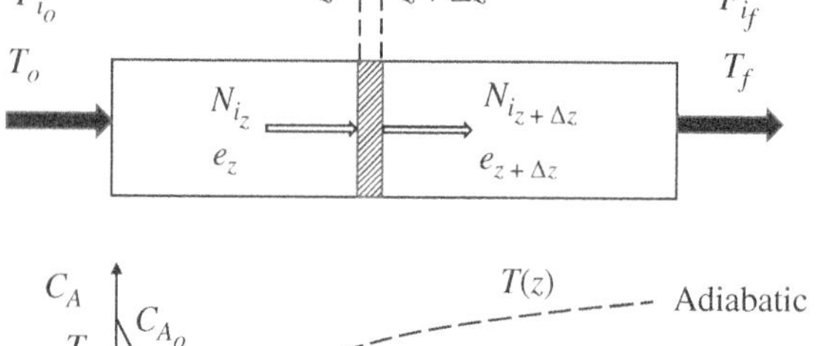

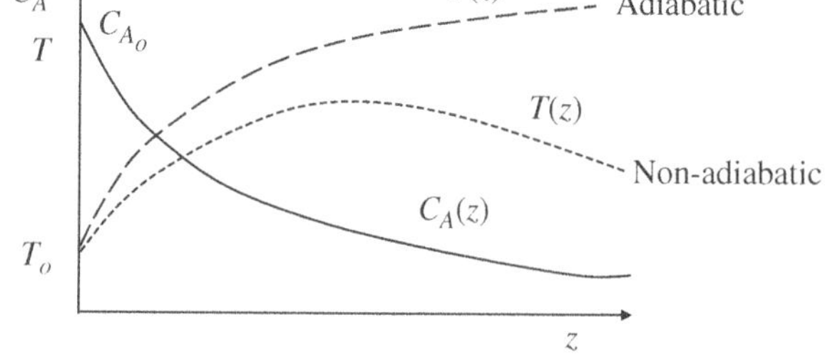

Figure 7.8 Temperature and concentration variations in a one-dimensional tubular flow reactor with an exothermic reaction.

Energy balance over the differential volume element between z and $z + \Delta z$ (Figure 7.8) can be expressed in terms of energy flux in the z-direction (e_z) and heat flux from the surface of the tubular reactor to the environment $\left(e_r|_{r=r_o}\right)$ as follows:

$$e_z|_z(\pi r_o^2) - e_z|_{z+\Delta z}(\pi r_o^2) - e_r|_{r=r_0} 2\pi r_o \Delta z = 0 \tag{7.27}$$

Dividing this equation by $\pi r_o^2 \Delta z$ and taking the limit as Δz approaching zero, the following differential energy balance equation is obtained:

$$-\frac{de_z}{dz} - \left(\frac{2}{r_o}\right) e_r|_{r=r_o} = 0 \tag{7.28}$$

Energy flux in the axial direction can be expressed as a summation of conduction heat flux and energy flux due to convective transport of species, as follows:

$$e_z = -\lambda_b \frac{dT}{dz} + \sum_i (N_{i_z} \hat{H}_i) \tag{7.29}$$

Here, λ_b is the thermal conductivity of the mixture in the tubular reactor. As for the heat loss term to the surroundings is concerned, it can be written in terms of an overall heat transfer coefficient and the temperature difference between the temperature of the reaction mixture and the temperature of the surroundings (T_a).

$$e_r|_{r=r_0} = U(T - T_a) \tag{7.30}$$

By substituting Eqs. (7.29) and (7.30) into Eq. (7.28), the following expression is obtained:

$$\lambda_b \frac{d^2T}{dz^2} - \frac{d}{dz}\left(\sum_i (N_{i_z} \hat{H}_i)\right) - \frac{2}{r_o} U(T - T_a) = 0 \tag{7.31}$$

Combining Eq. (7.26) with Eq. (7.31), the following expression can be obtained:

$$\lambda_b \frac{d^2T}{dz^2} - \sum_i \left(\hat{H}_i R_i + N_{i_z} \frac{d\hat{H}_i}{dz}\right) - \frac{2}{r_o} U(T - T_a) = 0 \tag{7.32}$$

Using the following relations between the rates of different species

$$\frac{R_A}{\nu_A} = \frac{R_B}{\nu_B} = \frac{R_i}{\nu_i} = R \tag{7.33}$$

and noting that $\sum_i \hat{H}_i \nu_i = \Delta H^o$, and $\frac{d\hat{H}_i}{dz} = C_{p_i} \frac{dT}{dz}$, differential energy balance equation reduces to

$$-\lambda_b \frac{d^2T}{dz^2} + \sum_i N_{i_z} C_{p_i} \frac{dT}{dz} + \frac{2}{r_0} U(T - T_a) = -R\Delta H^o \tag{7.34}$$

Note that ΔH^o is the heat of reaction at location z in the reactor, where the temperature is T. In the integration of this equation, temperature dependence of heat of reaction should be taken into consideration.

The right-hand side of Eq. (7.34) corresponds to the volumetric heat generation rate term in the reactor, while the terms on the left-hand side correspond to the volumetric heat removal rate terms. Among the heat removal terms, the order of magnitude of the conductive heat transfer rate term in the axial direction (the first term on the left-hand side of Eq. (7.34)) is generally much smaller than the energy transfer term due to convective transport of species in the reactor (the second term). Hence, Eq. (7.34) reduces to

$$\sum_i N_{i_z} C_{p_i} \frac{dT}{dz} + R\Delta H^o = -\frac{2}{r_0} U(T - T_a) \tag{7.35}$$

In the case of having a constant volumetric flow rate, constant density, and constant specific heat of the reaction mixture, Eq. (7.35) can be simplified, as shown below:

$$U_o \rho \bar{c} \frac{dT}{dz} + R\Delta H^o = -\frac{2}{r_o} U(T - T_a) \tag{7.36}$$

For an adiabatic reactor, the right-hand sides of Eqs. (7.35) and (7.36) become zero.

$$\sum_i N_{i_z} C_{p_i} \frac{dT}{dz} + R\Delta H^o = 0; \quad \text{(adiabatic tubular reactor)} \tag{7.37}$$

$$U_o \rho \bar{c} \frac{dT}{dz} + R\Delta H^o = 0; \quad \text{(adiabatic, constant density, and specific heat)} \tag{7.38}$$

The reaction rate term, which appears in the species conservation equation (Eq. (7.26)) and also in the energy equation (Eq. (7.35)), is a function of both temperature and concentrations of the reacting species. Hence, the simultaneous solution of these two equations is needed. For instance, for a simple first-order reaction, differential species conservation and energy balance equations can be written as follows:

$$-U_o \frac{dC_i}{dz} - k_o e^{-\frac{E_a}{R_g T}} C_i = 0 \tag{7.39}$$

$$\sum_i N_{i_z} C_{p_i} \frac{dT}{dz} + k_o e^{-\frac{E_a}{R_g T}} C_i \Delta H^o = -\frac{2}{r_o} U(T - T_a) \tag{7.40}$$

In writing Eq. (7.39), the molar flux of species i in the axial direction is expressed as the product of superficial velocity U_o and the concentration of the reactant i $\left(N_{i_z} = U_o C_i\right)$. A possible contribution of axial dispersion flux in the reactor is neglected in this relation, and a plug-flow assumption is made. Here, the velocity, which is the ratio of inlet volumetric flow rate to the cross-sectional area of the reactor $\left(U_o = \frac{Q}{\pi r_o^2}\right)$, is assumed constant.

By eliminating the reaction rate term from the energy and the species conservation equations, a relation can be obtained between the temperature and concentration of a reactant at any point within the reactor. For instance, for an adiabatic system, the following relation can be written:

$$R = \frac{\sum_i N_{i_z} C_{p_i}}{-\Delta H^o} \frac{dT}{dz} = -U_o \frac{dC_i}{dz} \tag{7.41}$$

The integration of Eq. (7.41) from the reactor inlet ($z = 0$) to an arbitrary location z in the reactor gives a relation between the temperature and the concentration of the reactant at that specific location.

$$\frac{\sum_i N_{i_z} C_{p_i}}{-\Delta H^o} (T - T_o) = U_o (C_{i_o} - C_i) \tag{7.42}$$

This relation can also be written in terms of molar flow rates of species flowing in the reactor (Eq. (7.43)) or in terms of average specific heat ($\bar{c}$, J/kg K) of the reaction mixture (Eq. (7.44)).

$$(T - T_o) = \frac{Q(-\Delta H^o)}{\sum_i F_i C_{p_i}}(C_{i_o} - C_i) \tag{7.43}$$

$$(T - T_o) = \frac{(-\Delta H^o)}{\rho \bar{c}}(C_{i_o} - C_i) \tag{7.44}$$

Here, ρ is the density of the reaction mixture flowing in the reactor. During the integration of Eq. (7.41), the heat of reaction and the molar heat capacities of the species $\left(C_{p_i}\right)$ are assumed constant.

Alternatively, the relation between the temperature and the fractional conversion of reactant i can be derived in an adiabatic reactor, following a similar procedure.

$$\sum_i F_i C_{p_i} \frac{dT}{dV} = -R\Delta H^o \tag{7.45}$$

$$-\frac{dF_i}{dV} = F_{i_o}\frac{dx_i}{dV} = R \tag{7.46}$$

$$F_{i_o}\frac{dx_i}{dV} = \left(\frac{\sum_i F_i C_{p_i}}{-\Delta H^o}\right)\frac{dT}{dV} \tag{7.47}$$

$$T - T_o = \frac{F_{i_o}(-\Delta H^o)}{\sum_i F_i C_{p_i}} x_i = \frac{F_{i_o}(-\Delta H^o)}{Q\rho\bar{c}} x_i \tag{7.48}$$

Substitution of the temperature vs. concentration relation (Eq. (7.44)) into Eq. (7.39) yields a differential equation, which does not contain the dependent variable T. This equation can be integrated numerically to predict the concentration profile within the reactor.

$$-U_o\frac{dC_i}{dz} - k_o \exp\left[-\frac{E_a}{R_g}\left(T_o + \frac{(-\Delta H^o)}{\rho\bar{c}}(C_{i_o} - C_i)\right)^{-1}\right]C_i = 0 \tag{7.49}$$

As noted before, the simultaneous solution of the species conservation equation and the energy equation is needed to design such a non-isothermal reactor. This can be achieved following a numerical procedure.

Alternatively, a macroscopic energy balance can be used to evaluate the volume of a non-isothermal tubular reactor. By rearrangement of Eq. (4.23), the following integral expression can be written to evaluate the volume of a tubular plug-flow reactor.

$$\frac{V}{F_{i_o}} = \int_0^{x_{i_f}} \frac{dx_i}{-R_i(T, x_i)} \tag{7.50}$$

Consider a macroscopic volume element of the reactor between $z = 0$ and an arbitrary axial position z (Figure 7.9). Temperature values at different locations along the reactor can be estimated using an energy balance around this macroscopic volume element.

$$F_{i_o}x_i(-\Delta H_T^o) = \sum_i (F_{i_o}C_{p_i})(T - T_o) + U\Delta T_{LM}2\pi r_o z \tag{7.51}$$

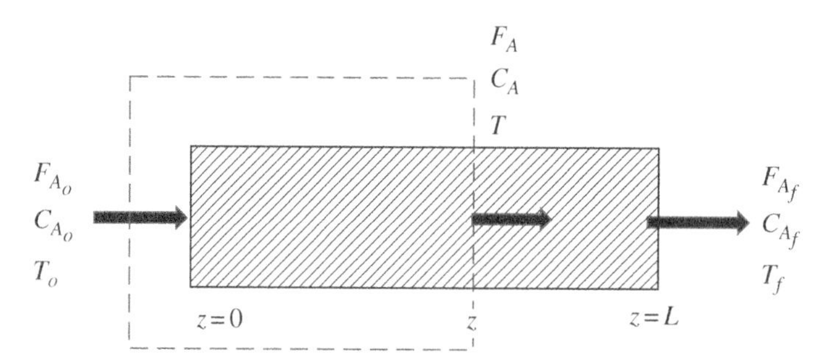

Figure 7.9 Schematic representation of a macroscopic volume element in a plug-flow reactor.

Here, ΔT_{LM} is the log-mean temperature difference between the reactor temperature and the temperature of the surrounding (cooling/heating) medium between $z = 0$ and an arbitrary axial position z. In the case of an adiabatic reactor, Eq. (7.51) reduces to

$$\sum_i (F_{i_o} C_{p_i})(T - T_o) + F_{i_o} x_i (\Delta H_T^o) = 0$$

This expression is equivalent to the application of the first law of thermodynamics to the section of the tubular reactor between $z = 0$ and an arbitrary position z. Derivation of macroscopic energy balance by using the first law of thermodynamics had already been discussed in Section 7.1 during the derivation of an energy balance around a CSTR.

In the design of a non-isothermal tubular reactor, temperatures at different conversion values (which correspond to different locations along the reactor) can be evaluated from the energy balance equation. The corresponding reaction rate values can then be evaluated using the temperature and fractional conversion values at different locations along the reactor. The reaction rate values evaluated along the reactor can then be used to evaluate the reactor volume by the numerical integration of Eq. (7.50). Note that the numerical integration of Eq. (7.50) (for instance, by using Simpson's rule) corresponds to the evaluation of the area under the $(1/-R_A)$ vs. fractional conversion curve, which gives the value of (V/F_{i_o}) (Figure 7.10).

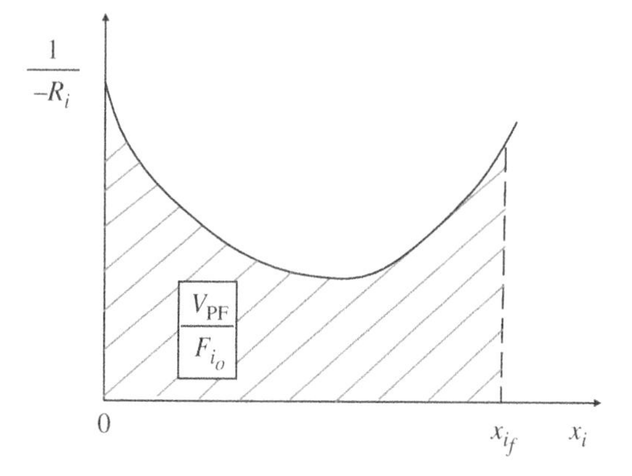

Figure 7.10 Schematic representation for the evaluation of the volume of an ideal tubular reactor.

7.4 Heat Effects in Multiple Reaction Systems

As discussed in Chapter 6, multiple reactions are involved in numerous processes. Heat effects of all of the reactions taking place in a reactor should be considered in writing the energy balance around the reactor. Heat effects in multiple reaction systems are illustrated in this section.

7.4.1 Heat Effects in a CSTR with Parallel Reactions

Let us consider two parallel reactions taking place in a CSTR (Figure 7.11).

$A + B \rightarrow R$; Specific Rate of Reaction 1: $R_{(1)}$; Heat of reaction: ΔH_1^o
$A \rightarrow S$; Specific Rate of Reaction 2: $R_{(2)}$; Heat of reaction: ΔH_2^o

For this system, the relations between the reaction rates of different species can be expressed as follows:

$$-R_A = R_{(1)} + R_{(2)}; \quad -R_B = R_{(1)}; \quad R_R = R_{(1)}; \quad R_s = R_{(2)}$$

Fractional conversion expressions of reactant A in Reactions 1 and 2, and its overall conversion can be defined as follows:

$$x_1 = \frac{F_R}{F_{A_o}}$$

$$x_2 = \frac{F_S}{F_{A_o}}$$

$$x_A = x_1 + x_2$$

If this reactor is operating adiabatically, the energy balance around the reactor can be written as follows:

$$\sum_i F_{i_o} C_{p_i} (T_R - T_o) + \Delta H^o_{1,T_R} F_{R_f} + \Delta H^o_{2,T_R} F_{S_f} + \sum_i F_{i_f} C_{p_i} (T_f - T_R) = 0 \tag{7.52}$$

$$\sum_i F_{i_o} C_{p_i} (T_R - T_o) + \Delta H^o_{1,T_R} F_{A_o} x_1 + \Delta H^o_{2,T_R} F_{A_o} x_2 + \sum_i F_{i_f} C_{p_i} (T_f - T_R) = 0 \tag{7.53}$$

$$\begin{aligned} &\left[F_{A_o} C_{p_A} + F_{B_o} C_{p_B}\right](T_R - T_o) + \Delta H^o_{1,T_R} F_{A_o} x_1 + \Delta H^o_{2,T_R} F_{A_o} x_2 \\ &\quad + \left[F_{A_o}(1 - x_1 - x_2) C_{p_A} + (F_{B_o} - F_{A_o} x_1) C_{p_B} + F_{A_o} x_1 C_{p_R} + F_{A_o} x_2 C_{p_S}\right](T_f - T_R) = 0 \end{aligned} \tag{7.54}$$

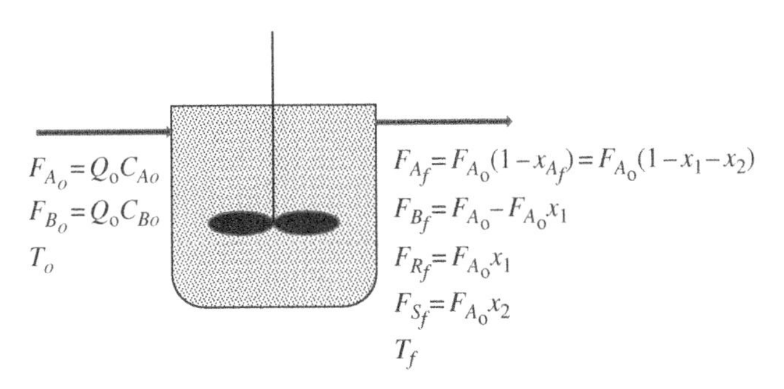

Figure 7.11 Schematic representation of a CSTR.

Molar heat capacity values are assumed as constant in writing these expressions.

Energy balance expression may also be expressed in terms of the rates of the reactions involved in this system. Species conservation equations for the products R and S around this CSTR are as follows:

$$-F_{R_f} + R_{R_f}V = 0; \quad -F_{R_f} + R_{(1)_f}V = 0 \tag{7.55}$$

$$-F_{S_f} + R_{S_f}V = 0; \quad -F_{S_f} + R_{(2)_f}V = 0 \tag{7.56}$$

Hence, energy balance expressed in Eq. (7.53) may also be written as

$$\sum_i F_{i_o}C_{p_i}(T_R - T_o) + \Delta H^o_{1,T_R}R_{(1)_f}V + \Delta H^o_{2,T_R}R_{(2)_f}V + \sum_i F_{i_f}C_{p_i}\left(T_f - T_R\right) = 0 \tag{7.57}$$

7.4.2 Heat Effects in a CSTR with Consecutive Reactions

Let us consider the following consecutive reaction system taking place in a CSTR.

$$A + B \rightarrow R \rightarrow S$$

Reaction rates of species involved in this system can be expressed in terms of the specific rates of the two consecutive reactions as follows:

$$-R_A = R_{(1)}; \quad -R_B = R_{(1)}; \quad R_R = R_{(1)} - R_{(2)}; \quad R_S = R_{(2)}$$

Considering an adiabatic CSTR, the energy balance around the reactor can be written as

$$\sum_i F_{i_o}C_{p_i}(T_R - T_0) + \Delta H^o_{1,T_R}\left(F_{A_o} - F_{A_f}\right) + \Delta H^o_{2,T_R}F_{S_f} + \sum_i F_{i_f}C_{p_i}\left(T_f - T_R\right) = 0 \tag{7.58}$$

Species conservation equations for reactant A and the product S can then be used to express the energy balance in terms of reaction rates.

$$\left(F_{A_o} - F_{A_f}\right) + R_{A_f}V = 0 \tag{7.59}$$

$$-F_{S_f} + R_{S_f}V = 0 \tag{7.60}$$

$$\sum_i F_{i_o}C_{p_i}(T_R - T_o) + \Delta H^o_{1,T_R}\left(-R_{A_f}V\right) + \Delta H^o_{2,T_R}\left(R_{S_f}V\right) + \sum_i F_{i_f}C_{p_i}\left(T_f - T_R\right) = 0 \tag{7.61}$$

One should be careful in defining fractional conversions in consecutive reaction systems. In this case, the overall fractional conversion of reactant A and the fractional conversion of A to S can be used as the two independent variables.

$$x_{A_f} = \frac{\left(F_{A_o} - F_{A_f}\right)}{F_{A_o}} \tag{7.62}$$

$$x_2 = \frac{F_{S_f}}{F_{A_o}} \tag{7.63}$$

With these conversion definitions, Eq. (7.58) becomes

$$\sum_i F_{i_o}C_{p_i}(T_R - T_o) + \Delta H^o_{1,T_R}F_{A_o}x_{A_f} + \Delta H^o_{2,T_R}F_{A_o}x_2 + \sum_i F_{i_f}C_{p_i}\left(T_f - T_R\right) = 0 \tag{7.64}$$

7.4.3 Energy Balance for a Plug-Flow Reactor with Multiple Reactions

For a multiple reaction system, the differential energy balance equation for a tubular plug-flow reactor can be expressed as follows:

$$\sum_i N_{i_z} C_{p_i} \frac{dT}{dz} + \Delta H_1^o R_{(1)} + \Delta H_2^o R_{(2)} + \cdots = -\frac{2}{r_o} U(T - T_a) \tag{7.65}$$

This equation can also be written in terms of molar flow rates of species along the reactor.

$$\sum_i F_i C_{p_i} \frac{dT}{dV} + \Delta H_1^o R_{(1)} + \Delta H_2^o R_{(2)} + \cdots = -\frac{2}{r_o} U(T - T_a) \tag{7.66}$$

In the case of an adiabatic reactor, the right-hand side of Eq. (7.66) becomes zero.

7.5 Heat Effects in Multiple Reactors and Reversible Reactions

7.5.1 Temperature Selection and Multiple Reactor Combinations

As discussed in Chapter 4, for an exothermic-reversible reaction, the temperature dependence of fractional conversion of the limiting reactant is expected to pass through a maximum for both flow and batch reactors. For instance, for a first-order reversible reaction ($A \rightleftarrows B$), fractional conversion expression for a CSTR is given by Eq. (4.49), as follows:

$$x_{A_f} = \frac{k_f \tau}{1 + \left(\frac{k_f \tau}{x_{Ae}}\right)} = \frac{k_f \tau}{1 + \left(\frac{1 + K_C}{K_C}\right) k_f \tau} \quad \text{(CSTR, first-order reversible reaction)} \tag{7.67}$$

The variation of fractional conversion as a function of temperature is illustrated in Figure 7.12 for an exothermic reaction taking place in a CSTR. Bell-shaped curves in Figure 7.12 correspond to the constant space-times ($\tau_3 > \tau_2 > \tau_1$).

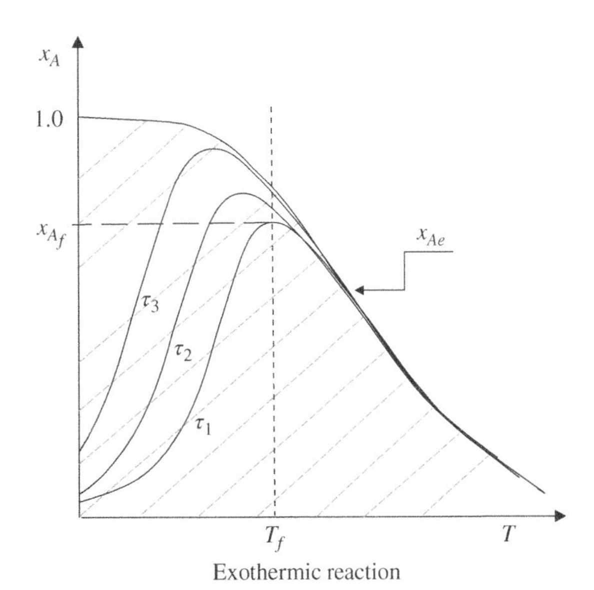

Figure 7.12 Variation of fractional conversion as a function of temperature for an exothermic reversible reaction at different space-times.

As illustrated in Figure 7.12, the best temperature (T_f) for a reactor operating at a particular space-time is the one corresponding to the maximum of the conversion vs. temperature curve. For reactor temperatures lower than this temperature, higher space-times are needed to reach the same fractional conversion. For reactor temperatures higher than the temperature corresponding to the maximum in the bell-shaped conversion curve, the increase of space-time will also cause some increase in fractional conversion until the equilibrium conversion. It is not possible to achieve a fractional conversion value higher than the equilibrium conversion. Equilibrium conversion can only be reached if the space-time in the reactor approaches infinity.

The feed stream temperature can be estimated from the energy balance around the reactor. Energy balance around a CSTR, operating adiabatically, is given by Eq. (7.68).

$$\sum_i F_{i_o} C_{p_i} (T_f - T_0) = -\Delta H^o_{T_f} F_{A_o} x_{A_f} \tag{7.68}$$

Rearrangement of this equation gives the following relation between fractional conversion and reactor temperature in a CSTR.

$$x_{A_f} = \frac{\sum_i F_{i_o} C_{p_i}}{-\Delta H^o_{T_f} F_{A_o}} (T_f - T_o) \tag{7.69}$$

The heat of reaction should be evaluated at the reactor temperature of T_f in Eq. (7.69). This expression can also be written as follows:

$$\begin{aligned} x_{A_f} &= \frac{\sum_i F_{i_o} C_{p_i}(T_R - T_o) + \sum_i F_{i_f} C_{p_i}(T_f - T_R)}{-\Delta H^o_{T_R} F_{A_o}} \\ &= \frac{\sum_i F_{i_o} C_{p_i}(T_R - T_o)}{-\Delta H^o_{T_R} F_{A_o}} + \frac{\sum_i F_{i_f} C_{p_j}}{-\Delta H^o_{T_R} F_{A_o}} (T_f - T_R) \end{aligned} \tag{7.70}$$

Here, F_{i_o} and F_{i_f} correspond to the molar flow rates of all the species present in the inlet and the outlet streams of the reactor, respectively. Hence, $\sum_i F_{i_o} C_{p_i}(T_o - T_R)$ and $\sum_i F_{i_f} C_{p_i}(T_f - T_R)$ represent the sensible enthalpies of the inlet and the outlet streams, with respect to the reference temperature T_R. The heat of reaction in Eq. (7.70) should be evaluated at the reference temperature of T_R. In the case of constant density, volumetric flow rate, and specific heat of the reaction mixture, Eq. (7.69) can also be simplified as

$$x_{A_f} = \left(\frac{Q\rho\bar{c}}{-\Delta H^o_{T_f} F_{A_o}} \right) (T_f - T_o) \tag{7.71}$$

This gives a linear relation between fractional conversion and the reactor temperature with a slope of $\left(\frac{Q\rho\bar{c}}{-\Delta H^o_{T_f} F_{A_o}} \right)$. For constant heat of reaction, the slope of this adiabatic line is constant. This is illustrated in Figure 7.13.

In the case of a non-adiabatic reactor with heat exchange with the surroundings, the energy balance around the reactor becomes

$$x_{A_f} = -\left(\frac{\sum_i F_{i_o} C_{p_i}(T_o) + UA_h T_a}{-\Delta H^o_{T_f} F_{A_o}} \right) + \frac{\sum_i F_{i_o} C_{p_i} + UA_h}{-\Delta H^o_{T_f} F_{A_o}} T_f \tag{7.72}$$

In some cases, multiple adiabatic reactors operating in series may be used to achieve higher conversions. For a two-adiabatic CSTR system operating in series (see Figure 7.14), the reaction

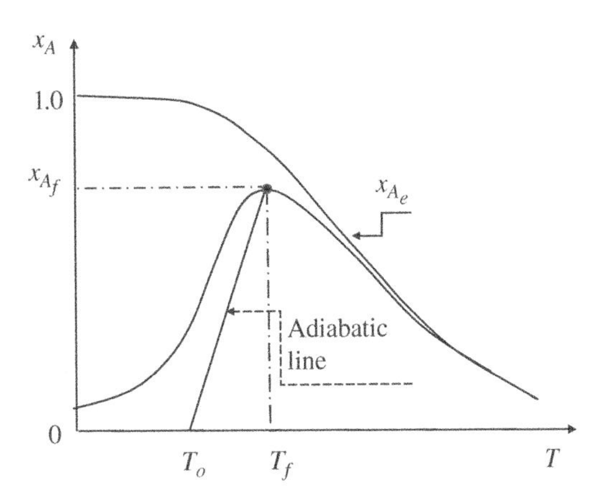

Figure 7.13 Prediction of reactor inlet and outlet temperatures of a CSTR-operating adiabatically (exothermic reaction).

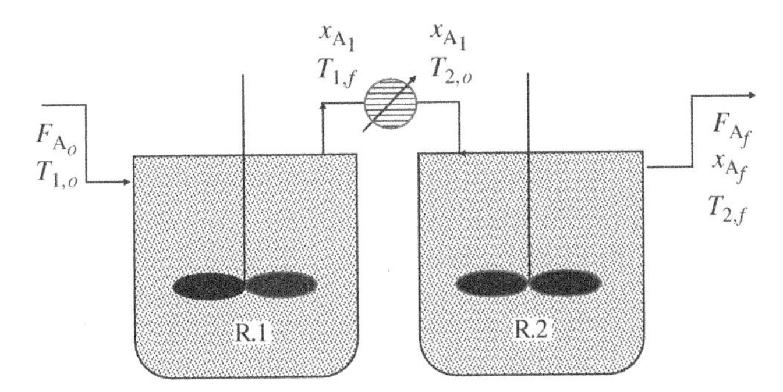

Figure 7.14 Two CSTRs operating in series with a heat exchanger between the reactors.

mixture leaving the first reactor may be cooled using a heat exchanger before entering the second reactor. In the case of the second CSTR, the material and the energy balance equations can be expressed as follows:

$$\frac{V_2}{F_{A_o}} = \frac{(x_{A_f} - x_{A_1})}{-R_{A_f}} \tag{7.73}$$

$$\sum_i F_{i_1} C_{p_i} (T_{2,f} - T_{2,o}) + \Delta H^o_{T_{2,f}} F_{A_o} (x_{A_f} - x_{A_1}) = 0 \tag{7.74}$$

Material and the energy balance equations for these two reactors can be rearranged as follows:

$$F_{A_o} x_{A_1} + R_{A_1} V_1 = 0 \quad \text{(Reactor 1)} \tag{7.75}$$

$$x_{A_1} = \frac{\sum_i F_{i_o} C_{p_i}}{-\Delta H^o_{T_{1,f}} F_{A_o}} (T_{1,f} - T_{1,o}) \quad \text{(Reactor 1)} \tag{7.76}$$

$$F_{A_o} (x_{A_f} - x_{A_1}) + R_{A_f} V_2 = 0 \quad \text{(Reactor 2)} \tag{7.77}$$

$$(x_{A_f} - x_{A_1}) = \frac{\sum_i F_{i_1} C_{p_i}}{-\Delta H^o_{T_{2,f}} F_{A_o}} (T_{2,f} - T_{2,o}) \quad \text{(Reactor 2)} \tag{7.78}$$

Here, $T_{1,f}$ and $T_{2,f}$ are the temperatures of the exit streams of reactors R.1 and R.2, respectively.

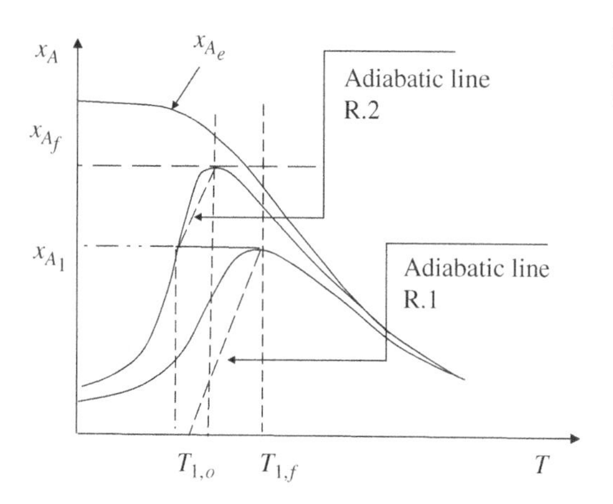

Figure 7.15 Conversion–temperature diagram for a two-CSTR system with cooling between the reactors.

Temperatures and conversions in this two-reactor system are illustrated in Figure 7.15. Constant space-time curves are also shown in Figure 7.15. All of these bell-shaped constant space-time curves have maximum values at specific temperatures. These stirred-tank reactors should operate at the temperatures corresponding to the maxima of the constant space-time curves. As illustrated in Figure 7.15, temperatures of the stirred-tank reactors can be selected at the intercepts of the adiabatic lines with the maxima of the constant space-time curves. The inlet temperatures of the reactors can be determined using the adiabatic energy balance expressions.

For some reversible-exothermic reactions with significant equilibrium limitations, multiple tubular reactors with interstage cooling between the reactors are used to achieve high conversions. A typical example is catalytic oxidation of SO_2 to SO_3 over the vanadium pentoxide catalyst.

$$SO_2 + \frac{1}{2}O_2 \rightleftarrows SO_3$$

A three-plug-flow reactor system with interstage cooling is illustrated in Figure 7.16. The species conservation equations and macroscopic energy balances between the inlets and the arbitrary positions in these plug-flow reactors (Figure 7.16) can be expressed as follows:

$$\frac{V_1}{F_{A_o}} = \int_0^{x_{A_1}} \frac{dx_A}{-R_A(T, x_A)} \quad \text{Reactor 1} \tag{7.79}$$

$$\frac{V_2}{F_{A_o}} = \int_{x_{A_1}}^{x_{A_2}} \frac{dx_A}{-R_A(T, x_A)} \quad \text{Reactor 2} \tag{7.80}$$

$$\frac{V_3}{F_{A_o}} = \int_{x_{A_2}}^{x_{A_f}} \frac{dx_A}{-R_A(T, x_A)} \quad \text{Reactor 3} \tag{7.81}$$

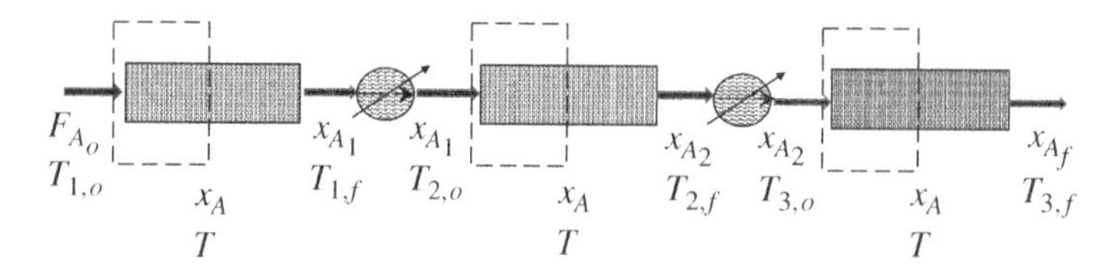

Figure 7.16 Three-plug-flow reactors, operating in series with cooling between the reactors.

$$\sum_i \int_{T_{1,o}}^{T_R} F_{i_o} C_{p_i} dT + \Delta H^o_{T_R} F_{A_0} x_A + \sum_i \int_{T_R}^{T} F_{i_z} C_{p_i} dT = 0; \quad \text{Reactor 1} \tag{7.82}$$

$$\sum_i \int_{T_{2,o}}^{T_R} F_{i_1} C_{p_i} \, dT + \Delta H^o_{T_R} F_{A_0} (x_A - x_{A_1}) + \sum_i \int_{T_R}^{T} F_{i_z} C_{p_i} dT = 0; \quad \text{Reactor 2} \tag{7.83}$$

$$\sum_i \int_{T_{3,o}}^{T_R} F_{i_2} C_{p_i} \, dT + \Delta H^o_{T_R} F_{A_0} (x_A - x_{A_2}) + \sum_i \int_{T_R}^{T} F_{i_z} C_{p_i} \, dT = 0; \quad \text{Reactor 3} \tag{7.84}$$

The inlet and outlet conditions of this system of reactors are illustrated in Figure 7.17. Bell-shaped constant rate curves are also shown in Figure 7.17. Constant rate curves are plotted by giving specific values to the reaction rate and plotting the variation of x_A as a function of temperature. During the interstage cooling between the reactors, it is usually preferred to decrease the temperature of the outlet stream of a reactor to a temperature at which the reaction rate is the same as its value at the exit conditions of the previous reactor. For instance, cooling of the outlet stream of the first reactor may be performed until a temperature such that the value of reaction rate at the inlet conditions of Reactor 2 $(x_{A_1}, \ T_{2,o})$ is about the same as the reaction rate at the outlet conditions of the first reactor $(x_{A_1}, \ T_{1,f})$. As illustrated in Figure 7.17, bell-shaped conversion vs. temperature curves (constant rate curves) pass through a maximum. In Figure 7.17, the rate values increase in the order of $R_3 > R_2 > R_1$.

For an arbitrary second-order liquid-phase reversible reaction ($A + B \rightleftarrows C$), species conservation and energy balance equations around the first and the second plug-flow reactors, which are illustrated in Figure 7.17, can be expressed for a constant volumetric flow rate case as follows:

$$-R_A = k\left(C_A C_B - \frac{1}{K} C_C\right) = k\left(C_{A_o}(1 - x_A)(C_{B_o} - C_{A_o} x_A) - \frac{1}{K} C_{A_o} x_A\right) \tag{7.85}$$

Species conservation equations for Reactors 1 and 2 are:

$$\frac{V_1}{F_{A_o}} = \int_0^{x_{A_1}} \frac{dx_A}{k\left(C_{A_o}(1 - x_A)(C_{B_o} - C_{A_o} x_A) - \frac{1}{K} C_{A_o} x_A\right)}; \quad \text{Reactor 1}$$

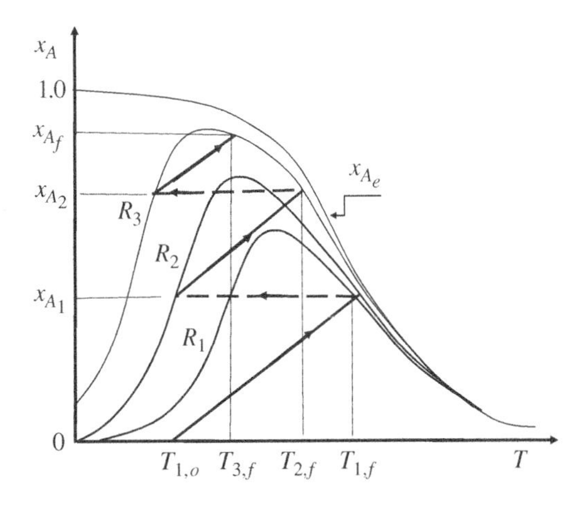

Figure 7.17 Conversion vs. temperature diagram showing the inlet and outlet conditions of a three-adiabatic reactor system with an exothermic-reversible reaction.

$$\frac{V_2}{F_{A_o}} = \int_{x_{A_1}}^{x_{A_2}} \frac{dx_A}{k\left(C_{A_o}(1-x_A)(C_{B_o}-C_{A_o}x_A) - \frac{1}{K}C_{A_o}x_A\right)}; \quad \text{Reactor 2}$$

Macroscopic energy balances between the inlets and arbitrary positions in the reactors are:

Reactor 1:

$$(F_{A_o}C_{P_A} + F_{B_o}C_{P_A})(T_R - T_{1,o}) + \Delta H^o_{T_R} F_{A_0} x_A$$
$$+ (F_{A_o}(1-x_A)C_{P_A} + (F_{B_o} - F_{A_o}x_A)C_{P_B} + F_{A_o}x_A C_{P_C})(T - T_R) = 0$$

Reactor 2:

$$(F_{A_o}(1-x_{A_1})C_{P_A} + (F_{B_o} - F_{A_o}x_{A_1})C_{P_B} + F_{A_o}x_{A_1}C_{P_C})(T_R - T_{2,0}) + \Delta H^o_{T_R} F_{A_o}(x_A - x_{A_1})$$
$$+ (F_{A_o}(1-x_A)C_{P_A} + (F_{B_o} - F_{A_o}x_A)C_{P_B} + F_{A_o}x_A C_{P_C})(T - T_R) = 0$$

Note that macroscopic energy balance equations are written between reactor inlets and the arbitrary locations within the reactors, where conversion and temperature are x_A and T. Both fractional conversion and temperature values are expected to change along the reactor. Hence, the reaction rate values are also expected to change along the reactors. Temperature values at different fractional conversions along the reactors can be estimated from the energy balance equations given above. Then, the evaluated conversion–temperature pairs can be used to determine the reaction rate values at different locations along the reactors, and the volumes of the reactors can be found by the numerical integration of the species conservation equations given above for Reactors 1 and 2. A similar analysis should be performed for the third reactor shown in Figure 7.17.

Example 7.2 Methanol Synthesis

Synthesis of methanol can be achieved by the reaction of CO and CO_2 with hydrogen, present in the synthesis gas.

$$CO + 2H_2 \rightleftarrows CH_3OH \quad \Delta H^o_{298} = -90.2\,\text{kJ/mol}$$

$$CO_2 + 3H_2 \rightleftarrows CH_3OH + H_2O \quad \Delta H^o_{298} = -49.0\,\text{kJ/mol}$$

We also expect the occurrence of a water–gas shift reaction in this reactor.

$$CO + H_2O \rightleftarrows CO_2 + H_2 \quad \Delta H^o_{298} = -41.2\,\text{kJ/mol}$$

However, only two of these reactions are independent.

Equilibrium conversion calculations of methanol synthesis are presented in Example 2.2, considering the reaction of CO with hydrogen over a Cu/ZnO/alumina-based solid catalyst. Temperature and the pressure dependences of the equilibrium conversion of CO are illustrated in that example in the absence of carbon dioxide. The feed stream composition was selected as 30% CO and 70% H_2. In the example given here, a reactor design model is illustrated for a non-isothermal reactor by taking the same feed composition as in Example 2.2. The design procedure for a methanol synthesis reactor with a feed stream containing both CO and CO_2 is given in Online Appendix O6.3.

Selection of Operating Conditions

Methanol synthesis is an exothermic reaction, and equilibrium conversion shows a decrease with an increase in temperature (Figure 2.5). However, an increase in equilibrium conversion is predicted with an increase in pressure due to the reaction stoichiometry.

The selection of the temperature and the pressure values depends upon the equilibrium and the reaction rate information. Equilibrium calculations indicate that low temperatures favor high conversions in methanol synthesis. However, the $Cu/ZnO/Al_2O_3$-type catalysts, which are generally used in methanol synthesis, show reasonable activity at temperatures higher than 500 K. On the other hand, catalyst deactivation due to sintering of the active metal copper was reported at temperatures over 600 K. Formation of undesired side products were also reported at such high temperatures. Hence, the best temperature range for this reaction was reported as between 523 and 573 K.

As indicated in Figure 2.5, the pressure of the reactor should be higher than 50 atm to achieve high conversions of CO to methanol in the temperature range of 523–573 K. However, the increase in pressure also increases the cost of the process. Also, safety issues become more critical in high-pressure operations. Equilibrium calculations also indicate that the effect of the increase of pressure on equilibrium conversion becomes less significant over 80 atm. Most methanol synthesis reactors are designed to operate at pressure values between 50 and 100 atm.

A number of rate expressions are reported in the literature for methanol synthesis [5, 6]. One of the rate expressions reported for methanol synthesis from CO and H_2 is of Leonov et al. [5]. A possible contribution of carbon dioxide to methanol synthesis is neglected in that work.

$$R_{\text{MetOH}} = k\left(\frac{P_{\text{CO}}^{1/2}P_{\text{H}_2}}{P_{\text{MetOH}}^{0.66}} - \frac{P_{\text{MetOH}}^{0.34}}{K_p P_{\text{CO}}^{0.5} P_{\text{H}_2}}\right)$$

If we denote CO, H_2, and methanol by A, B, and C, respectively

$$\text{A} + 2\text{B} \rightleftarrows \text{C}$$

the species conservation and the energy balance equations in a tubular reactor can be expressed as follows.

Species Conservation Equation for CO

$$-\frac{dF_A}{dV} + R_A = 0 \quad \Rightarrow \quad F_{A_o}\frac{dx_A}{dV} - k\left(\frac{P_A^{0.5}P_B}{P_C^{0.66}} - \frac{P_C^{0.34}}{K_P P_A^{0.5} P_B}\right) = 0 \tag{7.86}$$

Differential Energy Balance Equation in the Tubular Reactor

$$\sum_i F_i C_{P_i}\frac{dT}{dV} + \Delta H_T^o R = -\frac{2}{r_o}U(T - T_a) \tag{7.87}$$

In the case of adiabatic operation of the reactor, the heat transfer term with the surrounding becomes zero, and the differential energy balance can be expressed as follows:

$$\left[F_{A_o}(1-x_A)C_{P_A} + F_{A_o}x_A C_{P_C} + (F_{B_o} - 2F_{A_o}x_A)C_{P_B}\right]\frac{dT}{dV} + \Delta H_T^o k\left(\frac{P_A^{0.5}P_B}{P_C^{0.66}} - \frac{P_C^{0.34}}{K_P P_A^{0.5} P_B}\right) = 0 \tag{7.88}$$

Molar flow rates of A, B, and C can be expressed as follows:

$$F_A = F_{A_o}(1-x_A), \quad F_B = F_{B_o} - 2F_{A_o}x_A, \quad F_C = F_{A_o}x_A$$

The total molar flow rate at any fractional conversion is

$$F_T = F_{A_o} + F_{B_o} - 2F_{A_o}x_A = F_o - 2F_{A_o}x_A$$

The partial pressures of different species involved in this reaction can be expressed in terms of fractional conversion of CO as follows:

$$P_A = P\frac{F_A}{F_T} = P\frac{F_{A_o}(1-x_A)}{(F_o - 2F_{A_o}x_A)} \tag{7.89}$$

$$P_B = P\frac{(F_{B_o} - 2F_{A_o}x_A)}{(F_o - 2F_{A_o}x_A)} \tag{7.90}$$

$$P_C = P\frac{F_{A_o}x_A}{(F_o - 2F_{A_o}x_A)} \tag{7.91}$$

Hence, the reaction rate expression can be expressed in terms of fractional conversion values, as follows:

$$R_C = k\left(P^{0.84}\frac{(y_{A_o}(1-x_A))^{0.5}(y_{B_o} - 2y_{A_o}x_A)}{(1-2y_{A_o}x_A)^{0.84}(y_{A_o}x_A)^{0.66}} - \frac{(y_{A_o}x_A)^{0.34}(1-2y_{A_o}x_A)^{1.16}}{K_P P^{1.16}(y_{A_o}(1-x_A))^{0.5}(y_{B_o} - 2y_{A_o}x_A)}\right) \tag{7.92}$$

The rate of this reaction is generally reported in the literature in the units of moles of methanol produced per unit mass of the catalyst per unit time (see Online Appendix 6.3). However, in the design equation (Eq. 7.86) it should be in the units of moles of methanol produced per unit volume of the reactor per unit time. Therefore, the reaction rate constant given in the units of moles per catalyst mass per unit time should be multiplied by the density of the catalyst and the solid fraction of the bed as $[k = k'\rho_p(1 - \varepsilon_b)]$.

The following expression is given for the equilibrium constant of the methanol synthesis reaction.

$$K_p = 3.6 \times 10^{-12} \exp\left(\frac{90\,200}{R_g T}\right)$$

The equilibrium constant expression given above is written by assuming constant heat of reaction. A more precise evaluation of the temperature dependence of the equilibrium constant is illustrated in Example 2.2.

The reaction rate expression (Eq. (7.92)) depends upon temperature and fractional conversion of CO. Hence, it changes along the reactor. By inserting the reaction rate expression (Eq. (7.92)) to the material and the energy balance expressions (Eqs. (7.86) and (7.88)), two differential equations can be obtained for x_A and T. Hence, the simultaneous solution of these equations should be performed to evaluate the fractional conversion and the temperature values along the reactor.

Instead of writing a differential equation for the energy balance, a macroscopic balance may also be written between the inlet of the reactor and an arbitrary position along the reactor.

$$(F_{A_o}C_{P_A} + F_{B_o}C_{P_A})(T - T_o) + \Delta H_T^o F_{A_o}x_A = 0 \tag{7.93}$$

Temperature T and the fractional conversion x_A, which appear in this expression, correspond to their values at an arbitrary position within the reactor. Hence, Eq. (7.93) may be used to predict the temperature values corresponding to different fractional conversions along the reactor.

Temperature values corresponding to different fractional conversions should be used to evaluate the reaction rate values at different locations along the reactor. Then, the volume of the reactor can be evaluated by the numerical integration of Eq. (7.94).

$$\frac{V}{F_{A_o}} = \int_0^{x_A} \frac{dx_A}{R_C(T, x_A)} \tag{7.94}$$

The increase of temperature along the reactor should be controlled to achieve high fractional conversions of CO to methanol. This can be achieved either by cooling the reactor or by using multiple tubular reactors with interstage cooling or cold injection between the stages.

7.5.1.1 Endothermic-Reversible Reactions in a Multi-stage Reactor System

In the case of endothermic reactions, equilibrium conversion is expected to increase with an increase in temperature. Since the reaction rate also increases with an increase in temperature, high temperatures are usually preferred for endothermic reactions. However, there is a maximum temperature over which the catalyst loses its activity due to sintering, in solid-catalyzed reactions. Also, safety and economic considerations limit the maximum achievable temperature in the reactors.

In the case of an adiabatic operation, the outlet stream temperature is expected to be less than the temperature of the inlet stream for endothermic reactions. Hence, fractional conversion approaches to the equilibrium conversions toward the exit of such reactors. To reach a desired fractional conversion of x_{A_f}, it may be necessary to use multiple reactor systems with interstage heating between the reactors. This is illustrated in Figure 7.18 for a three-adiabatic reactor system shown in Figure 7.16.

Material and energy balance equations, which are written above (Eqs. (7.79)–(7.85)), can also be used to design multi-stage reactors for endothermic reactions. The only difference is that the heat of the reaction is positive for an endothermic reaction, and hence the temperature will decrease along the adiabatic reactor.

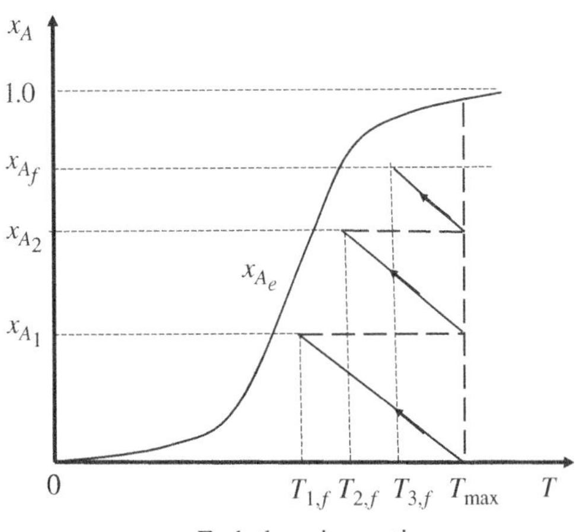

Figure 7.18 Schematic illustration of the conversion vs. temperature diagram for a three-adiabatic reactor system for an endothermic-reversible reaction.

Example 7.3 Ethanol Reforming

Fast depletion of oil reserves and related environmental considerations initiated significant research for the development of environmentally benign alternative fuels from non-fossil resources. Hydrogen, with its clean-burning properties and very-high-energy density, attracted significant attention as a fuel alternative for the future. Conventionally, hydrogen is produced by the steam-reforming reaction of methane, liquid fossil resources, and coal gasification. Production of hydrogen from a bio-mass-based resource is considered as an attractive alternative. Ethanol is one of the bio-based species, which contain quite a high number of hydrogen atoms. Up to 6 mols of hydrogen can, in principle, be produced from 1 mol of ethanol through a catalytic steam-reforming reaction. During the synthesis of bio-ethanol by a fermentation process, a large amount of water is also produced. Hence, some of the water present in the fermentation product may, in principle, be used during the steam-reforming process. Another advantage of using ethanol as a resource for hydrogen production is that its steam reforming takes place at much lower temperatures than the steam reforming of methane.

In the reforming reactor of ethanol, steam reforming and water–gas shift reactions simultaneously take place within the same system.

$$C_2H_5OH + H_2O \rightleftarrows 2CO + 4H_2 \quad \text{steam reforming}$$

$$CO + H_2O \rightleftarrows CO_2 + H_2 \quad \text{water - gas shift reaction}$$

The overall stoichiometry of the ethanol reforming can be expressed as follows:

$$C_2H_5OH + 3H_2O \rightleftarrows 2CO_2 + 6H_2 \quad \text{steam reforming}$$

In this example, it is asked to perform equilibrium calculations and to set up the design equations for an ethanol-reforming reactor. The effects of temperature and the feed composition on the equilibrium conversion of ethanol are analyzed, and operating conditions of the reactor are selected to achieve close to complete conversion. It is also asked to evaluate the equilibrium molar ratio of CO/CO_2 in the product stream at 523 K and 1 bar. The feed composition of the reactor can be taken as

$$\left(\frac{y_{H_2O}}{y_{C_2H_5OH}}\right)_{feed} = 3$$

Solution

This system of reactions involve five molecular species, namely C_2H_5OH, H_2O, CO, CO_2, and H_2. The relations between the rates of these species can be expressed by writing conservation relations for atomic species, namely C, O, and H.

$$C: \quad 2R_{C_2H_5OH} + R_{CO} + R_{CO_2} = 0$$

$$H: \quad 6R_{C_2H_5OH} + 2R_{H_2} + 2R_{H_2O} = 0$$

$$O: \quad R_{C_2H_5OH} + R_{CO} + 2R_{CO_2} + R_{H_2O} = 0$$

The difference between the number of rate values and the number of relations for the conservation of atomic species is 2. Since there are two independent rate values, only two of the reactions given above are independent.

By arbitrarily taking the steam-reforming reactions yielding CO and CO_2 as the two independent reactions, the following analysis is performed. In this analysis, ethanol is denoted as A, carbon monoxide as B, carbon dioxide as C, hydrogen as D, and water as E.

$$A + E \rightleftarrows 2B + 4D \tag{R.7.1}$$

$$A + 3E \rightleftarrows 2C + 6D \qquad (R.7.2)$$

By defining fractional conversion of ethanol to carbon monoxide and carbon dioxide as x_1 and x_2, respectively, mole fractions of all of the molecular species at any axial position in the reforming reactor can be expressed as follows:

$$x_1 = \frac{F_B}{2F_{A_0}}; \quad x_2 = \frac{F_C}{2F_{A_0}}$$

$$F_A = F_{A_0}(1 - x_1 - x_2)$$

$$F_B = 2F_{A_0}x_1$$

$$F_C = 2F_{A_0}x_2$$

$$F_E = F_{E_0} - F_{A_0}x_1 - 3F_{A_0}x_2$$

$$F_D = 4F_{A_0}x_1 + 6F_{A_0}x_2$$

$$\text{Total flow rate}: \ F_T = (F_{A_0} + F_{E_0}) + 4F_{A_0}x_1 + 4F_{A_0}x_2$$

Equilibrium constants for the two reactions ((R.7.1) and (R.7.2)) can then be expressed in terms of the partial pressures of the molecular species involved in the reaction. The partial pressures of species can be related to the two unknown conversion parameters x_{1_e} and x_{2_e}.

$$K_{p1} = \left(\frac{P_B^2 P_D^4}{P_A P_E}\right)_e; \quad K_{p2} = \left(\frac{P_C^2 P_D^6}{P_A P_E{}^3}\right)_e$$

$$P_{A_e} = P_T y_{A_e} = P_T \frac{F_{A_e}}{F_{T_f}} = P_T \frac{F_{A_0}(1 - x_{1_e} - x_{2_e})}{(F_{A_0} + F_{E_0}) + 4F_{A_0}x_{1_e} + 4F_{A_0}x_{2_e}} = P_T \frac{y_{A_0}(1 - x_{1_e} - x_{2_e})}{1 + 4y_{A_0}(x_{1_e} + x_{2_e})}$$

$$P_{B_e} = P_T y_{B_e} = P_T \frac{F_{B_e}}{F_{T_f}} = P_T \frac{2y_{A_0}x_{1_e}}{1 + 4y_{A_0}(x_{1_e} + x_{2_e})}$$

$$P_{C_e} = P_T y_{C_e} = P_T \frac{F_{C_e}}{F_{T_f}} = P_T \frac{2y_{A_0}x_{2_e}}{1 + 4y_{A_0}(x_{1_e} + x_{2_e})}$$

$$P_{D_e} = P_T y_{D_e} = P_T \frac{F_{D_e}}{F_{T_f}} = P_T \frac{y_{A_0}(4x_{1_e} + 6x_{2_e})}{1 + 4y_{A_0}(x_{1_e} + x_{2_e})}$$

$$P_{E_e} = P_T y_{E_e} = P_T \frac{F_{E_e}}{F_{T_f}} = P_T \frac{y_{E_0} - y_{A_0}(x_{1_e} + 3x_{2_e})}{1 + 4y_{A_0}(x_{1_e} + x_{2_e})}$$

The overall conversion of ethanol is:

$$x_A = x_1 + x_2$$

These partial pressure expressions can be substituted into the two equilibrium relations, and simultaneous solution of x_{1_e} and x_{2_e} can then be achieved using these two equations.

$$K_{p1} = \left(\frac{P_B^2 P_D^4}{P_A P_E}\right)_e = P_T{}^4 \frac{(2y_{A_0}x_{1_e})^2 [y_{A_0}(4x_{1_e} + 6x_{2_e})]^4}{[1 + 4y_{A_0}(x_{1_e} + x_{2_e})]^4 y_{A_0}(1 - x_{1_e} - x_{2_e})[y_{E_0} - y_{A_0}(x_{1_e} + 3x_{2_e})]} \qquad (7.95)$$

Table 7.2 Standard gas-phase Gibbs free energy and enthalpy of formation values at 298 K.

Species	C_2H_5OH	H_2O	CO	H_2	CO_2
ΔH_f^o (kJ/mol)	−235.1	−241.8	−110.5	0	−393.5
ΔG_f^o (kJ/mol)	−168.5	−228.6	−137.2	0	−394.4

Source: Sandler [7].

$$K_{p2} = \left(\frac{P_C^2 P_D^6}{P_A P_E{}^3}\right)_e = P_T{}^4 \frac{(2y_{A_0}x_{2_e})^2 [y_{A_0}(4x_{1_e} + 6x_{2_e})]^6}{[1 + 4y_{A_0}(x_{1_e} + x_{2_e})]^4 y_{A_0}(1 - x_{1_e} - x_{2_e})[y_{E_0} - y_{A_0}(x_{1_e} + 3x_{2_e})]^3} \tag{7.96}$$

To proceed with the solution of x_{1_e} and x_{2_e}, the values of the equilibrium constants are needed at the desired temperature. For this calculation, we need the free energies of formation and heat of formations of the species involved in the reaction. Also, we need the heat capacities of the species. All these data are given in Tables 7.2 and 7.3.

The heat of reaction and Gibbs' free energy of reaction values for reactions (R.7.1) and (R.7.2) are evaluated at 298 K, using the data reported in Tables 7.2 and 7.3.

$$\Delta G_1^o = \left[2\Delta G^o{}_{f,CO} + 4\Delta G^o{}_{f,H_2}\right] - \left[\Delta G^o{}_{f,C_2H_5OH} + \Delta G^o{}_{f,H_2O}\right] = 122.7 \text{ kJ/mol}$$

$$\Delta G_2^o = \left[2\Delta G^o{}_{f,CO_2} + 6\Delta G^o{}_{f,H_2}\right] - \left[\Delta G^o{}_{f,C_2H_5OH} + 3\Delta G^o{}_{f,H_2O}\right] = 65.5 \text{ kJ/mol}$$

Hence, the equilibrium constants for these two reactions at 298 K are

$$K_{1,298} = \exp\left(\frac{-122\,700}{(8.314)298}\right) = 3.10 \times 10^{-22}; \quad K_{2,298} = \exp\left(\frac{-65\,500}{(8.314)298}\right) = 3.30 \times 10^{-12}$$

To evaluate the equilibrium constant values at a temperature higher than 298 K, van't Hoff relation given in Eq. (2.35) is used together with Eq. (2.42) for the temperature dependence of heat of reaction.

$$\Delta H_{1,298}^o = \left[2\Delta H^o{}_{f,CO} + 4\Delta H^o{}_{f,H_2}\right] - \left[\Delta H^o{}_{f,C_2H_5OH} + \Delta H^o{}_{f,H_2O}\right] = 255.9 \text{ kJ/mol}$$

Table 7.3 The coefficients of gas-phase molar heat capacities of species.

		$C_p = a + bT + cT^2 + dT^3$		
Species	a	$b \times 10^2$	$c \times 10^5$	$d \times 10^9$
C_2H_5OH	19.875	20.946	−10.372	20.042
H_2O	32.218	0.192	1.055	−3.593
CO	28.142	0.167	0.537	−2.221
H_2	29.088	−0.192	0.400	−0.870
CO_2	22.243	5.977	−3.499	7.464

Source: Sandler [7].

$$\Delta H^o_{2,298} = \left[2\Delta H^o_{f,CO_2} + 6\Delta H^o_{f,H_2}\right] - \left[\Delta H^o_{f,C_2H_5OH} + 3\Delta H^o_{f,H_2O}\right] = 173.50\ \text{kJ/mol}$$

These results show that both of these reactions are endothermic. However, the heat of reaction of the water–gas shift reaction is −41.2 kJ/mol, indicating that it is an exothermic reaction. These factors have a significant effect on the product distributions at different temperatures. Temperature dependence of the heat of reaction of (R.7.1) and its equilibrium constant at 523 K can then be evaluated as follows:

$$\Delta H^o_{1,T} = \Delta H^o_{1,298} + \int_{298}^{T} \Delta C_p\, dT = \Delta H^o_{1,298} + \int_{298}^{T} (\Delta a_1 + \Delta b_1 T + \Delta c_1 T^2 + \Delta d_1 T^3)\, dT$$

$$\Delta H^o_{1,T} = 255\,900 + 120.54(T - 298) - 21.57 \times 10^{-2}\left(\frac{T^2 - 298^2}{2}\right) + 11.99 \times 10^{-5}\left(\frac{T^3 - 298^3}{3}\right) - 24.37 \times 10^{-9}\left(\frac{T^4 - 298^4}{4}\right)$$

$$K_{1,523} = K_{1,298} \exp\left(\int_{298}^{523} \frac{\Delta H^o_{1,T}}{R_g T^2}\, dT\right)$$

$$K_{1,523} = 0.015$$

Similarly, the equilibrium constant for (R.7.2) at 523 K is found as,

$$K_{2,523} = 115.9$$

Using these values of equilibrium constants in the equilibrium relations given by Eqs. (7.95) and (7.96), simultaneous solution of these equations gives

$$x_{1_e} = 0.055; \quad x_{2_e} = 0.74$$

The overall equilibrium conversion of ethanol, the ratio of CO to CO_2 in the product stream, and equilibrium mole fraction of hydrogen are then evaluated.

$$x_{A_e} = 0.055 + 0.74 = 0.795; \quad \frac{y_{CO}}{y_{CO_2}} = 0.074; \quad y_{H_2} = 0.65$$

Since both (R.7.1) and (R.7.2) are endothermic reactions, an increase in the equilibrium constant is expected with an increase in temperature. Also, an increase in the conversion of ethanol is expected, with an increase in the water to ethanol ratio in the feed stream (Figure 7.19). It may be proposed to operate the ethanol-reforming reactor at temperatures higher than 573 K and with a $\frac{y_{H_2O}}{y_{C_2H_5OH}}$ ratio higher than 5 in the feed stream, based on the equilibrium calculations. An increase in this ratio causes a significant increase in fractional conversion of ethanol. However, an increase in this ratio also increases the operating and fixed costs of the process.

Several studies are available in the literature for the kinetics of ethanol-reforming reaction. The forward rate expressions for reactions (R.7.1) and (R.7.2) were reported in the literature [8] in power-law form for the case of excess water in the feed stream:

$$R_{(1)} = k_{1_o} \exp\left(-\frac{E_{a,1}}{R_g T}\right) P_A^{0.75} \tag{7.97}$$

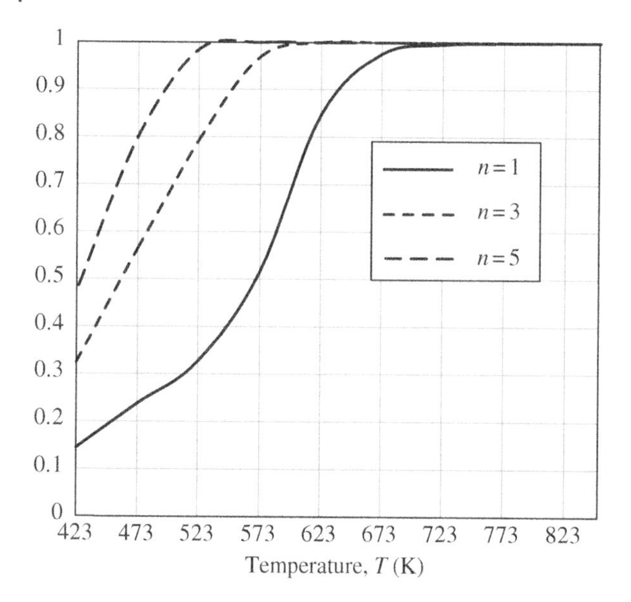

Figure 7.19 Temperature dependence of fractional conversion of ethanol at equilibrium with different $n = \frac{y_{H_2O}}{y_{C_2H_5OH}}$ ratios in the feed stream.

$$R_{(2)} = k_{2_0} \exp\left(-\frac{E_{a,2}}{R_g T}\right) P_A^{0.8} \tag{7.98}$$

As illustrated in Figure 7.19, for reactor temperatures higher than 573 K, together with water to ethanol ratios being higher than 5 in the feed stream, the reforming reactions can be assumed as irreversible. Rate expressions (7.97) and (7.98) may be used at these conditions. Due to the high water to ethanol ratio in the feed stream, the rates of reforming reactions were assumed zero order with respect to the concentration of H_2O in these expressions.

Since there are two independent reactions taking place in this system, we need to write down two species conservation equations and an energy balance equation to model this tubular reactor. Design equations for a plug-flow ethanol-reforming reactor are then expressed using Eqs. (7.97) and (7.98) as follows:

Species conservation equation for CO:

$$-F_{A_0} \frac{dx_1}{dV} + k_{1_0} \exp\left(-\frac{E_{a,1}}{R_g T}\right) P_A^{0.75} = 0$$

$$-F_{A_0} \frac{dx_1}{dV} + k_{1_0} \exp\left(-\frac{E_{a,1}}{R_g T}\right) P^{0.75} \left(\frac{y_{A_0}(1 - x_1 - x_2)}{1 + 4y_{A_0}(x_1 + x_2)}\right)^{0.75} = 0 \tag{7.99}$$

Species conservation equation for CO_2:

$$-F_{A_0} \frac{dx_2}{dV} + k_{2_0} \exp\left(-\frac{E_{a,2}}{R_g T}\right) P_A^{0.80} = 0$$

$$-F_{A_0} \frac{dx_2}{dV} + k_{2_0} \exp\left(-\frac{E_{a,2}}{R_g T}\right) P^{0.80} \left(\frac{y_{A_0}(1 - x_1 - x_2)}{1 + 4y_{A_0}(x_1 + x_2)}\right)^{0.80} = 0 \tag{7.100}$$

Differential energy balance:

$$\sum_i F_i C_{p_i} \frac{dT}{dV} + \Delta H^o_{1,T} R_1 + \Delta H^o_{2,T} R_2 = -\frac{2}{r_o} U(T - T_a) \tag{7.101}$$

Molar flow rates of all the species involved in this reaction should be expressed in terms of the fractional conversions x_1 and x_2. The reaction rate expressions $R_{(1)}$ and $R_{(2)}$ depend upon the fractional conversions and the reaction temperature. Simultaneous numerical solution of the species conservation equations (7.99) and (7.100), together with the energy balance equation (Eq. (7.101)), should be performed to design this reactor.

Further discussion of heat effects in chemical reactors may be found in Refs. [2–4, 9].

7.5.2 Cold Injection Between Reactors

For some exothermic-reversible reactions, a certain fraction of the cold feed stream is bypassed the first reactor and injected into the exit stream of this reactor to decrease its temperature and to get away from the equilibrium limitations [9]. A typical example is an ammonia synthesis reaction, which is a highly exothermic-reversible reaction.

$$N_2 + 3H_2 \rightleftarrows 2NH_3; \quad -\Delta H^o_{298} = 91.4\ \text{kJ/mol}$$

The equilibrium constant of this reaction decreases quite sharply with an increase in temperature. For instance, the equilibrium constant values for this reaction at 25, 200, 400, and 500 °C are 6.4×10^2, 4.4×10^{-1}, 1.64×10^{-4}, and 1.45×10^{-5}, respectively. Due to the thermodynamic limitations, lower temperatures are preferred to achieve higher conversions. However, from the reaction rate point of view, the reaction temperature should be higher than 400 °C. Usually, iron-based catalysts are used in this process, and these catalysts show activity only over this temperature. To achieve sufficiently high conversions at high temperatures, quite high pressures, over 100 atm, are used in the conventional ammonia synthesis reactors. Due to the stoichiometry of the ammonia synthesis reaction, higher pressures favor higher ammonia yields, according to Le Chatelier's principle. Conventionally, multi-stage tubular catalytic reactors are used for this reaction. Due to the high pressure of the reaction system, all stages of this multi-stage reactor system are placed into a single vessel, and cooling between the stages is achieved by the injection of some fractions of the cold reactant mixture into the reactor effluent streams after the first and the second stages.

A typical two-stage plug-flow reactor system with an injection of a fraction of cold feed stream between the reactors is illustrated in Figure 7.20.

Let us consider a first-order exothermic-reversible reaction taking place in this two-stage adiabatic reactor system.

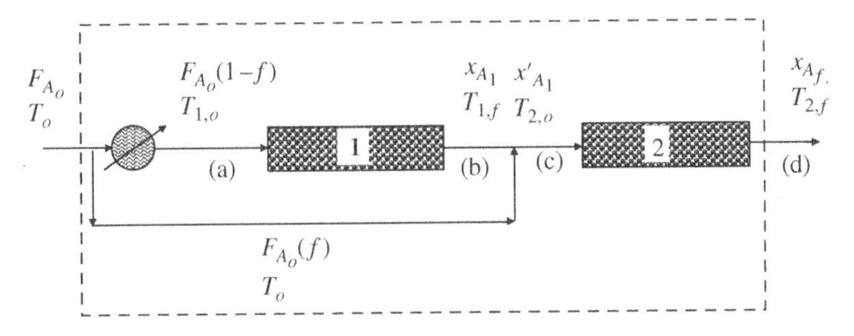

Figure 7.20 A two-stage reactor system with cold injection between the reactors.

$$\text{A} \rightleftarrows \text{B}; \quad -R_A = k\left[C_A - \frac{1}{K}C_B\right] = kC_{A_o}\left[(1-x_A) - \frac{1}{K}x_A\right]$$

In this system, the feed stream temperature is first increased, using a heat exchanger, from T_o to $T_{1,o}$, which is the required inlet temperature of the first stage of the reactor system. Due to the exothermic nature of the reaction, the temperature is expected to increase along the first reactor, from $T_{1,o}$ to the exit temperature of the reactor ($T_{1,f}$). A fraction of F_{A_o} is separated from the main feed stream before the heat exchanger and introduced into the effluent stream of the first reactor. As a result of this cold injection step, both temperature and the apparent conversion of reactant A decrease after the injection point. Points (a), (b), (c), and (d) in this system are shown in the conversion vs. temperature diagram, given in Figure 7.21.

Species conservation equations for this two-stage system can be expressed as follows:

$$\text{Reactor 1}: \quad \frac{V_1}{F_{A_o}(1-f)} = \int_0^{x_{A_1}} \frac{dx_A}{-R_A(T,x_A)} \tag{7.102}$$

$$\text{Reactor 2}: \quad \frac{V_2}{F_{A_o}} = \int_{x'_{A_1}}^{x_{A_f}} \frac{dx_A}{-R_A(T,x_A)} \tag{7.103}$$

Here, f is the fraction of F_{A_o} introduced into the exit stream of the first reactor. To evaluate x'_{A_1}, we should write a material balance for species A at the mixing point after Reactor 1.

$$F_{A_o}(1-f)(1-x_{A_1}) + F_{A_o}(f) = F_{A_o}\left(1-x'_{A_1}\right) \tag{7.104}$$

Hence, x'_{A_1} can be evaluated from

$$x'_{A_1} = x_{A_1}(1-f) \tag{7.105}$$

Numerical integration of Eqs. (7.102) and (7.103) requires information about the temperature variations along Reactors 1 and 2. Hence, energy balances should be written for these two reactors. For constant heat of reaction and constant heat capacity values of species, macroscopic energy balance equations between the inlets and arbitrary locations within Reactors 1 and 2 can be written for a feed stream containing only A, as follows:

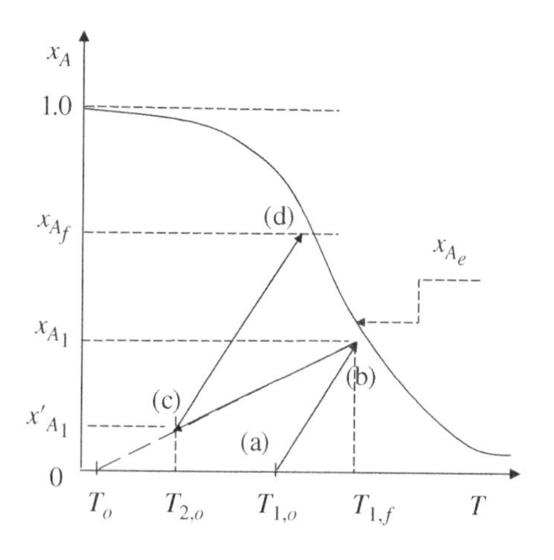

Figure 7.21 Illustration of the two-reactor system with a cold injection.

$$F_{A_o}(1-f)C_{p_A}(T-T_{1,o}) + \Delta H_T^o(F_{A_o}(1-f)x_A) = 0 \tag{7.106}$$

$$[(F_{A_o}(1-f)(1-x_{A_1}) + F_{A_o}(f))C_{p_A} + F_{A_o}(1-f)x_{A_1}C_{p_B}](T-T_{2,o}) + \Delta H_T^o F_{A_o}\left(x_A - x'_{A_1}\right) = 0 \tag{7.107}$$

The temperature of the feed stream entering the second reactor may be estimated by making an energy balance at the mixing point before the second reactor:

$$\begin{aligned}&\left[(F_{A_o}(1-f)(1-x_{A_1}))C_{p_A} + F_{A_o}(1-f)x_{A_1}C_{p_B}\right](T_{1f}-T_R) + F_{A_o}(f)C_{p_A}(T_o - T_R)\\&= \left[F_{A_o}\left(1-x'_{A_1}\right)C_{p_A} + F_{A_o}x'_{A_1}C_{p_B}\right](T_{2,o}-T_R)\end{aligned} \tag{7.108}$$

These energy balances yield the following relation for the slope of the line connecting the points b and c in Figure 7.21:

$$\frac{x_{A_1}}{T_{1f}-T_o} = \frac{x_{A_1}-x'_{A_1}}{T_{1f}-T_{2,o}}$$

Temperatures at different locations along the reactor can be evaluated at different fractional conversion values, using these energy balances. Numerical integration of Eqs. (7.102) and (7.103) can then be performed using these conversion–temperature pairs to evaluate the reaction rate values at different locations along the reactors.

7.5.3 Heat-Exchanger Reactors

Instead of using multiple tubular reactors with heat exchangers between the reactors (as illustrated in Figure 7.16), the use of shell and tube-type heat-exchanger reactors may also be suggested to control the temperature within the reactor. A schematic representation of such a reactor is illustrated in Figure 7.22. The catalyst may be packed into the tubes of a multi-tubular reactor. A cooling or a heating fluid may flow through the shell section, depending upon the exothermicity or endothermicity of the reaction, respectively. Depending upon the specific application, the shell section may be divided into multiple compartments or may be used as

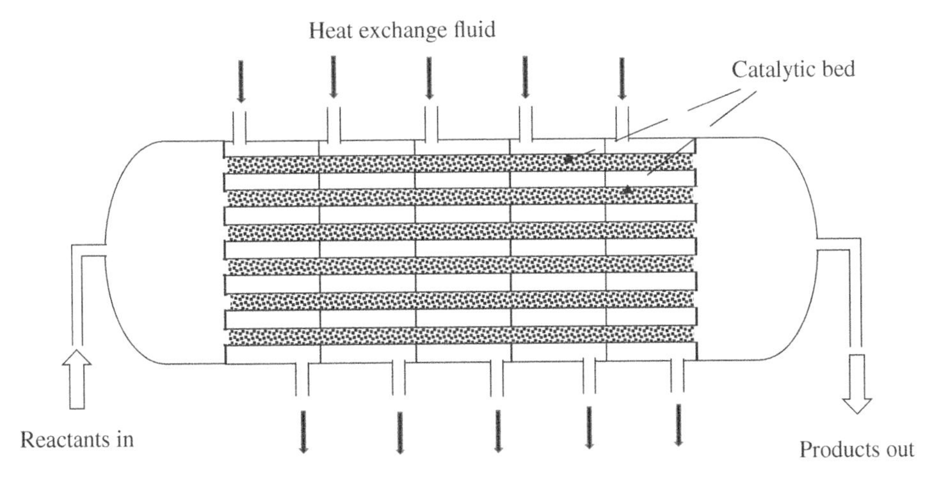

Figure 7.22 Schematic representation of a shell and tube reactor.

a single baffled section. Simultaneous solution of the species conservation (Eq. (7.39)) and energy balance equations (Eq. (7.40)) should be performed for each section of such a heat-exchanger reactor.

Problems and Questions

7.1 A first-order exothermic liquid-phase reaction is to be carried out in a CSTR. The volumetric feed rate of the inlet stream to the reactor is given as 120 l/min. Concentrations of the reactant in the reactor feed and exit streams are given as 0.8 and 0.05 mol/l, respectively. The temperature of the inlet stream to the reactor is given as 300 K. Some of the heat generated in the reactor is transferred to the fluid in the jacket of the reactor, which is at 293 K. The heat transfer surface area of the reactor is given as 12.0 m^2. The overall heat transfer coefficient is given as 3000 J/m^2 s K. Evaluate the volume and the reactor temperature, using the following data

$$k = 1.8 \times 10^7 \exp\left(-\frac{5600}{T}\right); \quad \min^{-1}$$

$\Delta H_R^o = -210\,000$ J/mol (given as independent of temperature)

$\bar{c} = 3.1$ J/g K (specific heat of the mixture is given as constant)

$\rho_{mix} = 1.1$ g/cm^3 (density of the mixture is given as constant).

7.2 Repeat Problem 7.1 for an adiabatic plug-flow reactor and evaluate the reactor volume. In this case, there is no heat transfer with the surroundings.

7.3 Following elementary liquid-phase reactions are taking place in a CSTR.

$$A \rightarrow B; \quad k_1 = 10^{10} \exp\left(-\frac{8100}{T}\right) \quad \text{(desired reaction)}$$

$$A \rightarrow C; \quad k_2 = 10^7 \exp\left(-\frac{5500}{T}\right)$$

It is desired to achieve a 90% conversion of reactant A in the reactor. Reactor volume, inlet volumetric flow rate, and inlet molar flow rate of A are given as 1895 l, 27 l/s, and 135 mol/s, respectively.

Data:
Specific heat of the reaction mixture: 2.5 J/g K
The density of the reaction mixture: 0.5 g/cm^3
The heat of reaction 1: $\Delta H_1^o = 34\,000$ J/mol
The heat of reaction 2: $\Delta H_2^o = -23\,000$ J/mol.

(a) What should be the reactor temperature and the conversion values achieved in the first and the second reactions?

(b) How much heat should be transferred to or from the reactor to have the reactor temperature being the same as the inlet temperature?

7.4 The following endothermic liquid-phase reaction proceeds in a steam-jacketed well-mixed reactor (CSTR). Calculate the reactor temperature and fractional conversion of reactant A in this reactor.

$$A + B \rightarrow C + D; \quad -R_A = 2.5 \exp\left(-\frac{4000}{T}\right) C_A C_B \ (\text{mol/cm}^3\ \text{s})$$

The feed stream's total molar flow rate is given as 30 mol/h, and its temperature is 30 °C. The reactor volume is 2 m^3. Concentrations of both A and B are 0.02 mol/cm^3 in the feed stream.

Data:
Heat of reaction: $\Delta H_1^o = 45\,000$ J/mol (at 25 °C)
Heat capacities of species: $C_{p_A} = 25$ (J/mol)/K ; $C_{p_B} = 15$ (J/mol)/K ; $C_{p_C} = C_{p_D} = 20$ (J/mol)/K
Overall heat transfer coefficient: $U = 60$ (cal/s m^2)/K
Heat transfer area: 1.5 m^2
The temperature of steam in the jacket: 150 °C.

7.5 Consider a first-order reversible exothermic reaction taking place in an adiabatic CSTR.

$$A \rightleftarrows B; \quad -R_A = k\left(C_A - \frac{1}{K} C_B\right)$$

The values of the forward rate constant are given as 3.6×10^{-3} and 6.0 s^{-1} at 320 and 460 K, respectively. The heat of reaction is given as $\Delta H_{\text{Ref}}^o = -18\,000$ J/mol. The specific heat and the density of the reaction mixture are given as 4 J/g K and 1.0 g/cm^3, respectively. The equilibrium constant of the reaction is given as 8.2, at 300 K. Volumetric flow rate and the concentration of reactant A in the feed stream are 1.5 m^3/h and 5 M, respectively.

(a) Find the equilibrium constant and the corresponding temperature, if equilibrium conversion is $x_{A_e} = 0.8$.

(b) If the reactor temperature is 15 °C less than the equilibrium temperature found in part (a), evaluate the corresponding fractional conversion of reactant A. Inlet temperature of the reactor is given as 300 K.

(c) Find the reactor volume for the fractional conversion and reactor temperature values found in part (b) of this problem.

7.6 A first-order gas-phase reaction is carried out in an adiabatic flow reactor.

$$A \rightarrow B + C; \quad k = 160 \exp\left(-\frac{25\,000}{R_g T}\right) \text{s}^{-1}$$

Feed stream contains 90% A and 10% inerts. The pressure and the temperature of the feed stream are 8 bar and 380 K, respectively. The volumetric flow rate of the feed stream is given as 10^5 l/h.

Evaluate the CSTR volume for a fractional conversion value of 0.85 for reactant A.

Data:
$C_{p_A} = 45$ J/mol K ; $C_{p_B} = 25$ J/mol K; $C_{p_C} = 20$ J/mol K ; $C_{p_{\text{inerts}}} = 30$ J/mol K;

$\Delta H_1^o = -25\,000\,\text{J/mol}$ (at 298 K).

7.7 Solve the same problem given in Problem 7.6 for a plug-flow reactor.

7.8 A first-order liquid-phase reaction,

$$\text{A} \rightarrow \text{B} \quad k = 4 \times 10^4 \exp\left(-\frac{3000}{T}\right) \text{h}^{-1}$$

is to be carried out in two-flow reactors operating in series. The first reactor is a CSTR, while the second reactor is an isothermal plug-flow reactor. Inlet and outlet temperatures of the first reactor are given as 475 and 700 K, respectively, while the second reactor operates at 700 K. Feed rate of the first reactor and concentration of A in this stream are given as 10.0 mol/min and 1 mol/l, respectively. The heat of reaction and the molar heat capacity of A are −30 000 J/mol and 70 J/mol K, respectively.

(a) Evaluate the volume of the first reactor if the fractional conversion at the exit of this reactor is 0.6.

(b) Find the amount of heat to be removed from (or introduced to) the first reactor to keep the temperature of this reactor at 700 K.

(c) If the volume of the second reactor (isothermal plug flow) is the same as the first reactor's volume, what will be fractional conversion at the exit of the second reactor?

(d) What is the amount of heat removed from the second reactor to keep it isothermal?

7.9 Consider the multi-stage reactor system given in Figure 7.23 for an exothermic-reversible reaction. There is a cold injection after the first reactor, and a heat exchanger is used between stages 2 and 3 to decrease the temperature of the effluent stream of Reactor 2. Show this system on a fractional conversion vs. temperature diagram indicating points (a), (b), (c) (d), and (e). Write down all the necessary energy and material balance equations required to design the reactors shown in Figure 7.23.

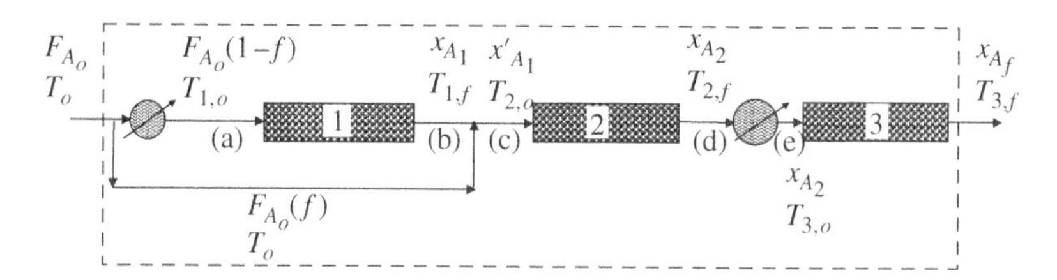

Figure 7.23 Reactor system for Problem 7.9.

7.10 The heat generation vs. reactor temperature curve is given in Figure 7.24 for an exothermic reaction taking place in an adiabatic CSTR. Pure A is fed to the reactor.

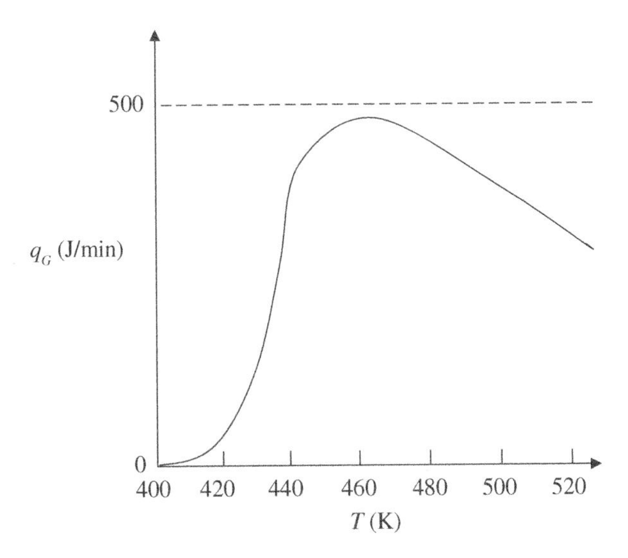

Figure 7.24 Heat generation curve for Problem 7.10.

(a) Explain the shape of the heat generation curve. What type of reaction would give this type of a heat generation curve? Justify your discussion with equations.

(b) Is it possible to have multiple steady-state temperatures for this system?

(c) What should be the temperature of the reactor to achieve the highest possible conversion? How do you decide about the inlet temperature without having steady-state multiplicity in this case?

7.11 Consider the following liquid-phase catalytic parallel reactions taking place in a fixed-bed plug-flow reactor. The desired product is C, while a parallel reaction produces the undesired product D.

$$\mathrm{A} + 1/2\mathrm{B} \rightarrow \mathrm{C}; \quad R_C = k_1 C_A C_B$$

$$\mathrm{A} + 3\mathrm{B} \rightarrow \mathrm{D}; \quad R_D = k_2 C_A C_B^{3/2}$$

Both of these reactions are highly exothermic, and the activation energy of the second reaction is higher than the activation energy of the first reaction. To control the temperature of the reactor, the reactor is cooled by a cooling fluid flowing in the jacket section of the reactor. The operation of the reactor is neither adiabatic nor isothermal. Answer the following questions:

(a) It is asked to set up all of the necessary differential material and energy balance equations, which are needed to design this reactor. State all of the assumptions, if any.

(b) Discuss the effects of reaction temperature and concentration of B on the selectivity of the desired product C with respect to A.

7.12 A second-order exothermic reaction between reactants A and B is carried out in an adiabatic CSTR, with a volume of 10 l.

$$A + B \rightarrow C; \quad -R_A = 3 \times 10^7 \exp\left(\frac{-14\,100}{T}\right) C_A C_B$$

The heat of reaction is given as constant, $\Delta H_R^o = -150$ kJ/mol. Molar heat capacities of A, B, and C can be taken as 60, 120, and 180 (J/mol)/K, respectively.

Concentrations of A and B in the feed stream are given as 1 and 2 M, respectively. Space-time in the reactor is given as 100 seconds.

(a) Write down the heat generation and the heat removal rate expressions as a function of reactor temperature.

(b) If the desired fractional conversion of A is 0.6, what should be the inlet and the outlet temperatures of the reactor?

7.13 A first-order irreversible reaction is taking place in an adiabatic CSTR. The feed rate to the reactor is 1.5 l/s. Inlet temperature and the inlet concentration of reactant A are 300 K and 1.8 M, respectively. The heat of reaction, the specific heat of the solution, and the feed stream's density are given as −40 J/mol, 3 J/g K, and 0.9 g/cm^3, respectively. The rate constant is given as

$$k = 2 \times 10^8 \exp\left(-\frac{7725}{T}\right) \text{s}^{-1} \quad (T \text{ in K})$$

Find the volume of the reactor for a conversion value of 0.85.

7.14 **(a)** A gas-phase exothermic reaction with an arbitrary rate expression is taking place in an adiabatic plug-flow reactor.

$$A \rightarrow B + 2C$$

The molar heat capacity of reactant A is given as constant. The feed stream contains only reactant A, and the feed temperature is T_o. Obtain an expression between the fractional conversion of reactant A and temperature at any axial position in the reactor.

(b) It is proposed to add some nitrogen to the inlet stream while keeping the flow rate of the feed stream the same as in the previous question. If the inlet stream contains 50% N_2 and 50% A in this second case, and if the molar heat capacity of nitrogen is twice the heat capacity of A, what would be the temperature at the position z, corresponding to the same conversion predicted in part (a)? Compare the temperatures that would be obtained at the same conversion levels in parts (a) and (b).

Case Studies

1 **Steam reforming of methane: (Online Appendix O6.4)** Steam reforming of methane is a conventional method for the production of hydrogen. You are asked to design a reactor (or reactor system) for the steam reforming of methane. During the steam-reforming reaction of methane, water–gas shift reaction (WGSR) also takes place.

$$CH_4 + H_2O \rightleftarrows CO + 3H_2 \quad \text{(SRM)}$$

$$CO + H_2O \rightleftarrows CO_2 + H_2 \quad \text{(WGSR)}$$

(a) Carry out a literature survey and obtain operating conditions of commercial SRM reactors. Also, obtain the thermodynamic and kinetic data for these reactions taking place within the same reactor.

(b) Perform equilibrium calculations and plot methane conversion, as well as product distributions at equilibrium, as a function of temperature, for the temperature and pressure ranges of commercial SRM reactors. In these calculations, you may use a feed stream composition of 27% CH_4, 65% H_2O, and 8% N_2.

(c) Decide about the temperature, pressure of the reactor. Also, decide about the type and operating mode of the reactor.

(d) Set up the design equations and perform the design calculations.

(e) Repeat the design calculations considering a membrane reactor, in which continuous removal of hydrogen is achieved from the reaction zone (Online Appendix O6.4).

(f) Repeat the design calculations considering a sorption enhanced reactor, in which the removal of produced carbon dioxide is achieved as a result of reaction with CaO, which is mixed with the catalyst.

2 **Methanol synthesis: (Online Appendix O6.3)** Methanol is considered a non-petroleum energy carrier and an intermediate to produce alternative motor vehicle fuels, fuel additives, and a number of petrochemicals. Methanol itself is an excellent motor vehicle fuel with an octane number of 116. Methanol-based fuel cells also have strong potential to be used for direct conversion of chemical energy to electrical energy. Conventionally, methanol is produced from synthesis gas in a high-pressure fixed-bed catalytic reactor. Synthesis gas, which is mainly composed of carbon monoxide, carbon dioxide, and hydrogen, may be obtained by gasification of coal, biomass, or by reforming of natural gas. Both carbon monoxide and carbon dioxide may react with hydrogen to yield methanol.

$$CO + 2H_2 \rightleftarrows CH_3OH$$

$$CO_2 + 3H_2 \rightleftarrows CH_3OH + H_2O$$

In this project, you are asked to design the reactor(s) to be used for methanol production at a rate of 500 tons/d.

(a) Perform a literature survey and obtain thermodynamic and kinetic data for this process.

(b) Considering a feed composition of 30% CO, 65% H_2, and 5% N_2, examine the thermodynamics of the methanol synthesis reaction to decide about the operating pressure. Plot equilibrium conversion vs. temperature graphs in a pressure range of 10–100 atm.

(c) Taking the pressure as 80 atm and considering a target conversion of CO as 40% decide about the reactor configurations, the operation mode of the reactors (adiabatic, non-adiabatic, isothermal), and reactor inlet temperature(s).

(d) Design the reactor(s) to find the catalyst volume, total reactor volume, and estimate the investment cost.

(e) Repeat the equilibrium and the design calculations with feed stream composition of 24% CO, 7% CO_2, 66% H_2, and 3% N_2 at 50 bar. Consider the simultaneous occurrence of the reactions of hydrogen with CO and CO_2 in the production of methanol.

3 **Dimethyl ether synthesis: (Online Appendix O6.5)** Due to the fast depletion of oil reserves, research activities related to finding alternative non-petroleum transportation fuels have been significantly accelerated. Having high cetane numbers (over 55), dimethyl ether (DME), which can be produced by dehydration of methanol, is considered as an excellent diesel fuel alternate.

Case 1: Steam reforming of methane (SRM): (Online Appendix O6.4) Conventionally, DME is produced by dehydration of methanol over a solid acid catalyst.

$$2CH_3OH \rightleftarrows CH_3OCH_3 + H_2O$$

Typically γ - Al_2O_3, some heteropolyacids or acidic zeolites may be used for this process. The design of a tubular catalytic reactor is asked in this case study.

(a) Perform a detailed literature survey to obtain kinetic and thermodynamic data for the dehydration of methanol. Also, obtain information about the properties and uses of DME.

(b) Perform equilibrium calculations. Decide about the operating temperature and pressure of the reactor. Also, decide about the operational mode of the reactor and the production rate of the DME synthesis plant.

(c) Perform the design calculations to decide about the dimensions of the reactor.

Case 2: Design of a DME reactor from synthesis gas:
Production of DME directly from synthesis gas is considered as an attractive and efficient process. By this method, synthesis gas, which may be produced by gasification of coal or biomass, can be converted to DME in a single reactor.

$$2CO + 4H_2 \rightarrow CH_3\text{-}O\text{-}CH_3 + H_2O$$

$$3CO + 3H_2 \rightarrow CH_3\text{-}O\text{-}CH_3 + CO_2$$

In this project, you are asked to design a reactor for DME synthesis from synthesis gas, having a composition of 10% CO_2, 40% CO, and 50% H_2.

(a) First, you should make a detailed and criticizing literature survey. Your survey should include properties, production methods, uses, characteristics of DME, and safety data.

(b) Obtain thermodynamic and kinetic data for DME synthesis from synthesis gas. Find out information about the catalysts used for this reaction. Decide about the plant capacity basing on the literature information. Decide about feed composition (of synthesis gas). Perform equilibrium conversion calculations at the selected pressure and temperature ranges.

(c) Basing on the thermodynamic and kinetic considerations, decide about the type and operating mode of the reactor and operating conditions (temperature, pressure).

(d) Set up all of the necessary design equations and describe the algorithm that you will use for the design. Perform the design calculations, evaluate the dimensions of the reactor.

(e) Discuss the limitations of your design calculations. Consider economic factors, environment, safety, and health issues.

4 **Ammonia synthesis:** The composition and feed flow rate of the reactant stream entering the first stage of the ammonia synthesis reactor are given as 21.5% N_2, 62.5% H_2, 5.5% NH_3, 10.5% inerts, and 10^5 mol/h, respectively.

$$N_2 + 3H_2 \rightleftarrows 2NH_3$$

(a) Plot equilibrium conversion vs. temperature diagrams in the pressure range of 50–150 bar for the feed composition of the first stage of the reactor.

(b) Decide about the reactor pressure and inlet temperature of the first stage of the reactor ($T > 700\ K$).

(c) Decide about the type and operating mode of the multi-stage reactor. Set up the design equations.

(d) If the production rate of ammonia is given as 10 000 m^3/h at standard conditions, evaluate the reactor volume.

5 **Ethylene by dehydrogenation of ethane:** Ethylene is one of the main chemicals used in the petrochemical industry. In this project, it is asked to design a reactor system to produce 100 tons/d ethylene with a purity of 98%, using a feedstock with a composition of 99% ethane and 1% methane.

(a) Collect information about ethylene production methods and the limitations of these methods.

(b) Obtain data for the thermodynamics and kinetics of the process.

(c) Decide about the type of the reactor, its operation mode, operating conditions.

(d) Set up the design equations and evaluate the volume and the dimensions of the reactor(s).

6 **Shift reactor:** The water–gas shift reaction is an important reaction to convert CO present in the reformer exit stream to CO_2 and H_2.

$$CO + H_2O \rightleftarrows CO_2 + H_2$$

You are asked to design a shift reactor to produce a gaseous stream containing less than 0.5% carbon monoxide. A gaseous stream with a composition of 15% CO, 58% H_2, 7% CO_2, 20% inerts is available from a steam-reforming process. This stream may be mixed with water vapor to adjust the CO to the H_2O ratio in the feed stream of the shift reactor.

(a) Collect information about the types of shift conversion reactors, catalysts to be used, and operating conditions. Also, obtain thermodynamic and kinetic data to be used in the design of this reactor.

(b) Plot equilibrium conversion of CO as a function of reaction temperature for different H_2O/CO ratios in the feed stream. Decide about the feed stream temperature, pressure, and the ratio of water vapor to carbon monoxide to be used in this process.

(c) Perform the design calculations, find the value of reactor volume, make some estimates about its dimensions, perform an economic analysis.

7 **Catalytic converter: (Online Appendix O6.6)**
Motor vehicles are responsible for a significant fraction of carbon monoxide, unburned hydrocarbons, and nitrogen oxide emissions to the atmosphere. Environmental regulations limit the concentrations of these pollutants in the exhaust streams of motor vehicles. Catalytic converters are generally used to reduce the concentrations of these pollutants to the emission standards. Monolithic catalytic converters causing low-pressure drop are generally preferred as catalytic converters. Surfaces of monolith channels are covered by a thin layer of wash-coated alumina, which is the support material for the active metals, like Pt, Pd, etc. Carbon monoxide and hydrocarbons are to be oxidized to CO_2

$$CO + 1/2O_2 \rightarrow CO_2$$

$$C_mH_n + \left(m + \frac{n}{4}\right)O_2 \rightarrow mCO_2 + \frac{n}{2}H_2O$$

while NO_x should be reduced to nitrogen in these converters. You are asked to design a monolithic catalytic converter for an automobile to meet European emission standards of CO, hydrocarbons, and NO_x in the exhaust stream. As a design basis, you may take a feed stream mass flow rate of exhaust gas as 100 g/s. You may also assume to have an exhaust stream composition of 1.8% CO, 0.05% C_3H_6, 0.05% NO, 9% CO_2, 9% H_2O, 3% O_2, the rest being nitrogen.

(a) Perform a detailed literature survey, obtain information about the types of these converters, catalysts used, kinetic, and thermodynamic data.

(b) Decide about the operating conditions of the reactor to be designed and set up the necessary equations to be used to design this converter.

(c) Evaluate the dimensions of such a converter, discuss the limitations of this design.

8 **From biogas to synthesis gas: (Online Appendix 06.2)**
Research activities related to using alternative resources, such as bio-waste and biogas, for the production of valuable chemicals and clean fuels have been accelerated during the last decades. Dry reforming of methane is an important reaction for converting biogas to synthesis gas, which may then be used to synthesize chemicals and alternative fuels through Fischer–Tropsch synthesis, methanol synthesis, dimethyl ether synthesis, etc.

$$CO_2 + CH_4 \rightleftarrows 2CO + 2H_2$$

The reverse water–gas shift reaction and dry reforming of methane cause some decrease in hydrogen yield.

$$CO_2 + H_2 \rightleftarrows CO + H_2O$$

You are asked to design a reactor for this purpose, considering a feed stream containing 45% CH_4, 40% CO_2, and 15% inerts. The feed rate of this gas stream is given as 0.5 kmol/h.

(a) Obtain information about the process, types of catalysts used, kinetic and thermodynamic data.

(b) Perform equilibrium calculations and plot CO_2 conversion and product compositions as a function of reaction temperature.

(c) Decide about the type and operating mode of the reactor, as well as the operating conditions (temperature, pressure) of the reactor inlet stream.

(d) Perform the design calculations and evaluate the dimensions of the reactor. Consider economic factors, environment, safety issues in the design procedure.

9 **Steam reforming of glycerol: (Online Appendix 6.1)**
One of the promising alternatives to transportation fuels is biodiesel, which can be produced through trans-esterification of oils with methanol or ethanol. Glycerol is the main byproduct of biodiesel production. Recent increasing trends of biodiesel production have the potential to create a significant surplus of glycerol. Steam reforming of glycerol to produce synthesis gas and hydrogen is considered an attractive method for its effective utilization. The overall stoichiometry of steam reforming of glycerol can be expressed as follows:

$$C_3H_8O_3 + 3H_2O \rightarrow 7H_2 + 3CO_2$$

This overall reaction is considered to be due to the combination of glycerol decomposition (R.1) and water–gas shift reaction (R.2).

$$C_3H_8O_3 \rightarrow 3CO + 4H_2 \qquad (R.1)$$

$$CO + H_2O \leftrightarrow CO_2 + H_2 \qquad (R.2)$$

You are asked to design a catalytic reactor to produce synthesis gas through steam reforming of glycerol. You can take the feed flow rate of glycerol as 10 tons/h.

(a) Perform a detailed literature survey for the steam reforming of glycerol. Obtain information about the thermodynamics, kinetics (rate data), types of reactors used, types of catalysts used for this process, operating conditions.

(b) Analyze the thermodynamics of this process, evaluate the equilibrium conversion of glycerol and product distributions as a function of reaction temperature, considering the simultaneous occurrence of (R.1) and (R.2). Perform equilibrium calculations for different water to glycerol ratios (between 3 and 10) in the feed stream.

(c) Decide about the reactor type, operation mode of the reactor, catalyst to be used, inlet temperature, pressure, and feed stream composition of the reactor.

(d) Set up and perform design calculations for the estimation of the reactor dimensions.

10 Autothermal reforming of methanol

Autothermal steam reforming of methanol was investigated by Lattner and Harold [10]. You are asked to review this article critically and answer the following questions.

(a) Describe the reactor type, discuss the reasons for operating the reactor auto-thermally.

(b) Set up the model equations to be solved for the design of such a reactor.

(c) Discuss the air feed system for this reactor.

(d) Discuss the effects of operating parameters of the reactor and select the feed stream temperature, pressure, and composition.

References

1. Slinko, M.G., Malinovskaja, O.A., and Beskov, V.S. (1967). *Chim. Prom.* **42**: 641.
2. Denbigh, F.R.S. and Turner, J.C.R. (1971). *Chemical Reactor Theory*. Cambridge: Cambridge University Press.
3. Smith, J.M. (1981). *Chemical Engineering Kinetics*. Singapore: McGraw Hill Co.
4. Fogler, H.S. (1992). *Elements of Chemical Reaction Engineering*. New Jersey: Prentice-Hall.
5. Leonov, V.E., Karavaev, M.M., Tsybina, E.N. et al. (1973). *Kinet. Katal.* **14**: 848.
6. Vanden Bussche, K.M. and Froment, G.F. (1996). *J. Catal.* **161**: 1–10.
7. Sandler, S.I. (1999). *Chemical and Engineering Thermodynamics*. New York: Wiley.
8. Mas, V., Baronetti, G., Amadeo, N. et al. (2008). *Chem. Eng. J.* **138**: 602–607.
9. Levenspiel, O. (1999). *Chemical Reaction Engineering*. New Jersey: Wiley.
10. Lattner, J.H. and Harold, M.P. (2007). *Catal. Today* **120**: 78–89.

8

Deviations from Ideal Reactor Performance

Perfectly mixed continuous stirred-tank flow reactor (CSTR) and ideal plug-flow reactor (PFR) are two of the most frequently used ideal reactor models, which involve several assumptions. In the case of the ideal PFR model, we assume no variation of concentrations and temperature in the radial direction and no dispersion of species in the axial direction. Hence, it is assumed that there is no distribution of residence times of molecules within the reactor. On the other hand, perfect mixing is assumed in a CSTR. These assumptions may not be valid for many of the real chemical reactors. There may be axial dispersion of species and radial temperature and concentration variations in a real tubular reactor. Deviations from the ideal reactor performance are discussed in this chapter.

8.1 Residence Time Distributions in Flow Reactors

Let us consider a residence time distribution experiment, as illustrated in Figure 8.1. Suppose we inject a small volume of an inert tracer A into the inlet stream of a tubular reactor in the form of an impulse (delta function). We would expect to observe the response curve of the tracer also in the form of a sharp delta function at the reactor exit if it had been an ideal PFR. In the case of a PFR, all of the tracer molecules would spend the same amount of time within the system, which would be $\tau = V/Q$ in a homogeneous system. However, if we perform the pulse-response experiment in a real tubular reactor, we would observe some distribution of residence times of the tracer molecules in the reactor exit stream. The tracer spreads away from the mean of the bell-shaped residence time distribution curve. Two critical factors responsible for this distribution of residence times are the hydrodynamic properties within the reactor and diffusion.

One of the critical factors causing the distribution of residence times of molecules in a tubular reactor is the velocity profile of the flowing fluid. For instance, if the flow regime in the reactor is laminar, a parabolic velocity distribution will be formed. Hence, the molecules near the centerline will move faster and spend less time in the reactor than the molecules near the reactor wall. However, in the case of turbulent flow, the velocity distribution is more like logarithmic. In the turbulent flow case, turbulent mixing in the axial direction contributes to the distribution of residence times. In the case of fixed-bed catalytic reactors, the velocity distribution is even more scattered than the homogeneous systems. Turbulent mixing around the particles is an important factor for the spread of the tracer in the fixed-bed catalytic reactors.

Fundamentals of Chemical Reactor Engineering: A Multi-Scale Approach, First Edition. Timur Doğu and Gülşen Doğu.

Companion website: www.wiley.com/go/dogu/chemreacengin

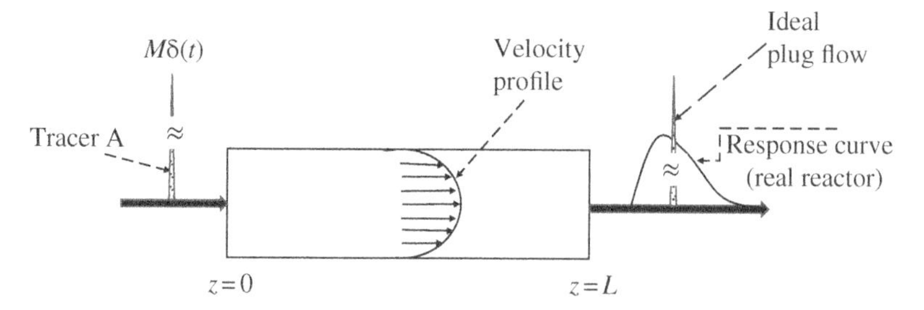

Figure 8.1 Pulse-response analysis of a real tubular reactor using an inert tracer.

Another important factor causing the distribution of residence times in a tubular reactor is the molecular diffusion of species both in the axial and the radial directions. The combined action of non-uniform velocity profiles, as well as molecular and turbulent diffusion, gives rise to the distribution of residence times of tracer molecules in a system, and this combined effect is described as dispersion. Hydrodynamics of the system and diffusional effects cause dispersion of molecules, both in the axial and in the radial directions.

The degree of dispersion in a flow reactor is an important factor for its performance. Fractional conversion of the reactant molecules in an axially dispersed tubular reactor differs from that of the corresponding conversion values predicted from the ideal PFR model. Deviations from the ideal plug-flow behavior become more significant as the dispersion effects gain importance. Dispersion effects cause a decrease in the efficiency of the ideal plug-flow behavior.

Ideal plug-flow and CSTR models are the two ideal limits of the flow reactors. For positive-order reactions, PFRs are more efficient than CSTR, giving higher fractional conversions than ideal stirred-tank reactors having the same volume. This is simply because of the higher concentration of reactants in a PFR than in a CSTR. Note that, according to the underlying assumption of the CSTR model, it is a perfectly mixed system, and the reactant concentrations within the reactor are uniform and assumed to be the same as the exit stream concentrations. However, for a PFR, we would have higher concentrations and higher reaction rates, especially close to the inlet of the reactor. A comparison of the performances of a CSTR and a PFR is discussed in Section 4.3.

Dispersion effects in a tubular reactor cause deviations from the ideal plug-flow behavior. Hence, for a tubular reactor with significant dispersion in the axial direction, fractional conversion values are expected to be less than the corresponding values in a PFR. In the case of the ideal PFR, we assume no distribution of residence times and hence no contribution of axial dispersion to the reactor performance. The degree of dispersion in a real flow reactor determines whether the performance of the reactor is close to an ideal PFR or a CSTR.

We would also expect some deviations from the ideal stirred-tank flow reactor model (CSTR) for real stirred-tank reactors. It may not be possible to achieve perfect mixing in the reaction vessel in a real stirred-tank reactor. If we perform a similar pulse-response experiment with an inert tracer in a stirred-tank reactor, we would expect an exponential decrease in the concentration of the tracer in the reactor exit stream with respect to time if that reactor had been an ideal CSTR (Figure 8.2).

$$C_{A_1} = C_{A_1}^{o} \exp\left(-\frac{t}{\tau}\right) \tag{8.1}$$

Here, $C_{A_1}^{o}$ is the initial concentration of the inert tracer A in the stirred-tank reactor, just after its injection into the reaction vessel. Experimental results usually indicate some deviations from the exponential decreasing trend of the concentration of the inert tracer in the exit stream of a stirred-tank reactor (Figure 8.2). The degree of spread of the response peak of the inert tracer in a flow

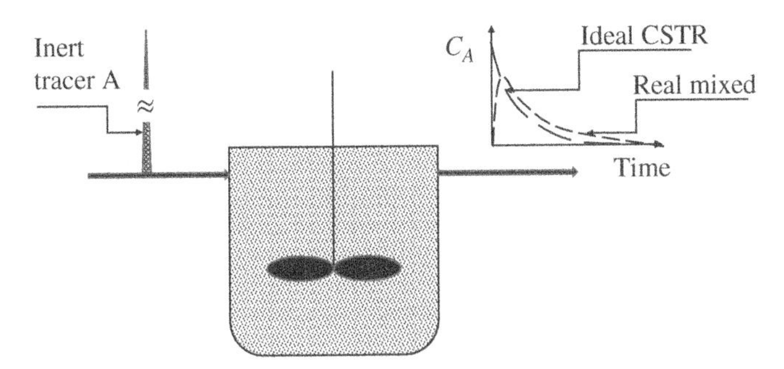

Figure 8.2 Pulse-response analysis of a real mixed reactor using an inert tracer.

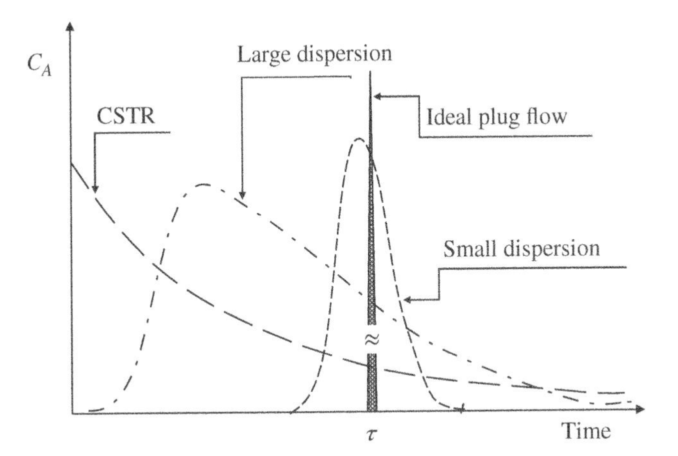

Figure 8.3 Typical residence time distributions curves for a tracer in flow reactors.

reactor gives information about the degree of dispersion in that system and the degree of deviation from the ideal reactor behavior (Figure 8.3).

If we perform the pulse-response experiments of an inert tracer in a system of two CSTRs operating in series, the equation of the response curve at the exit of the second reactor becomes

$$C_{A_2} = \frac{t}{\tau} C_{A_1}^{o} \exp\left(-\frac{t}{\tau}\right) \tag{8.2}$$

Equation (8.2) corresponds to a narrower response peak. If we increase the number of tanks in series, the shape of the response peak approaches to the behavior that would have been observed in a tubular reactor. In fact, for the case of an infinite number of CSTRs operating in series, the shape of the response peak becomes similar to the response curve that would have been obtained in an ideal PFR. These results indicate that as the number of CSTRs operating in series increases, the system approaches to a PFR model.

8.2 General Species Conservation Equation in a Reactor

Consider a reaction vessel with an arbitrary shape, in which the concentrations of the reacting species and the reaction temperature may vary in all three directions and also with time (Figure 8.4). Species conservation equation for component i can be written considering a differential volume element within this reactor (Figure 8.5).

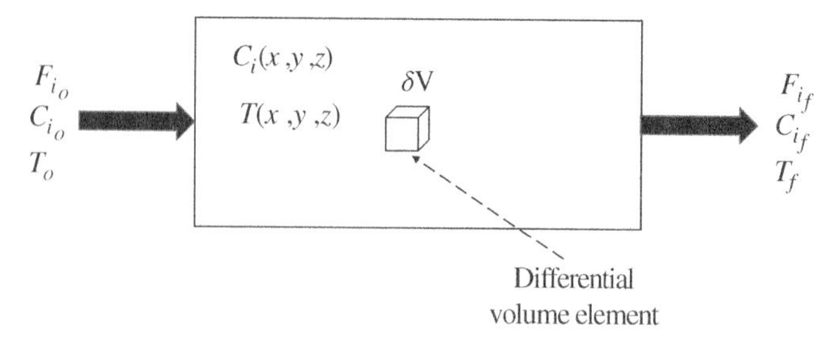

Figure 8.4 Schematic diagram of an arbitrary reaction vessel.

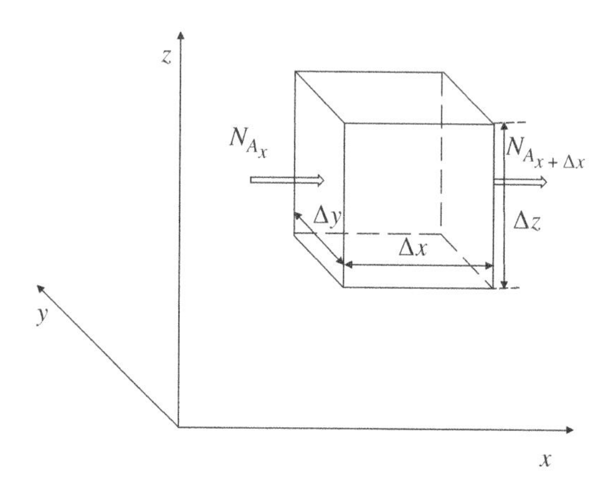

Figure 8.5 A cubical differential volume element for species conservation relation.

$$\left(N_{A_x}|_x - N_{A_x}|_{x+\Delta x}\right)\Delta y\Delta z + \left(N_{A_y}|_y - N_{A_y}|_{y+\Delta y}\right)\Delta x\Delta z + \left(N_{A_z}|_z - N_{A_z}|_{z+\Delta z}\right)\Delta y\Delta x$$
$$+ R_A\Delta x\Delta y\Delta z = \Delta x\Delta y\Delta z\frac{\partial C_A}{\partial t} \tag{8.3}$$

Here, N_{A_x}, N_{A_y}, and N_{A_z} are the fluxes of A in the x, y, and z directions, respectively. By dividing each term of Eq. (8.3) by $\Delta x\Delta y\Delta z$ and taking the limit as $\Delta x \to 0$, $\Delta y \to 0$, $\Delta z \to 0$ the following partial differential equation is obtained for the species conservation of A in this arbitrary reaction vessel, in rectangular coordinates:

$$-\left(\frac{\partial N_{A_x}}{\partial x} + \frac{\partial N_{A_y}}{\partial y} + \frac{\partial N_{A_z}}{\partial z}\right) + R_A = \frac{\partial C_A}{\partial t} \tag{8.4}$$

The general form of Eq. (8.4) can be expressed in terms of vectorial notation as follows:

$$-\nabla \cdot \boldsymbol{N_A} + R_A = \frac{\partial C_A}{\partial t} \tag{8.5}$$

Here, $\boldsymbol{N_A}$ and ∇ are the molar flux vector of species A and a differential operator (del operator), respectively.

$$N_A = \boldsymbol{i}N_{A_x} + \boldsymbol{j}N_{A_y} + \boldsymbol{k}N_{A_z} \tag{8.6}$$

$$\nabla = \boldsymbol{i}\frac{\partial}{\partial x} + \boldsymbol{j}\frac{\partial}{\partial y} + \boldsymbol{k}\frac{\partial}{\partial z} \tag{8.7}$$

Here, $\boldsymbol{i}, \boldsymbol{j}, \boldsymbol{k}$ are the unit vectors in the x, y, and z directions, respectively. Expansion of Eq. (8.5) in cylindrical and spherical coordinates is available in the online material of this book (Online Appendix 2).

By expressing the molar flux expressions of species A as a summation of convective and diffusive terms, the following relations can be written for constant total concentration:

$$N_{A_x} = C_A v_x - D_{AB}\frac{\partial C_A}{dx};\ N_{A_y} = C_A v_y - D_{AB}\frac{\partial C_A}{dy};\ N_{A_z} = C_A v_z - D_{AB}\frac{\partial C_A}{dz} \tag{8.8}$$

Here, v_x, v_y, and v_z are the velocity components in the x, y, and z directions. By substituting these flux expressions into Eq. (8.4) and assuming constant diffusion coefficient and constant density, the following steady-state general design equation can be expressed in the rectangular coordinates:

$$-\left[v_x\frac{\partial C_A}{\partial x} + v_y\frac{\partial C_A}{\partial y} + v_z\frac{\partial C_A}{\partial z}\right] + D_{AB}\left[\frac{\partial^2 C_A}{\partial x^2} + \frac{\partial^2 C_A}{\partial y^2} + \frac{\partial^2 C_A}{\partial z^2}\right] + R_A = 0 \tag{8.9}$$

For further details, the reader may also refer to any transport phenomena book [1].

The solution of Eq. (8.9) needs information about the hydrodynamics of the system, as well as temperature distributions. Variation of the velocity components with respect to x, y, and z coordinates should be determined from the differential momentum balance expressions (equation of motion). The rate term in Eq. (8.9) is a function of concentrations and the reaction temperature. Hence, the differential energy balance is also needed.

In many cases, this general species conservation equation is simplified according to the reactor geometry. In the case of a tubular reaction vessel, two-dimensional species conservation equation can be derived considering the general species conservation equation in cylindrical coordinates or by making a differential species material balance over the differential volume element illustrated in Figure 8.6.

For the steady-state operation of a tubular reactor, the species conservation equation over the differential volume element shown in Figure 8.6 can be written as follows:

$$(2\pi r\Delta r)\left(N_{A_z}\big|_z - N_{A_z}\big|_{z+\Delta z}\right) + (2\pi\Delta z)\left(rN_{A_r}\big|_r - rN_{A_r}\big|_{r+\Delta r}\right) + (2\pi r\Delta r\Delta z)R_A = 0 \tag{8.10}$$

This equation reduces to the following differential equation as $\Delta z \to 0$ and $\Delta r \to 0$.

$$-\frac{\partial N_{A_z}}{\partial z} - \frac{1}{r}\frac{\partial (rN_{A_r})}{\partial r} + R_A = 0 \tag{8.11}$$

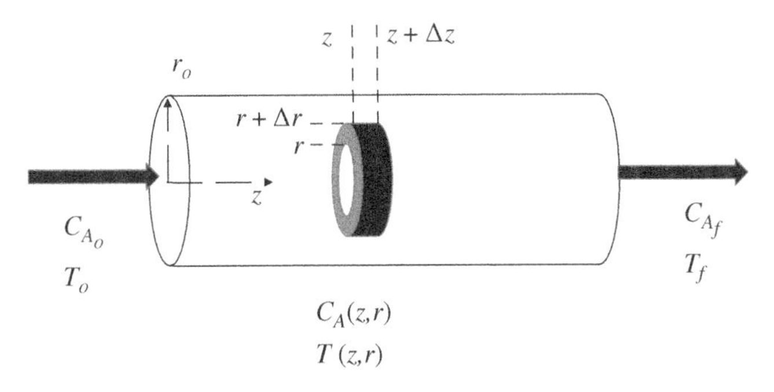

Figure 8.6 Illustration of a two-dimensional tubular reactor.

We consider only diffusion flux in the radial direction, while both convective and diffusive fluxes should be taken into account in the axial direction. With these considerations and assuming constant diffusion coefficient as well as constant density (no variation of v_z in the axial direction), Eq. (8.11) reduces to

$$-v_z\frac{\partial C_A}{\partial z} + D_{AB}\frac{\partial^2 C_A}{\partial z^2} + \frac{D_{AB}}{r}\frac{\partial}{\partial r}\left(r\frac{\partial C_A}{\partial r}\right) + R_A = 0 \tag{8.12}$$

For negligible radial variations of concentration of reactant A, this equation reduces to the one-dimensional design equation for a tubular reactor.

$$-v_z\frac{dC_A}{dz} + D_{AB}\frac{d^2 C_A}{dz^2} + R_A = 0 \tag{8.13}$$

If the diffusion term in the axial direction is also negligible, Eq. (8.13) reduces to

$$-v_z\frac{dC_A}{dz} + R_A = 0 \tag{8.14}$$

The velocity of the fluid in the z-direction may change in the radial direction. For instance, if the flow regime is laminar, the variation of velocity in the radial direction can be written as follows:

$$v_z = v_{max}\left(1 - \frac{r^2}{r_o^2}\right) \tag{8.15}$$

By making an additional assumption of constant v_z as

$$v_z = U_o = \frac{Q}{A} \tag{8.16}$$

we end up with the ideal PFR design equation for the constant volumetric flow rate case.

$$-U_o\frac{dC_A}{dz} + R_A = 0 \tag{8.17}$$

This analysis shows that the ideal PFR model involves the following assumptions:

- No variation of concentrations and temperature in the radial direction
- No diffusion/dispersion effects in the axial direction
- No variation of velocity in the radial direction, hence no distribution of residence times of molecules.

8.3 Laminar Flow Reactor Model

If the flow regime is laminar in a homogeneous tubular reactor, the substitution of Eq. (8.15) into Eq. (8.14) gives the following species conservation equation for the design of a laminar flow reactor:

$$-v_{max}\left(1 - \frac{r^2}{r_o^2}\right)\frac{dC_A}{dz} + R_A = 0 \tag{8.18}$$

Here, v_{max} is the maximum velocity of the fluid at the centerline of the reactor. In laminar flow, the average velocity is equal to half of the maximum velocity.

$$v_{max} = 2U_o \tag{8.19}$$

In this model, the contributions of diffusion flux in the axial and the radial directions are neglected. Due to the variation of velocity of the reacting fluid in the radial direction, residence times of the molecules are not the same in a laminar flow reactor. This causes a variation of concentration of reactant A in the radial direction. Fractional conversion of reactant A also changes in the radial direction. The molar flow rate of the reactant at the exit of this reactor can then be determined by the evaluation of the integral in Eq. (8.20).

$$F_{A_f} = F_{A_0}(1-\bar{x}_{A_f}) = \int_0^{r_0} (C_A|_{z=L}) v_z (2\pi r) dr \tag{8.20}$$

For an nth-order reaction, Eq. (8.18) becomes

$$-2U_o\left(1-\frac{r^2}{r_o^2}\right)\frac{dC_A}{dz} - kC_A^n = 0 \tag{8.21}$$

Rearrangement of this equation gives

$$\int_{C_{A_0}}^{C_{A_L}} \frac{dC_A}{C_A^n} = -\int_0^L \frac{k}{2U_o\left(1-\frac{r^2}{r_o^2}\right)} dz = -\frac{kL}{2U_o\left(1-\frac{r^2}{r_o^2}\right)} \tag{8.22}$$

For instance, for a second-order reaction ($n = 2$), integration of Eq. (8.22) gives a relation for the variation of concentration of reactant A in the radial direction at the reactor exit.

$$C_{A_L} = \frac{1}{\frac{1}{C_{A_0}} + \frac{kL}{2U_o\left(1-\frac{r^2}{r_o^2}\right)}} \tag{8.23}$$

If we substitute Eq. (8.23) into Eq. (8.20) and perform the integration, we can obtain the following expression for the average fractional conversion of A ($\bar{x}_{A_f}$) in the exit stream of this reactor:

$$\bar{x}_{A_f} = \frac{F_{A_0} - \int_0^{r_0}(C_{A_L} v_z)(2\pi r)dr}{F_{A_0}} = 1 - \frac{\int_0^{r_0} \frac{2U_o\left(1-\frac{r^2}{r_o^2}\right)}{\frac{1}{C_{A_0}} + \frac{kL}{2U_o\left(1-\frac{r^2}{r_o^2}\right)}}(2\pi r)dr}{U_o C_{A_0}}$$

$$\bar{x}_{A_f} = kC_{A_0}\tau\left[1 + \frac{kC_{A_0}\tau}{2}\ln\frac{kC_{A_0}\tau}{2 + kC_{A_0}\tau}\right] \tag{8.24}$$

This expression is derived initially by Denbigh [2], and analysis of other reaction orders is also reported by Levenspiel [3]. Corresponding fractional conversion equations for a second-order reaction in a CSTR and a PFR can be derived using Eqs. (4.3) and (4.24), respectively. The resultant expressions are

CSTR:

$$\frac{V}{F_{A_0}} = \frac{x_{A_f}}{kC_{A_0}^2(1-x_{A_f})^2} \tag{8.25}$$

$$x_{A_f} = \frac{(2k\tau C_{A_0} + 1) - \left((2k\tau C_{A_0} + 1)^2 - 4(k\tau C_{A_0})^2\right)^{1/2}}{2k\tau C_{A_0}} \tag{8.26}$$

PFR:

$$\frac{V}{F_{A_o}} = \int_0^{x_{A_f}} \frac{dx_A}{kC_{A_o}^2(1-x_A)^2} \tag{8.27}$$

$$x_{A_f} = \frac{k\tau C_{A_o}}{(1 + k\tau C_{A_o})} \tag{8.28}$$

The fractional conversion values obtained in the laminar flow reactor are in between the fractional conversion values obtained in plug-flow and perfectly stirred continuous tank reactors. For instance, for a $k\tau C_{A_o}$ value of 1.0, x_{A_f} values for a PFR, a laminar flow reactor, and a CSTR are 0.5, 0.45, and 0.38, respectively. As expected, fractional conversion values obtained in the PFR are higher than the fractional conversion values obtained in both laminar flow and perfectly mixed tank reactors, for positive-order reactions.

Parabolic velocity distribution of laminar flow causes some distribution of residence times. Hence, some dispersion of molecules in the laminar flow reactor. This dispersion phenomenon causes deviations from the plug-flow behavior. Residence time distributions for a laminar flow reactor are in between CSTR and PFR. Thus, fractional conversion of reactants is also expected to be in between the values that can be achieved in a CSTR and a PFR.

8.4 Dispersion Model for a Tubular Reactor

As discussed in Sections 8.1–8.3, the hydrodynamics of the reaction mixture and diffusional effects are two of the most critical factors causing the distribution of residence times in a tubular reactor. The spread of the residence times of the molecules in a tubular reactor causes deviations from the PFR performance. Such deviations usually lead to lower performance of a tubular reactor than the PFR performance for positive-order reactions. Effects of velocity distributions, the intensity of the intermixing of fluid packages, and diffusional effects cause dispersion both in radial and axial directions in tubular reactors.

Dispersion flux is generally expressed in a similar form as in Fick's law of diffusion by replacing the molecular diffusion coefficient with the dispersion coefficient. One-dimensional dispersion model is one of the most commonly used models to predict deviations from the plug-flow behavior. The axial molar flux of the species A is expressed as a summation of the convective flux and dispersion flux in the axial dispersion model.

$$N_{A_z} = U_o C_A - D_z \frac{dC_A}{dz} \tag{8.29}$$

The first term of the flux expression given in Eq. (8.29) corresponds to the molar flux in the case of the ideal plug-flow assumption. In this model, U_o is the average velocity of all of the species in the reactor $\left(U_o = \frac{Q_o}{A_c}\right)$. The second term on the right-hand side of Eq. (8.29) is the correction term to the ideal plug-flow assumption to describe the spread of residence times in a real tubular reactor. The axial dispersion coefficient D_z is expected to depend upon both hydrodynamics of the system and diffusion parameters. By substituting this flux expression into the differential species conservation equation (Eq. (8.30)), a second-order differential equation is obtained, describing the axial dispersion model for a tubular reactor (Eq. (8.31)). In writing this equation, U_o and D_z are taken as constant.

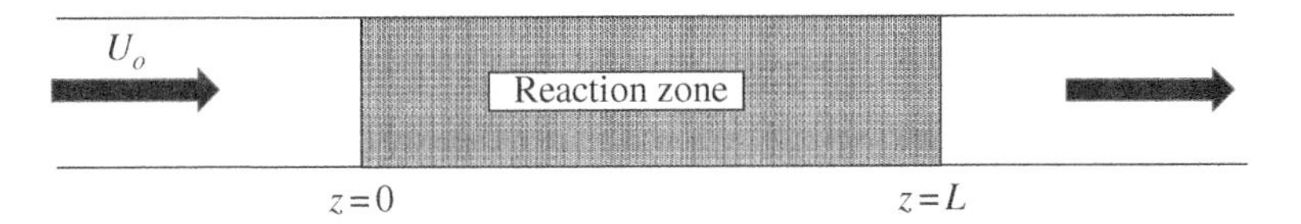

Figure 8.7 Schematic representation of a tubular reactor.

$$-\frac{dN_{A_z}}{dz} + R_A = 0 \tag{8.30}$$

$$D_z\frac{d^2C_A}{dz^2} - U_o\frac{dC_A}{dz} + R_A = 0 \tag{8.31}$$

The boundary conditions for the tubular reactor shown in Figure 8.7 were discussed in detail in the early literature [3–7]. The most straightforward boundary condition at the reactor inlet is $C_A|_{z=0} = C_{A_o}$. However, due to the dispersion effect within the reactor, this condition may not be exactly satisfied (Figure 8.7). Hence, the most commonly accepted boundary condition at the reactor inlet is the flux boundary condition [4].

$$\text{At } z = 0\,;\; U_oC_{A_o} = U_oC_A|_{z=0} - D_z\left.\frac{dC_A}{dz}\right|_{z=0} \tag{8.32}$$

Equation (8.32) is a flux boundary condition and considers the dispersion effects within the reactor at $z = 0$. The second boundary condition for Eq. (8.31) is generally written as

$$\text{At } z = L\,;\; \frac{dC_A}{dz} = 0 \tag{8.33}$$

Defining dimensionless concentration and dimensionless axial coordinate as

$$\psi = \frac{C_A}{C_{A_o}}\,;\; \varsigma = \frac{z}{L} \tag{8.34}$$

dimensionless species conservation equation becomes as follows for a first-order reaction rate expression.

$$\left(\frac{D_z}{U_oL}\right)\frac{d^2\psi}{d\varsigma^2} - \frac{d\psi}{d\varsigma} - k\tau\psi = 0 \tag{8.35}$$

The inverse of the dimensionless group, which appears in front of the first term of this equation, is called the axial Peclet number.

$$Pe_z = \frac{U_oL}{D_z} \tag{8.36}$$

The magnitude of the Peclet number is proportional to the ratio of convective and dispersive fluxes. It gives us information about the degree of deviation from the plug-flow behavior. At very large values of Peclet number, the performance of tubular reactor approaches to the plug-flow model. For small deviations from plug flow, $Pe_z^{-1} = \frac{D_z}{U_oL}$ must be small. On the other hand, as axial dispersion becomes important, the Peclet number approaches zero. In that case, the solution of Eq. (8.35) approaches the CSTR model. Hence, the value of the Peclet number gives us an indication of whether the performance of the reactor is close to a PFR or a CSTR.

The solution of Eq. (8.35) is expected to be a function of both $k\tau$ and the Peclet number. The solution of this equation with the dimensionless boundary conditions of

$$1 = \psi|_{\varsigma=0} - \frac{D_z}{U_oL}\frac{d\psi}{d\varsigma}\bigg|_{\varsigma=0} \text{ and } \frac{d\psi}{d\varsigma}\bigg|_{\varsigma=1} = 0 \tag{8.37}$$

is first derived by Wehner and Wihelm and reported in the early literature in several publications [3–7].

$$\psi = (1 - x_A) = \frac{4B\exp\left(\frac{Pe_z}{2}\right)}{(1+B)^2\exp\left(\frac{B}{2}Pe_z\right) - (1-B)^2\exp\left(-\frac{B}{2}Pe_z\right)} \tag{8.38}$$

Here, the parameter B is defined as follows:

$$B = \left(1 + 4k\tau\left(\frac{1}{Pe_z}\right)\right)^{1/2} \tag{8.39}$$

Note that, for an ideal PFR, the corresponding equation for ψ is

$$\psi = (1 - x_A) = \exp(-k\tau) \tag{8.40}$$

The effect of axial dispersion on the fractional conversion (which could be achieved in a tubular reactor for a first-order reaction) is illustrated in Figure 8.8, for constant space–time values (for constant $k\tau$ values). As shown in this figure, for quite low $k\tau$ values ($k\tau \ll 0.1$), the effect of $\frac{D_z}{U_oL}$ on fractional conversion is negligibly small. Such $k\tau$ values correspond to very small fractional conversions. Hence, the effect of axial dispersion is quite low at low fractional conversion values. On the other hand, in the case of very high space–times, ($k\tau > 100$) we expect very high conversions. Even with a CSTR, fractional conversion approaches to complete conversion for such $k\tau$ values. Hence, the fractional conversion values obtained in a PFR, in a CSTR, and in a tubular reactor with axial dispersion approach each other. On the other hand, the effect of axial dispersion on the conversion becomes highly significant for intermediate values of $k\tau$. As it is illustrated in Figure 8.8, for

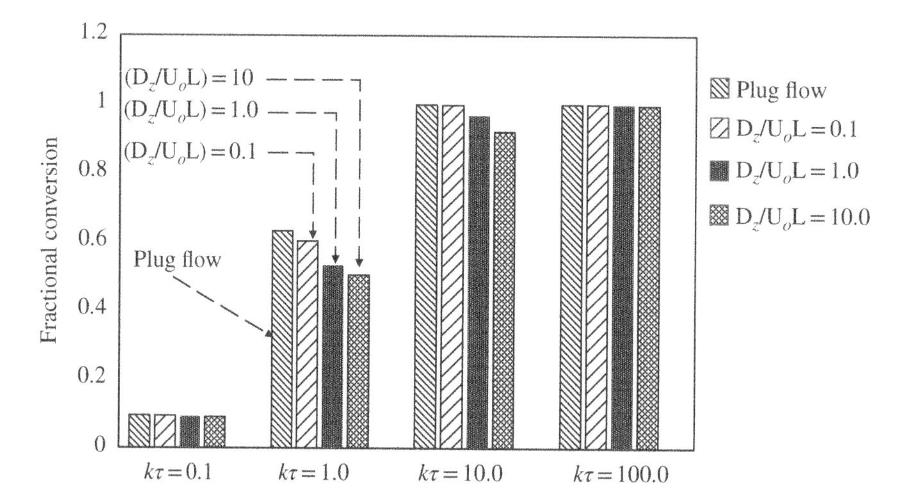

Figure 8.8 Effect of axial dispersion on fractional conversion for constant $k\tau$ values.

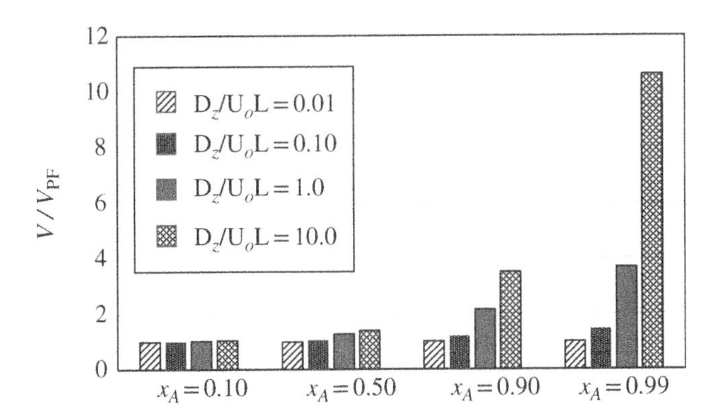

Figure 8.9 Effect of axial dispersion on the ratio of real and ideal tubular reactor volumes for different conversion levels.

$k\tau$ values of 1.0 and 10.0, a significant decrease of fractional conversion is expected with an increase in $\frac{D_z}{U_oL}$.

The effect of axial dispersion on reactor performance is also illustrated in Figure 8.9. In this figure, the effect of $\frac{D_z}{U_oL}$ on the ratio of real and PFR volumes is illustrated for different conversion levels. Different space–times are needed to reach a selected conversion level at different $\frac{D_z}{U_oL}$ values. As shown in Figure 8.9, we do not expect to have a significant effect of axial dispersion on the ratio of real to PFR volumes for low conversion levels. However, at high conversion levels, the ratio of real to ideal plug-flow tubular reactor volumes increases as $\frac{D_z}{U_oL}$ increases. For instance, if the desired fractional conversion level is selected as 0.99, the ratio of real to ideal tubular reactor volumes becomes over 10 at a $\frac{D_z}{U_oL}$ value of 10. As the value of $\frac{D_z}{U_oL}$ increases, the ratio of the volumes of the real and PFRs also increases. The limit of this ratio approaches the ratio of the volumes of the CSTR and the PFRs, as $\frac{D_z}{U_oL}$ approaches infinity. Hence, the dimensionless Peclet number can be considered as an important parameter, determining the degree of deviation from the ideal PFR performance.

$$V_{\text{real}} \Rightarrow V_{\text{plug flow}} \text{ as } Pe_z = \left(\frac{U_oL}{D_z}\right) \Rightarrow \infty \tag{8.41}$$

$$V_{\text{real}} \Rightarrow V_{\text{CSTR}} \text{ as } Pe_z = \left(\frac{U_oL}{D_z}\right) \Rightarrow 0 \tag{8.42}$$

The Peclet number can be considered as a correction factor for the ideal plug-flow performance.

As reported by Levenspiel [3], for small deviations from the ideal plug-flow model, the ratio of the volumes of the real and the PFRs to achieve the same conversion can be expressed as

$$\frac{V}{V_{\text{PF}}} \simeq 1 + \frac{k\tau}{Pe_z} \tag{8.43}$$

Hence, the following criterion should hold to have a maximum 1% deviation between the volumes of the real and the PFRs for a first-order reaction.

$$k\tau\left(\frac{D_z}{U_oL}\right) < 0.01 \tag{8.44a}$$

This criterion can also be expressed in terms of fractional conversion by combining Eq. (8.40) with Eq. (8.44a).

$$\left(\frac{D_z}{U_o L}\right) < \frac{0.01}{-\ln(1-x_A)} \tag{8.44b}$$

For instance, for fractional conversion values of 0.5 and 0.9, the right-hand side of Eq. (8.44b) becomes about 0.014 and 0.004, respectively.

8.5 Prediction of Axial Dispersion Coefficient

Analysis of the effect of axial dispersion on the performance of a tubular reactor depends upon the knowledge about the magnitude of the axial Peclet number. There have been many modeling and experimental studies published in the literature for the axial dispersion parameters in homogeneous tubular reactors, fixed-bed catalytic reactors, and other multiphase reactors, such as fluidized-bed reactors and bubbling-bed reactors. One of the earlier works on dispersion in empty tubes is of Taylor [8]. The axial dispersion coefficient is sometimes called the Taylor diffusion coefficient. According to Taylor's analysis, the axial dispersion coefficient in a tube can be expressed as in Eq. (8.45) in laminar pipe flow.

$$D_z = D_{AB} + \frac{U_o^2 d_t^2}{192 D_{AB}} \tag{8.45}$$

Here, d_t is the tube diameter. This equation can be written in dimensionless form for the axial Peclet number, which is defined using the tube diameter as the characteristic length (Pe_d).

$$Pe_d^{-1} = \left(\frac{D_z}{U_o d_t}\right) = \frac{1}{Re \cdot Sc} + \frac{ReSc}{192} \tag{8.46}$$

Here, Reynolds number and Schmidt number are defined as follows:

$$Re = \frac{d_t U_o \rho}{\mu} \tag{8.47}$$

$$Sc = \frac{\mu}{\rho D_{AB}} = \frac{\nu}{D_{AB}} \tag{8.48}$$

Taylor suggested that the axial dispersion coefficient should be proportional to the square root of friction factor and average velocity in the pipe for turbulent pipe flow (Re > 4500).

$$D_z = 3.57 U_o d_t (f)^{1/2} \tag{8.49}$$

Using the Blasius equation for the friction factor in turbulent pipe flow

$$f = 0.0791\, Re^{-0.25} \tag{8.50}$$

axial dispersion coefficient can be expressed as follows:

$$\frac{D_z}{U_o d_t} = Re^{-1/8} \tag{8.51}$$

The following empirical correlation (Eq. (8.52)) is then proposed for the Peclet number for turbulent pipe flow [9]. In this equation, the Peclet number is defined by using the tube diameter as the characteristic length.

$$Pe_d^{-1} = \frac{D_z}{U_o d_t} = \frac{3x10^7}{Re^{2.1}} + \frac{1.35}{Re^{1/8}} \tag{8.52}$$

Axial dispersion may also have a significant contribution to the performance of fixed-bed catalytic reactors. Turbulent mixing around the particles and molecular diffusion cause some distribution of residence times in such systems. Reynolds number and Peclet number are generally defined considering the particle diameter (d_p) as the characteristic length in fixed-bed systems.

$$Re_p = \frac{d_p U_o \rho}{\mu}\ ;\ Pe_p = \frac{U_o d_p}{D_z} \tag{8.53}$$

Sometimes, a dimensionless group, which is defined as $\frac{U_o d_p}{\varepsilon_b D_z}$, is used to predict the degree of axial dispersion in fixed-bed systems. This dimensionless group is called the Bodenstein number in some references. As shown in the early literature, gaseous flow Peclet number in a packed-bed reactor can be approximated as expressed in Eq. (8.54) [7, 10, 11].

$$\frac{Pe_p}{\varepsilon_b} = \frac{U_o d_p}{\varepsilon_b D_z} = 2 \tag{8.54}$$

This expression was shown to give satisfactory results, especially for particle Reynolds number values higher than 10.

We expect differences in axial Peclet numbers in gaseous and liquid flows through fixed-bed reactors. There are correlations for the Bodenstein number for liquid and gaseous flows. For most liquids, Bodenstein number is reported to be between 0.5 and 1.0, especially for particle Reynolds numbers being less than 100. A review of correlations for the axial dispersion coefficient in a fixed-bed column is reported in a recent publication [12]. One of the most commonly used correlations proposed for liquid dispersion in a packed bed is the one proposed by Chung and Wen [13].

$$Pe_p = 0.2 + 0.011\, Re_p^{0.48} \tag{8.55}$$

The value of particle Peclet number changes between 0.21 and 0.29, in the particle Reynolds number range of 1.0 and 100.

The relation between the Peclet number defined for a tubular reactor by taking the characteristic length as the length of the reactor (Eq. (8.36)) and the particle Peclet number defined for a fixed-bed reactor (Eq. (8.53)) can be expressed as follows:

$$Pe_z = \frac{U_o L}{D_z} = Pe_p \left(\frac{d_t}{d_p}\right)\left(\frac{L}{d_t}\right) \tag{8.56}$$

For a gas-phase reaction, Pe_p is approximately equal to one. As discussed in Section 8.4, the value of Pe_z should be very large to approach the ideal plug-flow model. Hence, as the ratios of L/d_t and d_t/d_p increase, we approach to a PFR.

8.6 Evaluation of Dispersion Coefficient by Moment Analysis

Dynamic techniques are frequently used for the evaluation of the rate and equilibrium parameters in chemical reactors. Moment technique is one of the most commonly used dynamic methods for the experimental evaluation of such parameters. This technique has been adopted to a fixed-bed tubular flow system in the early literature [14]. This technique is based on injecting a tracer A into the inlet stream of the vessel and analyzing the moments of the response curve. Those moment expressions are functions of the diffusion, adsorption, and reaction parameters of the system.

Consider the injection of a tracer A as an impulse into the inlet stream of a tubular vessel, which is shown in Figure 8.10. If the tracer is inert and no reaction is taking place in this homogeneous system, the shape of the response peak observed at the exit of the tube will be a function of the dispersion coefficient in the tube. The moments of the experimental response peaks can be evaluated, and these data can be used to determine the system parameters.

First, let us give a summary of the moment theory. The response peak of an injected tracer is simply the change of concentration of tracer A as a function of time at the exit of the system. The ordinary nth moment of this response curve is defined as follows [14, 15]:

$$m_n = \int_0^\infty t^n C_A(t)dt \tag{8.57}$$

The first absolute moment (μ_1), which is the mean of the distribution curve, is defined as the ratio of the first ordinary moment to zeroth ordinary moment and given by the following equation:

$$\mu_1 = \frac{m_1}{m_o} = \frac{\int_0^\infty tC_A(t)dt}{\int_0^\infty C_A(t)dt} \tag{8.58}$$

For the pulse-response experiment shown in Figure 8.10, the first absolute moment corresponds to the mean residence time of the tracer molecules in the vessel.

The second central moment, which is the second moment about the mean, is defined as

$$\mu_2' = \frac{\int_0^\infty (t-\mu_1)^2 C_A(t)dt}{\int_0^\infty C_A(t)dt} = \frac{m_2}{m_o} - \mu_1^2 \tag{8.59}$$

The second central moment is also called the variance of the response curve of the tracer, and it shows the deviation from the mean. Experimental values of the moments of the response curve can be evaluated using the definitions given above.

Derivation of the moment expressions can easily be performed using the following relation between moments and the Laplacian of the concentration of tracer at the exit of the system [14, 15]:

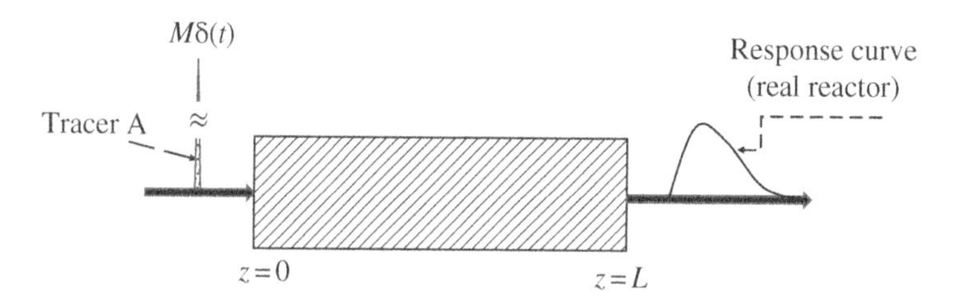

Figure 8.10 Pulse-response experiment of a tracer A in a tube.

$$m_n = \int_0^\infty t^n C_A(t)dt = (-1)^n \lim_{s \to 0} \frac{d^n \mathcal{L}(C_A)}{ds^n} \quad (8.60)$$

where s is the Laplace variable and $\mathcal{L}(C_A)$ is the Laplacian of C_A.

The unsteady-state differential conservation equation for the tracer should be solved in the Laplace domain to derive the moment expressions. Then, the moment expressions can be obtained using Eq. (8.60). The unsteady-state species conservation equation for an inert tracer, which is injected as an impulse into the inlet stream of a tubular vessel, can be written as follows:

$$\frac{\partial C_A}{\partial t} = -U_o \frac{\partial C_A}{\partial z} + D_z \frac{\partial^2 C_A}{\partial z^2} \quad (8.61)$$

Writing this partial differential equation in dimensionless form and taking the Laplacian of each term, the following ordinary differential equation is obtained:

$$\left(\frac{D_z}{U_o L}\right) \frac{d^2\overline{\psi}}{d\varsigma^2} - \frac{d\overline{\psi}}{d\varsigma} - \tau s \overline{\psi} = 0 \quad (8.62)$$

Here, $\overline{\psi}$ is the Laplacian of the dimensionless concentration ψ. The solution of this equation, with an inlet impulse boundary condition for the concentration of tracer A and using Eq. (8.60), moment expressions (m_n) can be derived. Then, the first absolute moment and the second central moment expressions are obtained by using their definitions given by Eqs. (8.58) and (8.59), respectively [14].

$$\mu_1 = \frac{L}{U_o} = \tau \quad (8.63)$$

$$\mu_2' = 2\left(\frac{D_z}{U_o L}\right)\tau^2 \quad (8.64)$$

Experimental evaluation of the second central moment values can be made using the response curve data by the numerical integration of the expression given in Eq. (8.59). The axial dispersion coefficient can then be evaluated by using the experimental μ_2' values in Eq. (8.64).

By considering an open vessel (flux boundary condition) with the undisturbed flow at the entrance and the exit of the reactor, the following expression for the second central moment was reported by Levenspiel [3].

$$\mu_2' = \left[2\left(\frac{D_z}{U_o L}\right) + 8\left(\frac{D_z}{U_o L}\right)^2\right]\tau^2 \quad (8.65)$$

If the degree of deviation from the ideal plug-flow performance is not high, the limit of Eq. (8.65) for small values of $\left(\frac{D_z}{U_o L}\right)$ becomes the same as expressed in Eq. (8.64).

8.7 Radial Temperature Variations in Tubular Reactors

One of the assumptions of plug-flow model is the neglect of possible variations of concentration and temperature in the radial direction. Modeling of a two-dimensional system considering radial and axial concentration variations is considered in Section 8.2. Temperature variations in the radial direction may also become quite significant, especially for systems with reactions having quite high heat of reactions [10]. In the case of an exothermic reaction taking place in a non-adiabatic tubular

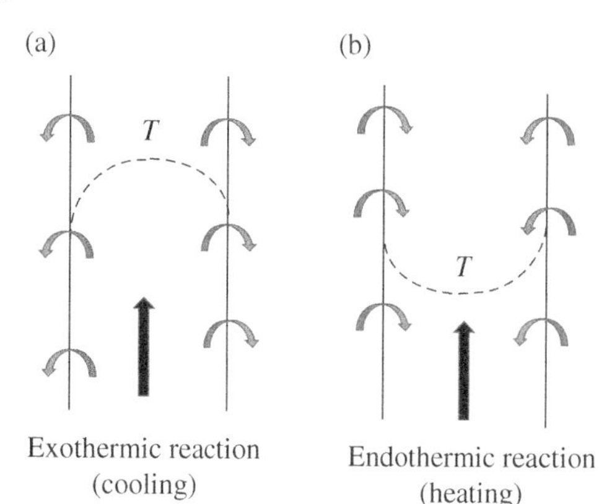

Figure 8.11 Temperature profiles in a non-adiabatic tubular reactor with exothermic and endothermic reactions.

reactor, a temperature profile having a maximum at the centerline of the reactor may develop within the reactor. On the other hand, in the case of an endothermic reaction taking place in a tubular reactor, which is heated from the walls, a temperature profile having a minimum at the center of the tube may be expected (Figure 8.11).

The velocity profile within the reactor is also expected to have a significant influence on the temperature distributions. For instance, molecules near the wall travel with a lower velocity than the molecules at the center of the tube in a laminar flow reactor. In that case, molecules near the wall will spend more time within the reactor than the molecules near the center, and they will have more time to react, causing more heat generation. If the reactor is perfectly insulated (adiabatic reactor), this may cause higher temperatures near the wall than at the center of the reactor for an exothermic reaction. In the case of a fixed-bed reactor, a more complex temperature distribution is expected. For a highly exothermic reaction in a non-adiabatic tubular reactor, transverse temperature profiles may develop within the reactor, causing the formation of a hot zone.

Let us consider an exothermic reaction with a typical activation energy of 120 kJ/mol taking place in a tubular reactor, having a wall temperature of 500 K. The ratio of the rate constant values at the hot point and the wall temperatures are evaluated using Eq. (8.66) and listed in Table 8.1 for different wall and hot point temperatures.

$$\frac{k_T}{k_{T_w}} \approx \frac{\exp\left(-\frac{120\,000}{R_g T}\right)}{\exp\left(-\frac{120\,000}{R_g T_w}\right)} \tag{8.66}$$

For the design of a tubular non-adiabatic jacketed reactor, we usually assume that the temperature variation is localized near the wall (Figure 8.12), and the temperature of the bulk fluid is

Table 8.1 The ratio of the rate constants evaluated for different temperature differences between the center and wall temperatures for the activation energy of 120 kJ/mol (T_w = 500 K).

$\Delta T = (T - T_w)$ (°C)	k_T/k_{T_w}
10	1.8
50	13.8
100	123.0

uniform. With this assumption, we can use the one-dimensional energy balance equation derived in Chapter 7.

$$\sum_i N_{i_z} C_{p_i} \frac{dT}{dz} + R\Delta H^o = -\frac{2}{r_o} U(T - T_a) \tag{8.67}$$

For the case of constant mass flow rate with constant specific heat for the fluid flowing in the reactor, Eq. (8.67) reduces to

$$U_o \rho \bar{c} \frac{dT}{dz} + R\Delta H^o = -\frac{2}{r_o} U(T - T_a) \tag{8.68}$$

If we consider temperature variations both in the axial and the radial directions in the tubular reactor, the two-dimensional energy balance equation given in Eq. (8.69) should be solved together with the species conservation equation discussed in the previous sections. The heat loss term to the surroundings appears in the boundary condition of this equation.

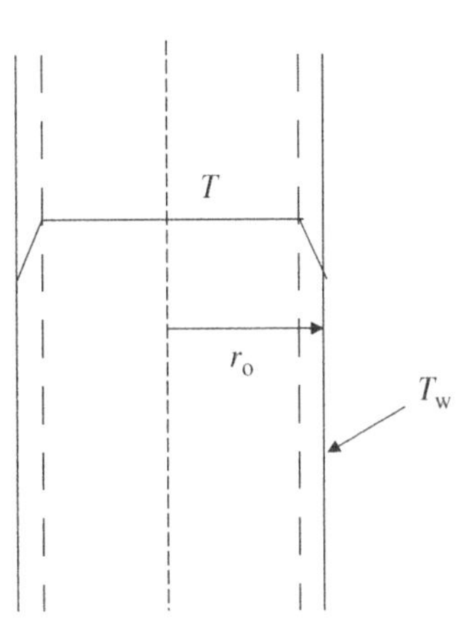

Figure 8.12 Schematic diagram of radial temperature variation for the one-dimensional model.

$$U_o \rho \bar{c} \frac{\partial T}{\partial z} - \frac{\partial}{\partial z}\left(\lambda_b \frac{\partial T}{\partial z}\right) - \frac{1}{r}\frac{\partial}{\partial r}\left(r\lambda_b \frac{\partial T}{\partial r}\right) + R\Delta H^o = 0 \tag{8.69}$$

Here, λ_b is the effective thermal conductivity of the reactor contents. In this equation, the axial heat conduction term (second term) is generally neglected by comparing its order of magnitude with the order of magnitude of the convective heat transfer term (first term of the equation), and Eq. (8.69) reduces to the following expression:

$$U_o \rho \bar{c} \frac{\partial T}{\partial z} - \frac{1}{r}\frac{\partial}{\partial r}\left(r\lambda_b \frac{\partial T}{\partial r}\right) + R\Delta H^o = 0 \tag{8.70}$$

For the solution of this equation, the following boundary conditions can be used for a non-adiabatic system:

$$\text{At } r = 0;\ \frac{\partial T}{\partial r} = 0 \text{ (symmetry)} \tag{8.71}$$

$$\text{At } r = r_o;\ -\lambda_b \left.\frac{\partial T}{\partial r}\right|_{r = r_o} = h_o\left(T|_{r = r_o} - T_a\right) \text{ (heat flux at the reactor wall)} \tag{8.72}$$

Here, h_o is the external heat transfer coefficient for heat transfer from the reactor wall to the surroundings and r_o is the radius of the tubular reactor. In writing the second boundary condition (Eq. (8.72)), heat conduction resistance through the reactor wall is neglected.

8.8 A Criterion for the Negligible Effect of Radial Temperature Variations on the Reaction Rate

The following criterion can be derived to test the justification of neglecting radial temperature variations in a tubular reactor. Let us consider Arrhenius-type temperature dependence of a reaction rate expression. By Taylor series expansion of the exponential temperature dependence term of this equation and by neglecting the higher-order terms of this expansion, the following approximate relations can be written:

$$R = k_o \exp\left(-\frac{E_a}{R_g T}\right) f(C_i)$$

$$R \approx k_o \exp\left(-\frac{E_a}{R_g T_w}\right) f(C_i)\left[1 + \frac{(T-T_w)}{T_w}\left(\frac{E_a}{R_g T_w}\right) + \cdots\right] = R_w\left[1 + \frac{(T-T_w)}{T_w}\left(\frac{E_a}{R_g T_w}\right)\right] \tag{8.73}$$

Here, T_w is the wall temperature. Neglecting the higher-order terms in the Taylor expansion can be justified for small variations of temperature.

For an exothermic reaction taking place in a tubular reactor with heat loss to the surroundings, a parabolic temperature profile can be assumed as a first approximation for small variations of temperature in the radial direction.

$$(T - T_w) \approx \Gamma\left(1 - \frac{r^2}{r_o^2}\right) \tag{8.74}$$

An expression for the parameter Γ is derived later in this section (Eq. (8.82)).

Considering a slit volume element (Figure 8.13) within the reactor, an average reaction rate within this element can be estimated as follows:

$$\pi r_o^2 \overline{R} = \int_0^{r_o} R 2\pi r dr \tag{8.75}$$

$$\pi r_o^2 \overline{R} = \int_0^{r_o} R_w\left[1 + \frac{(T-T_w)}{T_w}\left(\frac{E_a}{R_g T_w}\right)\right] 2\pi r dr \tag{8.76}$$

$$\pi r_o^2 \overline{R} = \int_0^{r_o} R_w\left[1 + \frac{\Gamma}{T_w}\left(1 - \frac{r^2}{r_o^2}\right)\left(\frac{E_a}{R_g T_w}\right)\right] 2\pi r dr \tag{8.77}$$

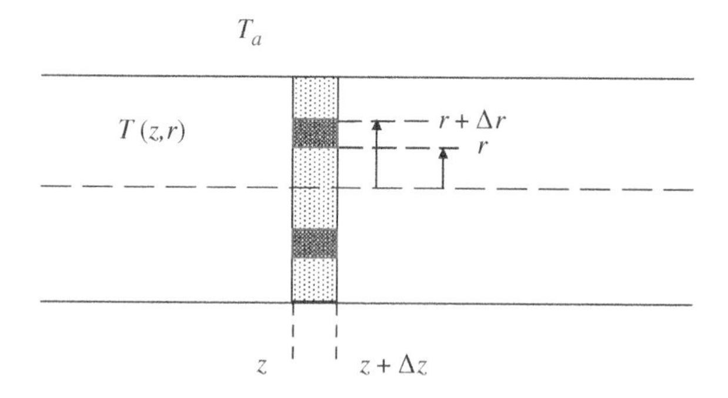

Figure 8.13 Schematic representation of the differential volume element in the tubular reactor for two-dimensional analysis.

Integration of this equation yields

$$\frac{\overline{R}}{R_w} = 1 + \frac{\Gamma E_a}{2R_g T_w^2} \tag{8.78}$$

Arbitrarily, if we select the criterion for neglect of the effects of radial temperature variations on the observed rate as

$$0.95 < \frac{\overline{R}}{R_w} < 1.05 \tag{8.79}$$

the following relation can be written:

$$\frac{\Gamma E_a}{R_g T_w^2} < 0.05 \tag{8.80}$$

Here, the parameter Γ can then be estimated by considering that all of the heat generated within the slit element is lost to the surroundings from the walls.

$$\overline{R}\Delta H^o\left(\pi r_o^2 \Delta z\right) = \left(2\pi r_o \Delta z\right)\lambda_b \left.\frac{dT}{dr}\right|_{r=r_o} \tag{8.81}$$

Using the temperature relation given by Eq. (8.74), the following expression can be obtained for the parameter Γ.

$$\Gamma = \frac{\left(-\Delta H^o r_o^2 \overline{R}\right)}{4\lambda_b} \tag{8.82}$$

The following criterion can then be obtained for the negligible effect of radial temperature variations on the observed rate by substituting the Γ expression given in Eq. (8.82) into Eq. (8.80). If this criterion is satisfied, we can neglect the radial temperature variations, and the one-dimensional model can be used.

$$\left(\frac{E_a}{R_g T_w^2}\right)\left(\frac{|-\Delta H^o|\overline{R} r_o^2}{\lambda_b}\right) < 0.4 \tag{8.83}$$

Testing the effects of radial temperature variations on the observed rate using this equation requires information about the effective thermal conductivity within the reactor. Some correlations are reported in the literature for effective thermal conductivity in a fixed-bed reactor. Heat transfer rate through a bed of solid catalysts is due to heat transfer through the void space as well as through the solid catalyst particles. Modeling of heat transfer in a packed bed is given by Smith [5]. For gaseous reaction mixtures in fixed-bed reactors, the value of λ_b was reported to be in the range of 0.17 to 0.52 J/m s K.

8.9 Effect of L/d_t Ratio on the Performance of a Tubular Reactor and Pressure Drop

As discussed in Sections 8.4 and 8.5, the value of the dimensionless group $\left(\frac{D_z}{U_o L}\right)$ should be very small to approach the ideal plug-flow model.

$$\text{As } \left(\frac{D_z}{U_o d_t}\right)\left(\frac{d_t}{L}\right) \rightarrow 0;\ V_{\text{Real}} \Rightarrow V_{\text{PF}} \tag{8.84}$$

It is possible to have different length to diameter ratios for a given value of reactor volume. As predicted by Eq. (8.84), an increase of the L/d_t ratio decreases the significance of axial dispersion and helps to approach plug-flow behavior. The selection of a higher L/d_t ratio for a tubular reactor of a given volume also helps to decrease the radial temperature gradients within the reactor. On the other hand, the increase of the L/d_t ratio, for a reactor of a given volume, causes a higher pressure drop in the reactor. Pressure drop in a homogeneous tubular reactor can be evaluated from the following expression [1]:

$$-\frac{\Delta P}{L} = \frac{2fLU_o^2}{d_t} \tag{8.85}$$

Here, f is the Fanning friction factor, which can be predicted using the following equations for laminar and turbulent tube flow, respectively:

$$f = \frac{16}{Re};\ \text{for laminar pipe flow} \tag{8.86}$$

$$f = \frac{0.079}{Re^{1/4}};\ \text{for turbulent flow (Blasius formula)} \tag{8.87}$$

In the case of turbulent flow, more complex expressions involving the contribution of the wall roughness to the friction factor are available in the literature.

In the case of fixed-bed catalytic reactors, the pressure drop in the reactor depends upon the catalyst pellet diameter, as well as superficial velocity within the bed and its void fraction ε_b. Ergun equation can be used for the prediction of pressure drop in a packed-bed catalytic reactor.

$$-\frac{\Delta P}{L} = 150\frac{\mu U_o}{d_p^2}\left(\frac{(1-\varepsilon_b)^2}{\varepsilon_b^3}\right) + 1.75\left(\frac{\rho U_o^2}{d_p}\right)\left(\frac{(1-\varepsilon_b)}{\varepsilon_b^3}\right) \tag{8.88}$$

Problems and Questions

8.1 A tubular reactor is designed with the assumption of plug flow. A first-order reaction is taking place in this reactor, which is operating isothermally. The fractional conversion of the reactant was calculated as 0.95 by the plug-flow model in a reactor of 5 cm inside diameter and 200 cm long. If the Reynolds number of the flowing fluid is 6000, what should be the volume of the real reactor to achieve the same conversion value?

8.2 Consider a gas-phase first-order reaction taking place in a tubular reactor. Reynolds and Schmidt numbers are given as 1000 and 2.5, respectively.

(a) Discuss the effect of axial dispersion on the reactor performance for the following cases of different L/d_t ratios:

$L/d_t = 2, 10, 100$

having the same space time ($k\tau = 3.0$.

(b) Estimate the ratio of volumes of real tubular and PFRs for the case of $L/d_t = 2$.

8.3 Residence time distribution obtained by an impulse injection of an inert tracer in a reactor vessel (R1) is given in the following table:

Time (s)	5	25	45	65	85	105	125	145
Tracer concentration (mg/l)	0.1	0.12	0.16	0.15	0.08	0.02	0.01	0

(a) Determine the space–time and axial Peclet number using these data.

(b) A first-order reaction was carried out in another PFR (R2) operating with the same residence time as reactor vessel R1. In this PFR, the fractional conversion of reactant A is found as 0.8. Calculate the rate constant.

(c) Determine the conversion if the same first-order reaction is carried out in the reaction vessel (R1).

8.4 The residence time distribution of an inert gaseous tracer in a reactor is given in the following table:

Time (s)	6	10	20	30	40	50	70	80
C_A/C_{A_0}	0	0.01	0.45	0.85	1.00	0.45	0.25	0.01

(a) Determine the Peclet number and space–time in this vessel.

(b) Estimate the fractional conversion if a first-order reaction with a rate constant of 0.2 min^{-1} takes place in this reactor.

(c) What would be the fractional conversion that would be reached if the plug-flow assumption had been satisfied for this reactor?

8.5 An inert liquid tracer is injected as an impulse into the liquid stream at the entrance of a packed-bed column filled with catalyst particles of 0.5 cm in diameter. The following residence time distribution data are obtained in a reactor having 1 m height and 0.25 m diameter.

t (min)	0	1	2	3	4	5	6
C_A/C_{A_0}	0	0.12	1.0	0.35	0.12	0.04	0.01

(a) Calculate the mean residence time and the variance using the residence time distribution data.

(b) Evaluate the value of the axial Peclet number.

(c) Discuss if the plug-flow model could be used for the design of this reactor.

Exercises

8.1 The volume of a tubular reactor is determined with the plug-flow assumption by taking the fractional conversion as 0.9. The reaction taking place in this reactor is first order. In a separate experiment, the value of the axial Peclet number is also determined.

(a) Do you expect to achieve the fractional conversion value of 0.9 if

$$\frac{U_oL}{D_z} = 100\,; \quad \frac{U_oL}{D_z} = 5\,; \quad \frac{U_oL}{D_z} = 0.1$$

(b) Discuss whether one would expect to achieve lower, higher, or equal fractional conversion in each case, without evaluating the numerical values of fractional conversion.

(c) Evaluate the fractional conversion of reactant A for the case of $\frac{U_oL}{D_z} = 5$.

8.2 Experiments have shown that the fractional conversion of limiting reactant A in a three CSTR system operating in series is equal to the fractional conversion reached in a tubular reactor. The rate expression is given as first order with respect to the concentration of reactant A.

(a) What should be the value of the axial Peclet number in the tubular reactor? In this analysis, assume that $k\tau = 2$ in the tubular reactor and also in the three-CSTR system (τ is the total space–time in the three-CSTR system). Discuss the effect of axial dispersion on the performance of this tubular reactor.

(b) Estimate the ratio of the volumes of the real and the PFRs for this problem.

8.3 Evaluate the fractional conversion values that can be achieved in a laminar flow reactor, PFR, CSTR, and tubular reactor with a $\frac{D_z}{U_oL}$ value of 2.0, for a first-order reaction with $k\tau = 1$. Compare your results and discuss the reasons for differences in fractional conversion values obtained in these reactors.

Can you model the tubular reactor with a $\frac{D_z}{U_oL}$ value of 2.0 as a multiple CSTR operating in series?

8.4 The volume of a tubular reactor is determined by the plug-flow assumption. The reaction is exothermic, homogeneous, and the reactor is cooled from the walls to approach isothermal operation. Two possible reactors of the same volume but different L/d_t ratios ($L/d_t = 20$ and $L/d_t = 2$) are available for this reaction. Do you expect to achieve the same fractional conversion in these reactors? Discuss the effect of the L/d_t ratio on the performance of a tubular reactor.

References

1 Bird, R.B., Stewart, W.E., and Lightfoot, E.N. (2002). *Transport Phenomena*. New York: Wiley.
2 Denbigh, K.G. (1951). *Appl. Chem.* 1: 227.
3 Levenspiel, O. (1999). *Chemical Reaction Engineering*. NJ: Wiley.
4 Danckwerts, P.V. (1953). *Chem. Eng. Sci.* 2: 1.
5 Smith, J.M. (1981). *Chemical Engineering Kinetics*. Singapore: McGraw Hill.
6 Wehner, J.F. and Wilhelm, R.H. (1959). *Chem. Eng. Sci.* 6: 89.
7 Levenspiel, O. and Bischoff, K.B. (1961). *Ind. Eng. Chem.* 53: 313.
8 Taylor, G.I. (1953). *Proc. R. Soc. A* 219: 186.
9 Wen, C.Y. and Fan, L.T. (1975). *Models for Flow Systems and Chemical Reactors*. New York: Marcel Dekker.
10 Denbigh, K.G. and Turner, J.C.R. (1971). *Chemical Reactor Theory*. Cambridge: Cambridge University Press.

11 Carberry, J.J. (1976). *Chemical and Catalytic Reactor Engineering*. New York: McGraw Hill.
12 Rastegar, S.O. and Tingyue, G. (2017). *J. Chromatogr.* 1490: 133–137.
13 Chung, S.F. and Wen, C.Y. (1968). *AIChE J.* 14: 857–866.
14 Schneider, P. and Smith, J.M. (1968). *AIChE J.* 14: 762.
15 Dogu, T. and Dogu, G. (2019). *Int. J. Chem. Reactor Eng.* 17: 5.

9

Fixed-Bed Reactors and Interphase Transport Effects

Solid catalysts promote the reaction rate in the desired direction by changing the reaction mechanism. The activation energy of a catalytic reaction path is generally lower than the fluid-phase reaction path. Many process improvements were achieved by the discovery of better routes involving new catalysts. Catalysts change the rates of reactions by involving in the reaction mechanism but remain unchanged at the end of the process. Catalysts change the reaction rate but do not have any effect on the equilibrium conversion of reactants, which is solely determined by the thermodynamics of the reaction. Catalysts may be either in fluid (homogeneous catalysts) or in solid (heterogeneous catalysts) forms.

Solid catalytic materials are used in many of the chemical conversions. Solid-catalyzed reactions proceed by the involvement of active sites on the catalyst surface. An introduction to catalysis and the involvement of active surface sites in the mechanisms of solid-catalyzed reactions are covered in Chapter 11. To have a large number of active sites per unit mass of the catalyst pellets, porous catalytic materials with quite a high pore surface area value are used in many reactions. Critical understanding of mass and heat transfer effects on the observed rates of the reactions catalyzed by porous catalytic materials is extremely important for the analysis and design of catalytic reactors [1–3]. Multi-scale nature of chemical reactor engineering involves mass, heat and momentum transfer concepts, as well as the chemistry and the thermodynamics of chemical conversions [4]. Observed rates of catalytic reactions are strongly influenced by the mass and heat transfer resistances over the external surface of the solid catalytic materials, as well as transport resistances within the porous catalyst pellets. Intrapellet diffusion and heat transfer effects on the observed rates of solid-catalyzed reactions are discussed in Chapter 10. Modeling of fixed-bed catalytic reactors and effects of interphase transport resistances on the observed rate are covered in this chapter.

9.1 Solid-Catalyzed Reactions and Transport Effects Within Reactors

For an arbitrary gas-phase reaction between reactants A and B, adsorption of both or one of these reactants is expected on the active sites over the catalyst surface, which will then be followed by a surface reaction step (Figure 9.1). The overall activation energy of a surface-catalyzed reaction is expected to be less than the activation energy of the gas-phase reaction. An increase in reaction rate is expected in a catalytic process due to a decrease of activation energy of the rate-limiting step. This is schematically illustrated in Figure 9.2 for an exothermic reaction.

Since solid-catalyzed reactions proceed by the involvement of active sites on the catalyst surface, porous catalysts with quite high surface area values are used in many of the industrially important

Fundamentals of Chemical Reactor Engineering: A Multi-Scale Approach, First Edition. Timur Doğu and Gülşen Doğu.

Companion website: www.wiley.com/go/dogu/chemreacengin

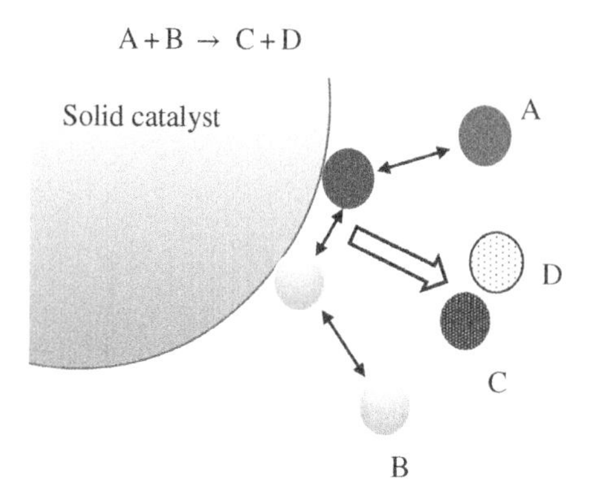

Figure 9.1 Schematic representation of a surface-catalyzed reaction.

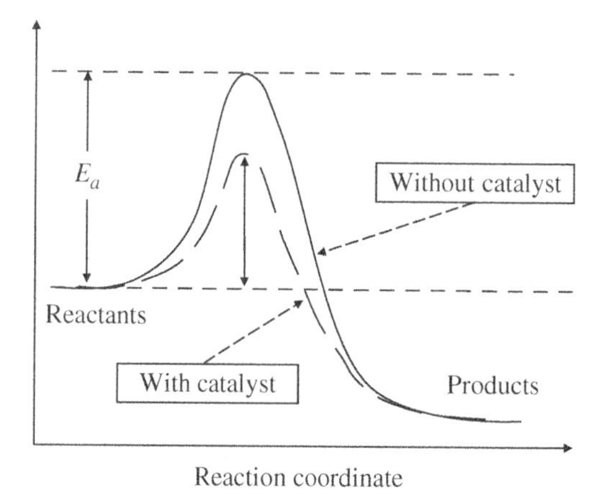

Figure 9.2 Decrease of activation energy by the use of a solid catalyst.

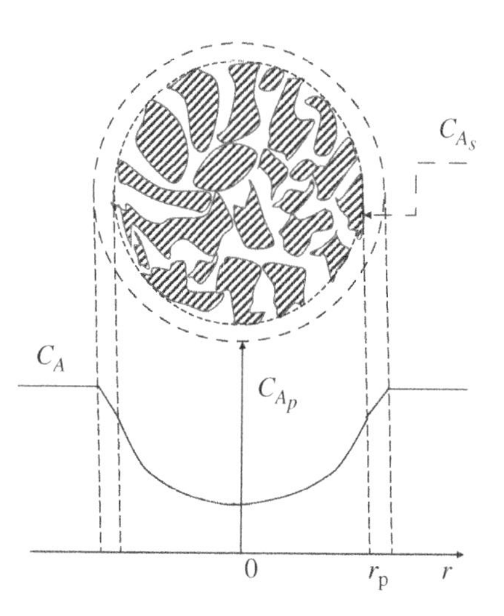

Figure 9.3 Concentration variations of reactant A due to the external mass transfer and pore diffusion resistances for a solid-catalyzed reaction.

reactions. Active sites on the pore surfaces of such catalytic materials are involved in the reaction mechanism. Pore surface area values reaching to 2000 m^2/g are reported in the literature for such porous catalytic materials. Transport of reactant molecules into the catalyst particle to reach the active sites on the pore surfaces involves pore diffusion and external mass transfer steps. Pore diffusion resistance, as well as the mass transfer resistance at the external surface of the catalyst, may cause lower concentrations of the reactants near the active sites within the pores than the bulk fluid concentrations. A schematic illustration of possible concentration variations over the external surface and within the porous catalyst pellet is given in Figure 9.3.

These transport resistances cause deviations of observed reaction rate (global reaction rate) from the ideal reaction rate, which would be predicted at the bulk fluid concentration and temperature. In most of the kinetic studies reported in the literature, the experimental data obtained are global (observed) reaction rates. To have information about the intrinsic rates, transport effects should be extracted from the global (observed) rate values. Heat effects within a catalyst pellet may also have a significant influence on the observed reaction rate values. This section will not consider heat effects and will formulate the mass transport effects on the observed rate for an isothermal catalytic reaction.

To express the observed reaction rate of a solid-catalyzed reaction, both intrapellet and interphase transport processes and the kinetics of surface reaction should be considered. Heat and mass transfer resistances on the external surface of a catalyst pellet, as well as pore diffusion effects, are expected to have significant influences on the observed rates. To express the observed reaction rate in terms of bulk concentrations of reactants, the following steps should be considered for a surface-catalyzed reaction:

1) *External mass transfer:* Transport of reactants from the bulk fluid to the external surface of the catalyst pellet.

2) *Intrapellet diffusion:* Pore diffusion of reactants into the catalyst pellet.
3) *Adsorption:* Adsorption of reactant(s) onto the surface sites within the pores and migration over the pore surfaces.
4) *Surface reaction:* Reaction of reactants by the involvement of active sites on the pore surfaces.
5) *Desorption:* Desorption of products from the pore surfaces to the porous space within the catalyst pellet.
6) *Pore diffusion of products:* Transport of desorbed product molecules to the external surface of the pellet.
7) *External mass transfer of products:* Mass transfer of product molecules from the external surface of the catalyst pellet to the bulk fluid phase within the reactor.

9.2 Observed Reaction Rate and Fixed-Bed Reactors

In many of the catalytic processes, tubular fixed-bed catalytic reactors are used. For the general case, variations of concentration and temperature are expected both in the axial and radial directions within these tubular reactors. Velocity profiles and diffusional effects may cause a distribution of residence times within such reactors. These effects and dispersion models are discussed in Chapter 8. In the design of a fixed-bed reactor, radial variations of concentration and temperature are frequently neglected, and one-dimensional design equations are used. For a fixed-bed reactor illustrated in Figure 9.4, the species conservation equation for the differential volume element within the reactor can be written following the general conservation equation, which is expressed as follows:

(Input Rate) – (Output Rate) + (Generation Rate) = (Accumulation Rate)

Differential species conservation equation for the reactant A can be written at steady state in terms of the molar flow rate of A (F_A) and the observed reaction rate, which is based on the unit volume of the reactor (R_A, mol/m^3 s).

$$F_A\big|_z - F_A\big|_{z+\Delta z} + R_A \Delta V = 0 \tag{9.1}$$

Equation (9.1) is a pseudo-homogeneous species conservation equation written for the fixed-bed reactor. Derivation of pseudo-homogeneous species conservation equations is given in Appendix O.3 of the online material.

In reality, the bed is composed of two zones, namely fluid phase and solid catalyst particles. The volume fraction of the bed, which is occupied by the fluid, is called the void fraction of the bed (ε_b), while the volume fraction occupied by the solid catalyst particles is $(1 - \varepsilon_b)$. In some cases, the rate of a catalytic reaction is expressed based on the volume of the catalyst pellet ($R_{A,p}$, mol/m^3 s) instead of reactor volume. In some other cases, the rate of a catalytic reaction may be expressed based on the mass of the catalyst ($R_{A,w}$, mol/g s) or based on the surface area of the catalyst ($R_{A,s}$, mol/m^2 s). The relations between R_A and $R_{A,p}$, $R_{A,w}$, and $R_{A,s}$ can be expressed as follows:

$$R_A = R_{A,p}(1 - \varepsilon_b) \tag{9.2}$$

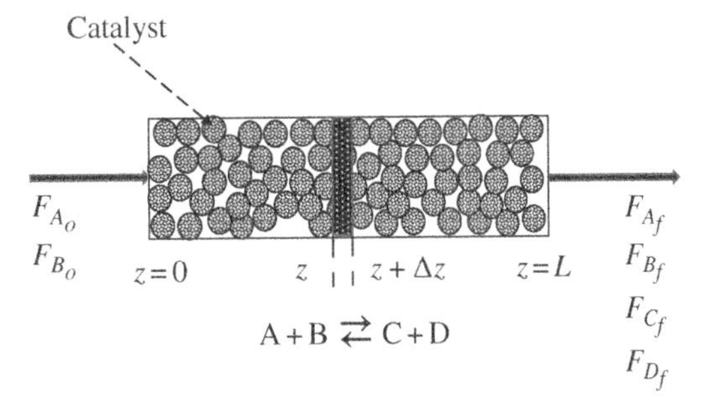

Figure 9.4 Schematic diagram of a fixed-bed catalytic reactor.

$$R_A = R_{A,w}\rho_p(1-\varepsilon_b) \tag{9.3}$$

$$R_A = R_{A,s}S_g\rho_p(1-\varepsilon_b) \tag{9.4}$$

Here, ρ_p and S_g are the apparent density (g/m^3) and the total active surface area (m^2/g) of the catalyst pellet, respectively.

By dividing each term of Eq. (9.1) by ΔV and by taking the limit as ΔV approaching zero, differential pseudo-homogeneous species conservation equation is obtained.

$$\lim_{\Delta V \to 0}\left(\frac{F_A|_z - F_A|_{z+\Delta z}}{\Delta V}\right) + R_A = 0 \tag{9.5}$$

$$-\frac{dF_A}{dV} + R_A = 0 \tag{9.6}$$

Here, the molar flow rate of reactant A can be expressed either in terms of its fractional conversion $F_A = F_{A_o}(1-x_A)$ or in terms of volumetric flow rate Q and concentration of A at a location z within the reactor ($F_A = QC_A$). Then, for an arbitrary nth-order reaction, Eq. (9.6) becomes as

$$F_{A_o}\frac{dx_A}{dV} - kC_A^n = 0 \tag{9.7}$$

If the rate of the catalytic reaction is given in terms of the catalyst mass $\left(R_{A,w} = k_wC_A^n\right)$, the species conservation equation should be written as follows:

$$F_{A_o}\frac{dx_A}{dV} - k_w\rho_p(1-\varepsilon_b)C_A^n = 0 \tag{9.8}$$

In this equation, C_A is the concentration of reactant A in the bulk fluid phase at a location z within the reactor. However, the catalytic reaction is expected to take place within the catalyst pellet on the pore surfaces. Suppose the external mass transfer resistance over the catalyst surface or pore diffusion resistance to transport the reactants to the active sites within the catalyst pellet is not negligible. In that case, a significantly lower concentration of reactant A is expected within the catalyst pores. Transport resistances may cause a significant decrease in the observed reaction rate for positive-order reactions. Hence, the observed reaction rate becomes lower than the ideal reaction rate, which is evaluated at the bulk fluid concentration (and temperature) at the same location. Hence, we define the effectiveness factor (η), which is the ratio of observed reaction rate to the reaction rate evaluated at the bulk concentration of the reactants and products, which is called the ideal reaction rate.

$$\eta = \frac{\text{Observed reaction rate}}{\text{Ideal Reaction rate}} = \frac{\left(R_{A,p}\right)_{\text{obs}}}{\left(R_{A,p}\right)_{\text{ideal}}} \tag{9.9}$$

Effectiveness factor may be considered as a correction factor for the ideal reaction rate, to take into account the external mass transfer and pore diffusion resistances. The effectiveness factor concept and its evaluation procedure are discussed in detail in Chapter 10.

By combining the intrinsic kinetics with the effects of transport resistances, observed (global) rate expressions can be obtained, which should be used in the analysis and design of reactors containing solid catalysts. Hence, for the catalytic reactions, which are influenced by the transport resistances, Eq. (9.6) should be expressed as follows:

$$-\frac{dF_A}{dV} + (1-\varepsilon_b)\eta R_{A,p} = 0 \tag{9.10}$$

This equation can be expressed in terms of fractional conversion of reactant A as follows:

$$F_{A_o}\frac{dx_A}{dV} + (1-\varepsilon_b)\eta R_{A,p} = 0. \tag{9.11}$$

In the case of an nth-order reaction rate, expressed based on the catalyst mass, Eq. (9.11) can be expressed as

$$F_{A_o}\frac{dx_A}{dV} - \eta k_w \rho_p (1-\varepsilon_b) C_A^n = 0 \tag{9.12}$$

The concentration of reactant A within the bulk fluid in the reactor can be expressed in terms of its fractional conversion and the volume expansion factor for a gas-phase reaction, as discussed in Chapter 4.

$$F_{A_o}\frac{dx_A}{dV} - \eta k_w \rho_p (1-\varepsilon_b) C_{A_o}^n \frac{(1-x_A)^n}{(1+\varepsilon x_A)^n} = 0 \tag{9.13}$$

In the case of a constant volumetric flow rate, this equation reduces to

$$F_{A_o}\frac{dx_A}{dV} - \eta k_w \rho_p (1-\varepsilon_b) C_{A_o}^n (1-x_A)^n = 0 \tag{9.14}$$

For the constant volumetric flow rate case, the species conservation equation can also be expressed in terms of concentration of reactant A, instead of its fractional conversion.

$$-Q\frac{dC_A}{dV} + (1-\varepsilon_b)\eta R_{A,p} = 0 \tag{9.15}$$

Another way of writing this equation is

$$-U_o\frac{dC_A}{dz} + (1-\varepsilon_b)\eta R_{A,p} = 0 \tag{9.16}$$

Here, U_o is the superficial velocity in the fixed-bed reactor ($U_o = Q/A_c$). Equation (9.16) corresponds to the plug-flow model for a fixed-bed reactor with a constant volumetric flow rate. In the case of having significant dispersion in the axial direction, the one-dimensional steady-state design equation should be written considering the dispersion term, as follows:

$$-U_o\frac{dC_A}{dz} + D_z\frac{d^2C_A}{dz^2} + (1-\varepsilon_b)\eta R_{A,p} = 0 \tag{9.17}$$

9.3 Significance of Film Mass Transfer Resistance in Catalytic Reactions

In a catalytic reactor, external film mass transfer resistance over the catalyst pellets might be quite significant in some cases. Mass transfer resistance through the boundary layer over the external surface of a catalyst pellet may give rise to differences between the bulk and the surface values of the concentration of reactants and products. Such differences may cause appreciable variations between the observed reaction rate and the ideal reaction rate, which should have been evaluated at the bulk fluid conditions.

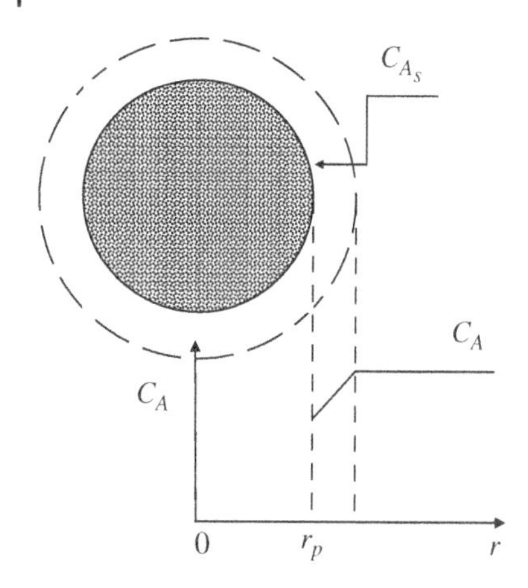

Figure 9.5 Schematic representation of external film resistance.

The importance of film mass transfer resistance on the observed reaction rate is illustrated here for a simple first-order surface reaction in a fixed-bed catalytic reactor. If we consider a case in which a non-porous solid catalyst is used, pore diffusion effects can be eliminated. A typical example of such a reaction is the oxidation of ammonia in a nitric acid plant. For this reaction, non-porous Pt gauze is generally used as the catalyst. This reaction is very fast, and the transport of reactants to the external surface of the catalyst is the controlling resistance.

For a simple first-order surface reaction taking place at the external surface of a non-porous catalyst (Figure 9.5), the reaction rate should be evaluated at the surface concentration where the reaction is taking place.

$$\mathrm{A} \rightarrow \mathrm{B} \quad -R_{A_s} = k_s C_{A_s}; \ \text{mol/m}^2\,\text{s}$$

To express the surface concentration of reactant A in terms of its bulk concentration, the rate of surface reaction can be equated to the flux of A to the external surface of the catalyst pellet.

$$k_m(C_A - C_{A_s}) = k_s C_{A_s} \tag{9.18}$$

Here, k_m is the film mass transfer coefficient over the surface of the catalyst pellet. Re-arrangement of Eq. (9.18) gives a relation between the surface and bulk concentrations of reactant A.

$$C_{A_s} = C_A\left(\frac{k_m}{k_m + k_s}\right) \tag{9.19}$$

The observed surface reaction rate can then be expressed as follows:

$$-R_{A_s} = k_s C_{A_s} = \left(\frac{k_s k_m}{k_s + k_m}\right) C_A = \frac{1}{\left(\dfrac{1}{k_s} + \dfrac{1}{k_m}\right)} C_A; \ \text{mol/m}^2\,\text{s} \tag{9.20}$$

Using Eq. (9.20), the fixed-bed reactor design equation for the constant volumetric flow rate case becomes

$$Q\frac{dC_A}{dV} - \rho_p S_g(1-\varepsilon_b)\left(\frac{1}{k_s} + \frac{1}{k_m}\right)^{-1} C_A = 0 \tag{9.21}$$

As shown in this equation, the observed reaction rate constant depends both on the surface reaction resistance ($1/k_s$) and also on the film mass transfer resistance ($1/k_m$). In the case of a very fast surface reaction, the surface concentration of A approaches zero, and film mass transfer resistance controls the reaction rate.

$$\text{If } \left(\frac{1}{k_s}\right) \ll \left(\frac{1}{k_m}\right); \quad -R_A = \rho_p S_g(1-\varepsilon_b) k_m C_A \quad \text{mass transfer controls} \tag{9.22}$$

$$\text{If } \left(\frac{1}{k_m}\right) \ll \left(\frac{1}{k_s}\right); \quad -R_A = \rho_p S_g(1-\varepsilon_b) k_s C_A \quad \text{surface reaction controls} \tag{9.23}$$

For the film mass transfer control case, we only need to know the mass transfer coefficient, rather than the surface reaction rate constant, for the design of this reactor.

The value of the mass transfer coefficient over the external surface of the catalyst pellet can be predicted from the correlations given in the literature. Such correlations are generally presented in terms of a J_D factor for mass transfer. Fixed-bed mass transfer data reported in the literature are correlated in terms of the J_D factor and particle Reynolds number [1].

$$J_D = \frac{k_m}{U_o}\varepsilon_b (Sc)^{2/3} = \frac{1.15}{Re_p^{1/2}} \tag{9.24}$$

Here, Schmidt number (Sc) is defined as the ratio of viscous diffusivity (kinematic viscosity) to the molecular diffusion coefficient of A in the bulk fluid D_{AB}. Schmidt number and particle Reynolds number can be predicted using their definitions.

$$Sc = \frac{\mu}{\rho D_{AB}} \tag{9.25}$$

$$Re_p = \frac{d_p U_o \rho}{\mu \varepsilon_b} \tag{9.26}$$

In these expressions, U_o is the superficial velocity in the reactor $\left(U_o = \frac{Q}{A_c}\right)$, ρ is the density of the fluid, μ is the viscosity of the fluid, d_p is the particle diameter of the catalyst, and A_c is the cross-sectional area of the tubular reactor. As can be predicted from Eq. (9.24), the mass transfer coefficient is a function of superficial velocity in the bed. Mass transfer resistance may decrease by increasing the superficial velocity in the reactor. Keeping the reactor volume and the total volumetric flow rate constant, if we decrease the diameter of the tubular reactor, we expect an increase in the superficial velocity. Hence, this will also increase the mass transfer coefficient, which may cause an increase in the fractional conversion of reactant A if mass transfer resistance controls the observed reaction rate.

The surface reaction rate term may also be rearranged in the following form:

$$-R_{A_s} = k_s C_{A_s} = \left(\frac{k_s k_m}{k_s + k_m}\right) C_A = \left(1 + \frac{k_s}{k_m}\right)^{-1} k_s C_A = \left(\frac{1}{1 + Da_s}\right) k_s C_A \tag{9.27}$$

Here, Da_s is the surface Damköhler number defined as the ratio of mass transfer resistance to surface reaction resistance.

$$Da_s = \frac{k_s}{k_m} \tag{9.28}$$

For the case of a very small Damköhler number $Da_s \rightarrow 0$, film mass transfer resistance becomes negligible, and the observed reaction rate constant becomes the same as the ideal reaction rate constant. However, in the case of very large values of the Damköhler number ($Da_s \rightarrow \infty$), the observed reaction rate constant approaches the value of the mass transfer coefficient. In this case, the activation energy of the observed rate constant approaches zero since the temperature dependence of the mass transfer coefficient is quite weak.

9.4 Tubular Reactors with Catalytic Walls

Let us consider a tubular reactor, in which there is no fluid-phase reaction. However, the walls of the tubular reactor are covered with a thin layer of catalyst, over which reactants are converted to the products (Figure 9.6). This catalyst layer may be porous or nonporous. If the catalyst layer over

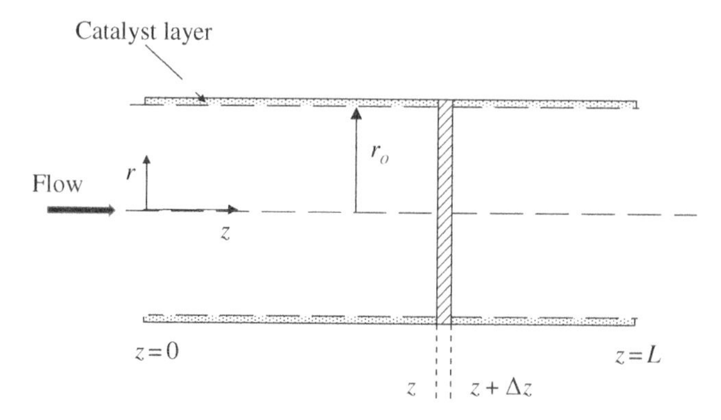

Figure 9.6 A tubular reactor with catalytic walls.

the wall surface of the tubular reactor is not porous, the reaction takes place on the external surface of the catalyst. Non-porous and porous catalyst layer cases are considered in this section.

Modeling of a wall-catalyzed reactor can be made either by considering the variation of concentration of reactant A only in the z-direction (one-dimensional model) or considering concentration variations both in the z and r directions (two-dimensional model). Model equations of these two cases are given below.

9.4.1 One-Dimensional Model

$$\pi r_o^2\left(N_{A_z}\big|_z - N_{A_z}\big|_{z+\Delta z}\right) - 2\pi r_o \Delta z\left(N_{A_r}\big|_{r_o}\right) = 0 \tag{9.29}$$

$$\left(\pi r_o^2\right)\frac{dN_{A_z}}{dz} + (2\pi r_o)\left(N_{A_r}\big|_{r_o}\right) = 0 \tag{9.30}$$

$$N_{A_r}\big|_{r_o} = k_m(C_A - C_{A_s}) = k_s C_{A_s} \tag{9.31}$$

Here, k_m and k_s are the mass transfer coefficient and the surface reaction rate constant, respectively. C_{A_s} is the concentration of reactant A at the surface of the catalyst layer, which is not necessarily the same as the bulk concentration C_A. In this model, a nonporous catalyst layer is considered.

The flux of reactant A in the axial direction may, in general, be written as follows:

$$N_{A_z} = U_o C_A - D_z \frac{dC_A}{dz} \tag{9.32}$$

With the plug-flow reactor assumption, the second term in Eq. (9.32) becomes negligible, and Eq. (9.30) reduces to the following expression:

$$U_o \frac{dC_A}{dz} + \left(\frac{2}{r_o}\right)(k_s C_{A_s}) = 0 \tag{9.33}$$

Using Eq. (9.31), the surface concentration of A can be expressed in terms of bulk concentration, as follows:

$$C_{A_s} = \left(\frac{k_m}{k_m + k_s}\right) C_A \tag{9.34}$$

Hence, the one-dimensional model equation reduces to

$$U_o\frac{dC_A}{dz} + \left(\frac{2}{r_o}\right)\left(\frac{k_s k_m}{k_m + k_s}\right)C_A = 0 \tag{9.35}$$

We can write the following limiting cases of this equation:

$$k_m \gg k_s;\quad U_o\frac{dC_A}{dz} + \left(\frac{2}{r_o}\right)(k_s)C_A = 0 \quad \text{surface reaction controls} \tag{9.36}$$

$$k_s \gg k_m;\quad U_o\frac{dC_A}{dz} + \left(\frac{2}{r_o}\right)(k_m)C_A = 0 \quad \text{mass transfer controls} \tag{9.37}$$

The modeling equations shown above correspond to a wall-catalyzed reactor, for which the catalyst layer was non-porous. In the case of having a porous catalyst layer on the surface of the reactor wall, pore diffusion resistance in this catalyst layer may also be significant. For a problem with significant pore diffusion resistance in the product layer, Eq. (9.31) should be expressed as follows:

$$N_{A_r}|_{r_o} = -\delta\eta R_{A,p} \tag{9.38}$$

Here, δ is the thickness of the porous catalyst layer, $R_{A,p}$ is the reaction rate per unit volume of the catalyst layer, and η is the effectiveness factor. For this case, the species conservation equation becomes

$$\frac{dN_{A_z}}{dz} - \left(\frac{2}{r_o}\right)\delta\eta R_{A,p} = 0 \tag{9.39}$$

For an nth-order catalytic reaction (based on catalyst volume), with the constant volumetric flow rate case, Eq. (9.39) reduces to

$$U_o\frac{dC_A}{dz} + \left(\frac{2}{r_o}\right)\delta\eta k C_A^n = 0 \tag{9.40}$$

Prediction of the effectiveness factor for different geometries is discussed in Chapter 10. Also, a more detailed presentation of wall-catalyzed reactions is given in Chapter 13 in relation to the modeling of monolithic reactors.

9.4.2 Two-Dimensional Model

Species conservation equation for a tubular reactor, in which there is no homogeneous reaction taking place, can be expressed as follows:

$$-\frac{\partial N_{A_z}}{\partial z} - \frac{1}{r}\frac{\partial (rN_{A_r})}{\partial r} = 0 \tag{9.41}$$

$$-v_z\frac{\partial C_A}{\partial z} + D_{AB}\frac{\partial^2 C_A}{\partial z^2} + \frac{D_{AB}}{r}\frac{\partial}{\partial r}\left(r\frac{\partial C_A}{\partial r}\right) = 0 \tag{9.42}$$

Diffusion flux in the axial direction is generally much smaller than the convective flux. Hence, Eq. (9.42) reduces to the following form:

$$-v_z\frac{\partial C_A}{\partial z} + \frac{D_{AB}}{r}\frac{\partial}{\partial r}\left(r\frac{\partial C_A}{\partial r}\right) = 0 \tag{9.43}$$

In this case, the reaction term appears in the boundary condition rather than being within the species conservation equation written for the bulk fluid phase.

Boundary conditions:

$$\text{At } r = r_o; \quad -D_{AB}\left.\frac{\partial C_A}{\partial r}\right|_{r=r_o} = k_s C_{A_s} \tag{9.44}$$

$$\text{At } r = 0; \quad \left.\frac{\partial C_A}{\partial r}\right|_{r=0} = 0 \tag{9.45}$$

In the case of a porous catalytic layer on the surfaces of the tubular reactor, the boundary condition expressed in Eq. (9.44) should be written as follows:

$$D_{AB}\left.\frac{\partial C_A}{\partial r}\right|_{r=r_o} = \delta\eta R_{A,p} \tag{9.46}$$

If the variation of the velocity in the radial direction is neglected, Eq. (9.43) can be simplified as follows:

$$-U_o\frac{\partial C_A}{\partial z} + \frac{D_{AB}}{r}\frac{\partial}{\partial r}\left(r\frac{\partial C_A}{\partial r}\right) = 0 \tag{9.47}$$

9.5 Modeling of a Non-isothermal Fixed-Bed Reactor

Modeling of a non-isothermal tubular reactor is discussed in Section 7.3, considering temperature variation only in the axial direction. In many cases, temperature and concentration variations in the radial direction in a tubular reactor may be quite significant, as well as their variations in the axial direction. In this case, two-dimensional species conservation and energy balance equations should be solved to predict product distributions and design such a reactor. For this case, dispersion and heat conduction terms in the radial direction should be included into the pseudo-homogeneous species conservation and energy balance equations for the fixed-bed reactor. In this section, first, two-dimensional steady-state model equations for a non-isothermal fixed-bed reactor are derived. Then, these equations are simplified to obtain one-dimensional design equations for the non-isothermal reactor.

Two-dimensional steady-state model equations for the conservation of reactant species A and the energy balance equation can be expressed as follows:

$$-U_o\frac{\partial C_A}{\partial z} + D_z\frac{\partial^2 C_A}{\partial z^2} + D_r\left(\frac{\partial^2 C_A}{\partial r^2} + \frac{1}{r}\frac{\partial C_A}{\partial r}\right) + (1-\varepsilon_b)\eta R_{A,p} = 0 \tag{9.48}$$

$$-\sum_i N_{i_z} C_{p_i}\frac{\partial T}{\partial z} + \lambda_e\frac{\partial^2 T}{\partial z^2} + \lambda_e\left(\frac{\partial^2 T}{\partial r^2} + \frac{1}{r}\frac{\partial T}{\partial r}\right) + (1-\varepsilon_b)\eta R_{A,p}(\Delta H^o) = 0 \tag{9.49}$$

The following boundary conditions can be used for the simultaneous solution of these pseudo-homogeneous species conservation and the energy balance equations:

$$C_A(r,0) = C_{A_o}; \quad T(r,0) = T_o \quad \text{at } z = 0 \tag{9.50}$$

$$\left[\frac{\partial C_A}{\partial z}\right]_{z=L} = 0; \quad \left[\frac{\partial T}{\partial z}\right]_{z=L} = 0 \tag{9.51}$$

$$\left[\frac{\partial C_A}{\partial r}\right]_{r=0} = 0; \quad \left[\frac{\partial T}{\partial r}\right]_{r=0} = 0 \text{ (symmetry condition)} \tag{9.52}$$

$$\lambda_e \frac{\partial T}{\partial r}\bigg|_{r=r_o} = -U\left(T|_{r=r_o} - T_a\right) \tag{9.53}$$

No flux condition is generally used as the boundary condition of Eq. (9.48) at $r = r_o$. The parameter D_r, which appears in Eq. (9.48), is the radial dispersion coefficient, and its value can be estimated for a packed-bed reactor from the following relation [2]:

$$\frac{U_o d_p}{\varepsilon_b D_r} \cong 11 \tag{9.54}$$

In the case of constant specific heat and constant density for the reacting fluid flowing in the reactor, Eq. (9.49) may also be expressed as follows:

$$-U_o \rho \bar{c} \frac{\partial T}{\partial z} + \lambda_e \frac{\partial^2 T}{\partial z^2} + \lambda_e \left(\frac{\partial^2 T}{\partial r^2} + \frac{1}{r}\frac{\partial T}{\partial r}\right) + (1-\varepsilon_b)\eta R_{A,p}(\Delta H^o) = 0 \tag{9.55}$$

In many cases, the order of magnitude of heat conduction term in the axial direction is much less than the order of magnitude of the convective heat flow term. In this case, Eq. (9.55) reduces to:

$$-U_o \rho \bar{c} \frac{\partial T}{\partial z} + \lambda_e \left(\frac{\partial^2 T}{\partial r^2} + \frac{1}{r}\frac{\partial T}{\partial r}\right) + (1-\varepsilon_b)\eta R_{A,p}(\Delta H^o) = 0 \tag{9.56}$$

The coupling term of the species conservation and the energy balance equations is the reaction rate, which depends on both temperature and the concentrations of the species. In the case of having more than one independent reaction in the reactor, more than one species conservation equation should be written and solved together with the energy balance equation. For instance, if we have n number of independent reactions, we should consider n steady-state species conservation equations to model a fixed-bed tubular reactor. Also, the energy balance equation should contain heat generation terms of all of the independent reactions.

$$-U_o \rho \bar{c} \frac{\partial T}{\partial z} + \lambda_e \left(\frac{\partial^2 T}{\partial r^2} + \frac{1}{r}\frac{\partial T}{\partial r}\right) - (1-\varepsilon_b)\sum_j \eta R_{(j),p} \Delta H^o_{(j)} = 0 \tag{9.57}$$

Here, $\Delta H^o_{(j)}$ and $R_{(j),p}$ are the heat of reaction and the intrinsic rate (per unit volume of the catalyst) of reaction j.

In the case of negligible radial temperature and concentration variations, Eqs. (9.48) and (9.57) may be simplified and the following one-dimensional energy and species conservation equations may be simultaneously solved to design a non-isothermal fixed-bed reactor.

$$-U_o \rho \bar{c} \frac{dT}{dz} + (1-\varepsilon_b)\eta R_{A,p}(\Delta H^o) = 0 \tag{9.58}$$

$$-U_o \frac{dC_A}{dz} + D_z \frac{d^2 C_A}{dz^2} + (1-\varepsilon_b)\eta R_{A,p} = 0 \tag{9.59}$$

9.6 Steady-State Multiplicity on the Surface of a Catalyst Pellet

Steady-state multiplicity in a CSTR is discussed in Section 7.2. It is also possible to have multiple steady-state surface temperatures of catalysts in catalytic reactions under certain conditions [3]. Let us consider an exothermic first-order surface-catalyzed reaction taking place on the external surface of a non-porous catalyst pellet (Figure 9.5). Due to film mass transfer resistance on the catalyst surface, the surface concentration of reactant A is expected to be less than its bulk concentration. We also expect to have some differences in the surface and the bulk fluid temperatures. The heat generated due to the chemical reaction on the catalyst surface should be transferred to the bulk fluid to keep the surface temperature at a steady-state value. The steady-state material and the energy balance equations describing this system can be written as follows:

$$k_m(C_A - C_{A_s}) = k_s C_{A_s} \tag{9.60}$$

$$h(T_s - T) = \left(-\Delta H^o_{T_s}\right) k_s C_{A_s} \tag{9.61}$$

Here, k_m and h are the mass transfer and heat transfer coefficients at the catalyst pellet external surface. C_A and T correspond to the concentration and the temperature of the bulk fluid, respectively. Following relation can then be written by combining these two relations:

$$h(T_s - T) = \left(-\Delta H^o_{T_s}\right)\left(\frac{k_s k_m}{k_m + k_s}\right) C_A \tag{9.62}$$

The right- and the left-hand sides of Eq. (9.62) are proportional to the heat generation and heat removal rates, respectively.

$$q_G = A_e\left(-\Delta H^o_{T_s}\right)\left(\frac{k_s k_m}{k_m + k_s}\right) C_A = A_e\left(-\Delta H^o_{T_s}\right)\frac{k_m k_o \exp\left(-\frac{E_a}{R_g T_s}\right)}{k_m + k_o \exp\left(-\frac{E_a}{R_g T_s}\right)} C_A; \ \ (\text{J/s}) \tag{9.63}$$

$$q_R = A_e h(T_s - T) \tag{9.64}$$

Here, A_e is the external surface area of the catalyst pellet. Temperature dependence of q_G gives an S-shaped curve, while the temperature dependence of q_R is linear, with a slope of $A_e h$. As illustrated in Figure 9.7, three intersections of the q_G curve and the q_R line are possible, corresponding to the

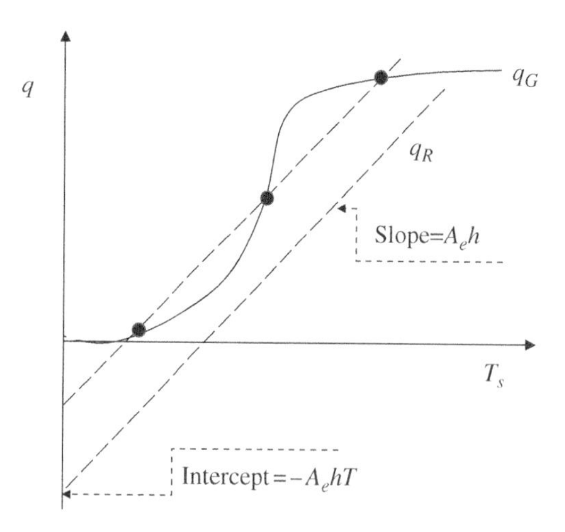

Figure 9.7 Schematic representation of heat generation and heat removal rate curves as a function of the catalyst surface temperature.

three steady-state surface temperatures. To not have multiple steady-state surface temperatures, either the slope of the heat removal line should be higher than the slope of the heat generation curve at its inflection point, or the bulk temperature of the fluid should be such that q_R line intercepts the q_G curve only at a single point.

Exercises

9.1 Consider a cylindrical radial flow reactor shown in Figure 9.8. The annular space between r_1 and r_2 is packed with spherical catalyst pellets having particle radii of r_p. Transport of species takes place in the radial direction through the catalyst bed, and the axial variation of concentrations (and temperature) is negligible. Consider a first-order catalytic reaction taking place in this reactor. Set up the necessary model equations and boundary conditions for the following cases. Make the equations dimensionless and discuss the possible effects of dimensionless groups on the overall conversion.

(a) Consider an isothermal reactor with negligible pore diffusion and film mass transfer effects on the observed rate. Both radial flow and dispersion in the radial direction through the catalyst layer are significant in this reactor.

(b) Consider an isothermal reactor with a negligible pore diffusion effect on the observed rate. External film mass transfer resistance over the catalyst surface may be significant. Both radial flow and dispersion in the radial direction through the catalyst layer are significant in this reactor.

(c) Consider an exothermic reaction taking place in this reactor, which operates adiabatically. Transport effects on the observed rate of the reaction are given as negligible. Also, neglect the radial dispersion through the catalyst layer.

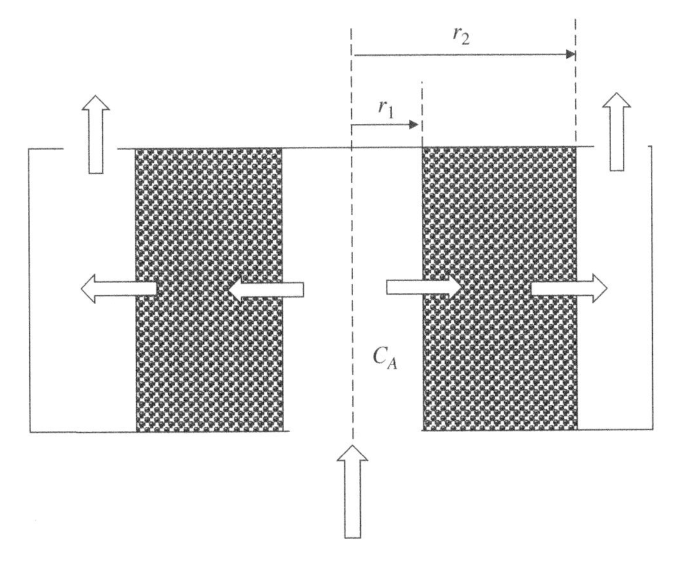

Figure 9.8 Schematic representation of a radial flow reactor.

9.2 Discuss the following limits of observed activation energy of a surface-catalyzed reaction:

$$Da_s \rightarrow 0$$

$$Da_s \rightarrow \infty$$

Also, qualitatively plot the observed activation energy of a first-order surface reaction as a function of $1/T$.

References

1 Carberry, H.J. (1976). *Chemical and Catalytic Reaction Engineering*. New York: McGraw Hill.

2 Denbigh, K.G. and Turner, J.C.R. (1971). *Chemical Reactor Theory: An Introduction*. Cambridge: Cambridge University Press.

3 Luss, D. (1977). Steady-state and dynamic behavior of a single catalyst pellet. In: *Chemical Reactor Theory: A Review* (eds. L.L. Lapidus and N.R. Amundsen), 191–265. London: Prentice-Hall.

4 Lerou, J.J. and Ng, K.M. (1996). *Chem. Eng. Sci.* 51: 1595–1614.

10

Transport Effects and Effectiveness Factor for Reactions in Porous Catalysts

Pore-diffusion resistance may cause concentration variations of reactants and products within a porous catalyst pellet. Depending upon the relative magnitudes of characteristic diffusion and reaction times, concentration profiles may be established within the catalyst pellets. If the reaction rate is much slower than the diffusion rate, reactant species may quickly diffuse into the catalyst pores to reach the active sites. In that case, we can neglect the concentration variations within the catalyst pellet. However, if the reaction rate is much faster than the diffusion rate of reactants into the pores, significant concentration variations may develop within a porous catalyst (Figure 10.1). As a result, the observed reaction rate may become much less than the reaction rate evaluated at the bulk fluid concentrations for positive-order reactions. This phenomenon was initially considered by Thiele, and an effectiveness factor was defined as a correction factor for the ideal reaction rate [1–7]. In the initial treatment of this phenomenon, diffusion and reaction in a capillary were considered, and an effectiveness factor definition was developed as the ratio of observed reaction rate to the ideal reaction rate, which was evaluated at the external surface concentrations and the temperature of the catalyst pellet.

10.1 Effectiveness Factor Expressions in an Isothermal Catalyst Pellet

Observed reaction rate per unit volume of the catalyst pellet can be evaluated either by integrating the point reaction rate with respect to the catalyst pellet volume or evaluating the total molar flow rate of reactant A diffusing into the pellet from the external surface of the catalyst.

$$\left(R_{A,p}\right)_{\text{obs}} = \frac{1}{V_p}\int_{V_p} R_{A,p}\left(C_{A_p}, T_p\right)dV = \frac{1}{V_p}\int_{A_e} D_e\left(\nabla C_{A_p}\cdot \boldsymbol{n}\right)dA_e \tag{10.1}$$

Here, V_p is the volume of the catalyst pellet. C_{A_p} and T_p are the concentration and temperature within the catalyst pellet, which is considered as a pseudo-homogeneous medium. A_e is the external surface area of the catalyst pellet, and $\boldsymbol{n}$ is the unit normal vector at the external surface. The effective diffusion coefficient D_e within the porous catalyst pellet depends upon the pore structure of the catalyst, temperature, and the chemical nature of the diffusing molecules (Chapter 12).

The effectiveness factor is defined as the ratio of the observed reaction rate to the ideal reaction rate, which is evaluated at the bulk fluid properties at the external surface of the pellet.

Fundamentals of Chemical Reactor Engineering: A Multi-Scale Approach, First Edition. Timur Doğu and Gülşen Doğu.

Companion website: www.wiley.com/go/dogu/chemreacengin

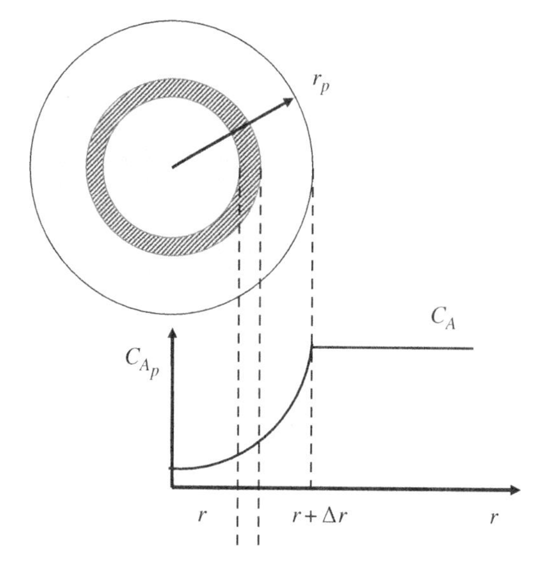

Figure 10.1 Schematic representation of concentration variation within a porous spherical catalyst pellet.

$$\eta = \frac{\text{Observed reaction rate}}{\text{Ideal reaction rate}} = \frac{\int_{V_P} R_{A,p}(C_{A_p}, T_p)dV}{V_p[R_{A,p}(C_A, T)]_{\text{bulk}}} \tag{10.2}$$

Here, C_A and T are the concentration and temperature within the bulk fluid surrounding the catalyst pellet, respectively. In the following derivation, possible film mass transfer resistance at the external surface of the catalyst pellets is neglected. Hence, external surface concentration and the temperature values are assumed to be the same as the corresponding bulk fluid values. For a spherical catalyst pellet, Eq. (10.2) can be written as follows:

$$\eta = \frac{\int_0^{r_p} R_{A,p}(C_{A_p}, T_p)4\pi r^2 dr}{\frac{4}{3}\pi r_p{}^3[R_{A,p}(C_A, T)]_{\text{bulk}}} = \frac{-D_e\left.\frac{dC_{A_p}}{dr}\right|_{r=r_p} 4\pi r_p^2}{\frac{4}{3}\pi r_p{}^3[R_{A,p}(C_A, T)]_{\text{bulk}}} \tag{10.3a}$$

As illustrated in Eq. (10.3a), the observed reaction rate for a catalyst pellet can be estimated either by integrating the reaction rate through the volume of the catalyst pellet or by evaluating the net flow rate of limiting reactant A into the catalyst pellet from its external surface. The rate of transport of reactant A into the catalyst pellet can be written as a product of the molecular flux of A at the catalyst external surface and the pellet external surface area. For an nth-order catalytic reaction, Eq. (10.3a) reduces to

$$\eta = \frac{-D_e\left.\frac{dC_{A_p}}{dr}\right|_{r=r_p} 4\pi r_p^2}{-\frac{4}{3}\pi r_p{}^3 k C_A^n} = \left.\frac{3D_e}{r_p k C_A^n}\frac{dC_{A_p}}{dr}\right|_{r=r_p} \tag{10.3b}$$

In many of the chemical engineering applications, dimensionless variables and dimensionless numbers are frequently used. For this system, dimensionless concentration and dimensionless radial coordinate can be expressed by dividing C_{A_p} and r with the characteristic bulk concentration C_A and the characteristic length, respectively. As the characteristic length, a typical selection is the radius of the catalyst pellet r_p. An alternative characteristic length can be defined as the ratio of pellet volume (V_p) to its external surface area (A_e). This is a more general characteristic length definition, which can be used for any catalyst pellet shape [5]. Hence, dimensionless concentration and radial coordinate within a spherical catalyst pellet are defined in this book as follows:

$$\psi = \frac{C_{A_p}}{C_A}; \quad \zeta = \frac{r}{\frac{V_p}{A_e}} = \frac{\left(4\pi r_p^2\right) r}{\frac{4}{3}\pi r_p^3} = \frac{3r}{r_p} \tag{10.4}$$

The substitution of these dimensionless variable definitions into Eq. (10.3b) gives a dimensionless expression for the effectiveness factor in a spherical catalyst pellet.

$$\eta = \frac{1}{{\varphi_n}^2}\left(\frac{d\psi}{d\zeta}\right)_{\zeta = 3} \tag{10.5}$$

The dimensionless parameter φ_n in Eq. (10.5) is the shape-generalized Thiele modulus, which is defined for an nth-order reaction taking place in a spherical catalyst pellet, as follows:

$$\varphi_n = \frac{r_p}{3}\left(\frac{kC_A^{n-1}}{D_e}\right)^{1/2}; \text{ spherical pellet} \tag{10.6}$$

For a catalyst pellet having a general shape, the definition of Thiele modulus can be expressed as follows:

$$\varphi_n = \frac{V_p}{A_e}\left(\frac{kC_A^{n-1}}{D_e}\right)^{1/2} \tag{10.7}$$

For the cylindrical (infinite cylinder) and slab-shaped catalyst pellets, characteristic length V_p/A_e can be expressed in terms of the radius of the cylinder and half-thickness of the slab, as $r_{cy}/2$ and L_s, respectively.

$$\varphi_n = \frac{r_{cy}}{2}\left(\frac{kC_A^{n-1}}{D_e}\right)^{1/2}; \text{(cylindrical pellet with a radius of } r_{cy}) \tag{10.8}$$

$$\varphi_n = L_s\left(\frac{kC_A^{n-1}}{D_e}\right)^{1/2}; \text{(slab-shaped pellet with a half thickness of } L_s) \tag{10.9}$$

Thiele modulus is proportional to the ratio of diffusion time to reaction time within a catalyst pellet. In the case of a first-order reaction taking place in a spherical catalyst pellet, it can be expressed as

$$\varphi = \frac{r_p}{3}\left(\frac{k}{D_e}\right)^{1/2} = \frac{1}{3}\left(\frac{k}{D_e/r_p^2}\right)^{1/2} \propto \left(\frac{\text{Diffusion time}}{\text{Reaction time}}\right) \tag{10.10}$$

For large values of Thiele modulus, the reaction rate is much higher than the diffusion rate into the pellet (diffusion control), while for small values of Thiele modulus, the reaction rate is much lower than the diffusion rate (reaction control).

If $\varphi \gg 1$ (diffusion-control region) (10.11a)

If $\varphi \ll 1$ (reaction-control region) (10.11b)

If the value of Thiele modulus is much less than unity, the effectiveness factor is expected to approach 1. However, for Thiele modulus values higher than unity, the observed reaction rate is expected to be less than the ideal reaction rate, which is evaluated at the bulk fluid concentration and temperature values.

We need to know the concentration profile within the catalyst pellet to derive the effectiveness factor expression. Steady-state species conservation equation written for the differential volume element between r and $r + \Delta r$ (Figure 10.1) yields the differential pseudo-homogeneous species conservation equation for reactant A.

$$(4\pi r^2 N_{A_r})_r - (4\pi r^2 N_{A_r})_{r+\Delta r} + 4\pi r^2 \Delta r R_{A,p} = 0 \tag{10.12}$$

$$\lim_{\Delta r \to 0}\left(\frac{(r^2 N_{A_r})_r - (r^2 N_{A_r})_{r+\Delta r}}{\Delta r}\right) + r^2 R_{A,p} = 0 \tag{10.13}$$

$$-\frac{1}{r^2}\frac{d(r^2 N_{A_r})}{dr} + R_{A,p} = 0 \tag{10.14}$$

The radial flux of reactant A into the pellet can be expressed using Fick's law of diffusion:

$$N_{A_r} = -D_e \frac{dC_{A,p}}{dr} \tag{10.15}$$

Here, D_e is the effective diffusion coefficient within the porous catalyst pellet. The values and the prediction methods of the effective diffusion coefficient are discussed in Chapter 12. Using Fick's law flux expression for the diffusion of reactant A into the pellet, the species conservation equation for a first-order catalytic reaction becomes as follows:

$$\frac{D_e}{r^2}\left(\frac{d}{dr}\left(r^2 \frac{dC_{A_p}}{dr}\right)\right) - kC_{A_p} = 0$$

$$\frac{D_e}{r^2}\left(r^2 \frac{d^2 C_{A_p}}{dr^2} + 2r\frac{dC_{A_p}}{dr}\right) - kC_{A_p} = 0 \tag{10.16}$$

In this equation, D_e is the effective diffusivity in the porous pellet. The value of D_e is assumed as constant in this pseudo-homogeneous species conservation equation. Derivation of pseudo-homogeneous transport equations and predictive models of the effective diffusion coefficient are discussed in the Online Appendix O3 of this book and in Chapter 12, respectively.

The boundary conditions for Eq. (10.16) can be expressed as follows:

$$\text{At } r = 0; \quad \frac{dC_{A_p}}{dr} = 0 \quad \text{or} \quad C_{A_p} \to \text{finite} \tag{10.17}$$

$$\text{At } r = r_p; \quad C_{A_p}\big|_{r=r_p} = C_A \tag{10.18}$$

In writing the second boundary condition (Eq. (10.18)), film mass transfer resistance over the external surface of the catalyst pellet is neglected. Hence, the external surface concentration is considered equal to the bulk concentration of reactant A. Solution of Eq. (10.16) with these boundary conditions can be achieved by making a substitution as $u = rC_A$. With this substitution, Eq. (10.16) becomes

$$\frac{d^2u}{dr^2} - \frac{k}{D_e}u = 0 \tag{10.19}$$

Using the dimensionless radial coordinate definition given in Eq. (10.4), Eq. (10.19) reduces to the following expression:

$$\frac{d^2u}{d\zeta^2} - \varphi^2 u = 0 \tag{10.20}$$

Equation (10.20) is a second-order homogeneous differential equation with constant coefficients. The solution of this equation can be expressed as follows:

$$u = C_1 \sinh(\varphi\zeta) + C_2 \cosh(\varphi\zeta) \tag{10.21}$$

Integration constants C_1 and C_2 can be evaluated using the boundary conditions expressed in Eqs. (10.17) and (10.18). The concentration profile expression within the spherical catalyst pellet can then be obtained by the solution of the dimensionless species conservation equation, as follows:

$$\psi = \frac{C_{A_p}}{C_A} = \frac{3\sinh(\varphi\zeta)}{\zeta \sinh(3\varphi)} \tag{10.22}$$

This expression shows that concentration variation in the catalyst pellet is a strong function of Thiele modulus. The effectiveness factor expression for a first-order reaction is then derived by substituting Eq. (10.22) into the definition of effectiveness factor (Eq. (10.5)). The effectiveness factor expression for a first-order reaction in a spherical pellet can then be obtained as

$$\eta = \frac{1}{\varphi}\left(\coth(3\varphi) - \frac{1}{3\varphi}\right) \tag{10.23}$$

The effectiveness factor is a function of Thiele modulus. As shown in Figure 10.2, effectiveness factor approaches 1.0 in the reaction-control region (for $\varphi \ll 1$). However, the limit of the effectiveness factor at large values of Thiele modulus approaches

$$\eta \to \frac{1}{\varphi} \quad \text{for } \varphi \gg 1 \tag{10.24}$$

Derivation of effectiveness factor expression for other geometries yields similar conclusions. The shape-generalized pseudo-homogeneous species conservation equation within a catalyst pellet can be written as follows [2]:

$$\frac{1}{x^a}\frac{d}{dx}\left(x^a D_e \frac{dC_{A_p}}{dx}\right) + R_{A,p} = 0 \tag{10.25}$$

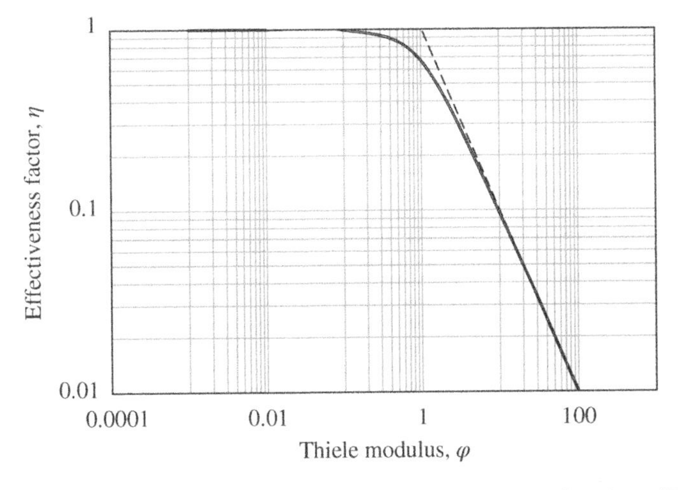

Figure 10.2 The variation of the effectiveness factor as a function of Thiele modulus for a first-order reaction in a spherical catalyst pellet.

$$\text{Slab geometry: } a = 0$$

$$\text{Cylindrical geometry: } a = 1; \quad x \to r$$

$$\text{Spherical geometry: } a = 2; \quad x \to r$$

For a first-order reaction, concentration profiles within the slab and the cylindrical catalyst pellets can then be obtained by solving Eq. (10.25).

$$\psi = \frac{C_{A_p}}{C_A} = \frac{\cosh(\varphi\zeta)}{\cosh\varphi}; \quad \varphi = L\left(\frac{k}{D_e}\right)^{1/2} \quad \text{slab geometry} \tag{10.26}$$

$$\psi = \frac{C_{A_p}}{C_A} = \frac{I_0(\varphi\zeta)}{I_0(2\varphi)}; \quad \varphi = \frac{r_{cy}}{2}\left(\frac{k}{D_e}\right)^{1/2} \quad \text{cylindrical geometry} \tag{10.27}$$

Here, L is the half-thickness of the slab-shaped catalyst pellet. In the case of cylindrical pellets, it is assumed that the length to diameter ratio of the pellet was sufficiently high so that the transport of reactants to the pellet is mainly from the cylindrical wall.

The effectiveness factor expressions for the slab and the cylindrical pellet geometries can then be derived, as expressed in Eqs. (10.28) and (10.29), respectively.

$$\eta = \frac{\tanh(\varphi)}{\varphi}; \text{slab geometry} \tag{10.28}$$

$$\eta = \frac{I_1(2\varphi)}{\varphi I_o(2\varphi)}; \text{cylindrical geometry} \tag{10.29}$$

Here, I_o and I_1 correspond to the Bessel functions. As shown in Figure 10.3, the variation of effectiveness factor with Thiele modulus is quite similar for the slab, cylindrical, and spherical geometries. For all geometries, the limiting forms of effectiveness factor are as follows [2, 5]:

$$\eta \to 1 \quad \text{for} \quad \varphi \ll 1 \tag{10.30a}$$

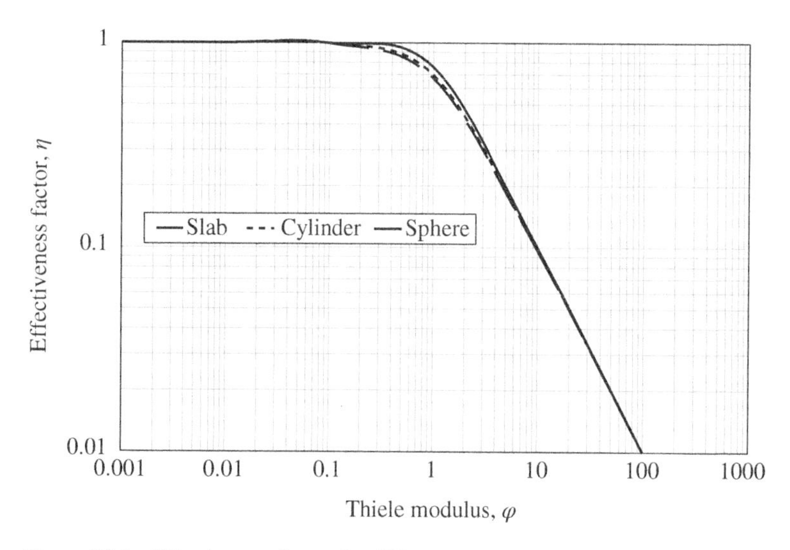

Figure 10.3 Effectiveness factor for different geometries (first-order reaction).

$$\eta \rightarrow \frac{1}{\varphi} \quad \text{for} \quad \varphi \gg 1 \tag{10.30b}$$

An order and shape-generalized Thiele modulus can also be defined as

$$\hat{\varphi}_n = \left(\frac{n+1}{2}\right)^{1/2} \frac{V_p}{A_e} \left(\frac{kC_A^{n-1}}{D_e}\right)^{1/2} \tag{10.31}$$

Analysis of the effectiveness factor expressions for any positive nth -order reaction yields the following limits in terms of the order and the shape-generalized Thiele modulus:

$$\eta \rightarrow 1 \quad \text{for} \quad \hat{\varphi}_n \ll 1 \tag{10.32a}$$

$$\eta \rightarrow \frac{1}{\hat{\varphi}_n} \quad \text{for} \quad \hat{\varphi}_n \gg 1 \tag{10.32b}$$

For a number of catalytic reactions, Langmuir–Hinshelwood (L-H) type rate expressions describe the reaction rates (Chapter 11). The intrinsic rate expression for a simple mono-molecular L-H type catalytic reaction may be expressed in the following form:

$$R_p = \frac{kC_A}{(1 + KC_A)^2} \tag{10.33}$$

For a catalytic reaction having a rate expression given in Eq. (10.33), the rate of the reaction approaches a negative first-order reaction for very large values of the adsorption term (KC_A).

$$\lim_{KC_A \gg 1} R_p = \left(\frac{k}{K^2}\right) C_A^{-1} \tag{10.34}$$

For such a case, a decrease in the concentration of reactant A within the catalyst pellet will increase the reaction rate. Hence, pore-diffusion resistance, causing a lower concentration of the limiting reactant within the catalyst pellet than its external surface concentration, may increase the observed rate at specific concentration ranges for this kind of a rate expression. Hence, the effectiveness factor values higher than 1 may be observed under certain conditions for certain rate expressions. However, in the case of power-law type rate expressions with positive orders, pore-diffusion resistance is expected to decrease the observed reaction rate compared to the ideal reaction rate.

10.2 Observed Activation Energy and Observed Reaction Order

The observed rate of a catalytic reaction can be estimated by multiplying the ideal reaction rate with the effectiveness factor. In the case of pore-diffusion-control region (for $\hat{\varphi}_n \gg 1$), observed reaction rate expression for an nth -order reaction can be written as

$$\left(R_{A,p}\right)_{obs} = \eta k C_A^n = \frac{1}{\hat{\varphi}_n} k C_A^n \simeq \frac{A_e}{V_p} \left(\frac{2kD_e}{n+1}\right)^{1/2} C_A^{(n+1)/2} \tag{10.35}$$

As shown in Eq. (10.35), the observed reaction order of an nth-order reaction is $(n + 1)/2$ in the diffusion-control region. For a first-order reaction, the observed rate is also first order. However, for higher reaction orders, the observed order is less than the actual reaction order. For instance, for a second-order reaction, the observed reaction order becomes 3/2 in the diffusion-control region.

Another important conclusion that we get from Eq. (10.35) is the temperature dependence of the observed reaction rate constant. Incorporation of the Arrhenius expression for the reaction rate constant into Eq. (10.35) yields the temperature dependence of the observed reaction rate constant. For a spherical catalyst pellet, this relation becomes as follows:

$$\left(R_{A,p}\right)_{obs} = \frac{3}{r_p}\left(\frac{2k_oD_e}{n+1}\right)^{1/2} \exp\left(-\frac{E_a}{2R_gT}\right)C_A^{\frac{(n+1)}{2}} = k_{o_{obs}} \exp\left(-\frac{E_{a_{obs}}}{R_gT}\right)C_A^{n_{obs}} \quad (10.36)$$

As shown in Eq. (10.36), the observed activation energy of a diffusion-controlled catalytic reaction is half of its actual activation energy.

$$E_{a_{obs}} = \frac{E_a}{2}; \text{ for the diffusion-control region} \quad (10.37)$$

This is illustrated in Figure 10.4 for the temperature dependence of the observed rate constant of a catalytic reaction. According to the Arrhenius law, the slope of the ln k vs. $1/T$ curve gives the activation energy $\left[\text{slope} = -\left(E_{a_{obs}}/R_g\right)\right]$. As the temperature increases, the reaction rate constant, and hence the Thiele modulus also increases. If the increase in Thiele modulus with an increase in temperature is quite high, the reaction may pass to the diffusion-control region. Hence, a decrease in the observed activation energy may be observed. For Thiele modulus values being much higher than 1, observed activation energy becomes half of the actual activation energy, causing a decrease in the slope of ln(k) vs. $1/T$ curve at higher temperatures. The ratio of the observed and the actual activation energy values can be related to the Thiele modulus as follows [2, 8]:

$$\frac{E_{a_{obs}}}{E_a} = 1 + \frac{1}{2}\frac{d\ln(\eta)}{d\ln(\varphi)}$$

Evaluation of the value of the Thiele modulus requires information for the value of the ideal reaction rate constant and actual reaction order. However, these may not be available in many cases. Instead, we may have information about the observed reaction rate. Hence, another dimensionless group, namely observable modulus, is defined as follows:

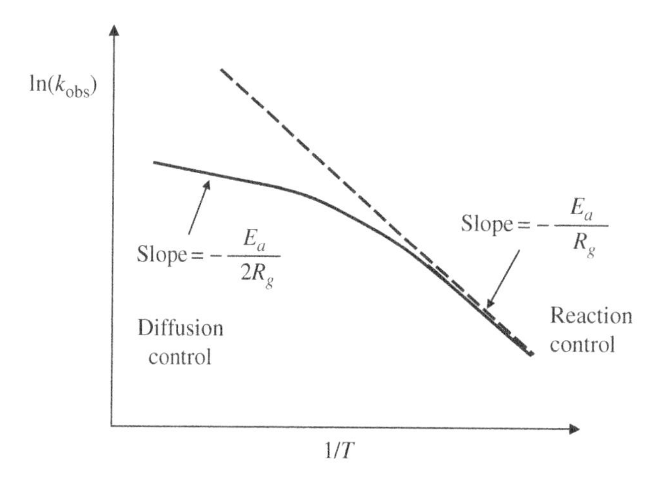

Figure 10.4 Effect of pore-diffusion resistance on the observed activation energy.

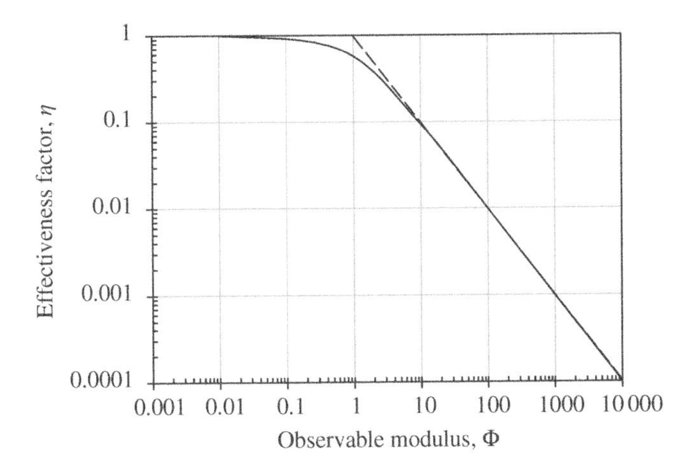

Figure 10.5 Variation of effectiveness factor as a function of observable modulus for a first-order reaction in a spherical catalyst pellet.

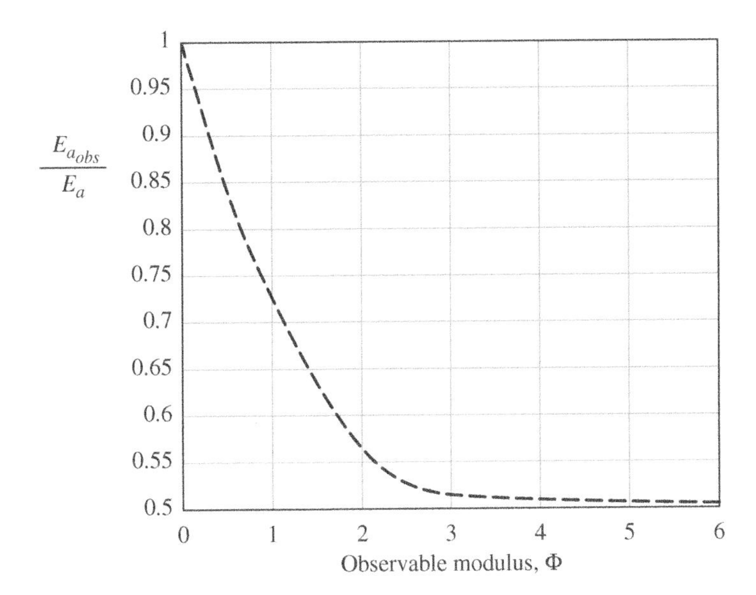

Figure 10.6 Effect of pore-diffusion resistance on observed activation energy.

$$\Phi = \eta\varphi^2 = \frac{\left(\frac{V_p}{A_e}\right)^2 R_{p_{\text{obs}}}}{D_e C_{A_s}}; \text{(observable modulus)} \tag{10.38}$$

Variation of effectiveness factor as a function of observable modulus for a spherical catalyst pellet is shown in Figure 10.5 for a first-order reaction.

The ratio of the observed and the actual activation energy values of a reaction can also be related to the observable modulus. The variation of $E_{a_{\text{obs}}}/E_a$ as a function of observable modulus is illustrated in Figure 10.6 for a first-order reaction in a slab-shaped catalyst pellet.

10.3 Effectiveness Factor in the Presence of Pore-Diffusion and Film Mass Transfer Resistances

In Section 10.1, only the effect of intrapellet diffusion resistance was considered in the prediction of the effectiveness factor and the observed reaction rate. In a catalytic reactor, mass transfer resistance between the bulk fluid phase and the external surface of the porous catalyst may also have some contribution to the observed rate [1]. Mass and heat transfer resistances through the boundary layer developed over the catalyst pellet external surface may cause differences between the bulk and the surface concentration and temperature values. In the case of a significant concentration gradient across this boundary layer, the boundary condition at $r = r_p$ given by Eq. (10.18) should be replaced by Eq. (10.39).

$$k_m\left(C_A - C_{A_p}\big|_{r=r_p}\right) = D_e \frac{dC_{A_p}}{dr}\bigg|_{r=r_p} \tag{10.39}$$

The dimensionless form of this equation can be written as follows:

$$Bi_m\left(1 - \psi|_{\zeta=3}\right) = \frac{d\psi}{d\zeta}\bigg|_{\zeta=3} \tag{10.40}$$

Here, Bi_m is the Biot number for mass transfer in spherical geometry. The definition of Biot number for mass transfer is given by Eq. (10.41), using the characteristic length as $(r_p/3)$ for spherical geometry.

$$Bi_m = \frac{k_m r_p}{3D_e} \tag{10.41}$$

Definition of mass transfer Biot number for a general catalyst shape can be expressed as follows:

$$Bi_m = \frac{k_m\left(\frac{V_p}{A_e}\right)}{D_e} \tag{10.42}$$

Physically, the Biot number is proportional to the ratio of pore diffusion to external mass transfer resistances. In the case of large Biot number values, external mass transfer resistance becomes negligible, and Eq. (10.18) may be used as the boundary condition at $r = r_p$ in the derivation of the effectiveness factor expression.

The solution of the pseudo-homogeneous species conservation equation (Eq. (10.16)), using the boundary conditions given in Eqs. (10.17) and (10.40), the effectiveness factor expression can be derived for the case of having both pore-diffusion resistance as well as external mass transfer resistance. The effectiveness factor expression for a first-order reaction taking place in a spherical catalyst pellet can then be derived for this case as

$$\eta = \frac{3\varphi - \tanh(3\varphi)}{3\varphi^2 \tanh(3\varphi) + Da(3\varphi - \tanh(3\varphi))} \tag{10.43}$$

In this equation Da is the Damköhler number, which is defined as the ratio of the square of Thiele modulus to Biot number. For the case of a first-order reaction in a spherical catalyst pellet, its definition is as given in Eq. (10.44).

$$Da = \frac{\varphi^2}{Bi_m} = \frac{r_p}{3}\left(\frac{k}{k_m}\right) \tag{10.44}$$

Depending upon the geometry of the catalyst pellet and the reaction order, some modifications should be made in the definition of Damköhler number. Damköhler number is proportional to the ratio of reaction rate to external mass transfer rate. The magnitude of the Damköhler number gives information about the significance of external mass transfer resistance as compared to the reaction resistance. Its large values indicate that film mass transfer resistance is quite significant, and the effect of external mass transfer resistance on the observed rate should not be neglected.

The concentration profiles within the pellets of different geometries can be obtained by solving the species conservation equations, with the boundary conditions taking the film mass transfer resistance at the external surface of the catalyst pellet into account. The corresponding effectiveness factor expressions for the cylindrical and slab geometries are given in Eqs. (10.45) and (10.46), respectively, for the case of having both pore-diffusion and external mass transfer resistances.

$$\eta = \frac{I_1(2\varphi)}{[\varphi I_o(2\varphi) + Da\, I_1(2\varphi)]} \text{ (cylindrical geometry)} \tag{10.45}$$

$$\eta = \frac{\sinh(\varphi)}{\varphi \cosh(\varphi) + Da \sinh(\varphi)} \text{ (slab geometry)} \tag{10.46}$$

Definitions of Biot and Damköhler numbers for the infinite cylinder and the slab geometries are as follows:

$$Da = \frac{r_p}{2}\left(\frac{k}{k_m}\right); Bi_m = \frac{r_p}{2}\left(\frac{k_m}{D_e}\right) \text{ (infinite cylinder)} \tag{10.47}$$

$$Da = L\left(\frac{k}{k_m}\right); Bi_m = L\left(\frac{k_m}{D_e}\right) \text{ (slab geometry)} \tag{10.48}$$

The variation of the effectiveness factor with Thiele modulus for different Damköhler number values is illustrated in Figure 10.7 for spherical geometry. For large values of Damköhler number, the limit of effectiveness factor for small values of Thiele modulus approaches to

$$\lim_{Da \to \infty} \eta = \frac{1}{Da} \tag{10.49}$$

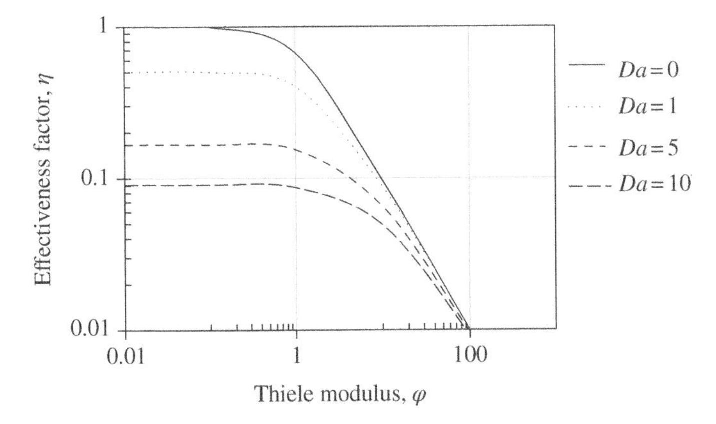

Figure 10.7 Effectiveness factor for various Damköhler numbers for a first-order reaction (spherical catalyst pellet). *Source:* Dogu [2], Reproduced with permission of Taylor & Francis.

instead of approaching 1. This is due to the significant resistance of external mass transfer for large values of the Damköhler number.

10.4 Thermal Effects in Porous Catalyst Pellets

In addition to concentration gradients, significant temperature gradients may also develop within porous catalyst pellets due to the heat effects of surface-catalyzed reactions. A variation of temperature could arise across the pellet, especially for reactions with a high heat of reaction and low thermal conductivity of the catalyst pellet. For highly exothermic reactions, the temperature at the center of the catalyst pellet may reach values higher than the bulk temperature of the fluid surrounding the pellet. On the other hand, some decrease in temperature from the external surface to the center of the catalyst pellet may be expected for highly endothermic reactions. Analysis of this behavior and prediction of the observed reaction rate in a nonisothermal catalyst pellet require the simultaneous solution of the species conservation equation and the energy balance equation within the catalyst pellet.

For a spherical catalyst pellet, steady-state differential energy balance equation over the volume element between r and $r + \Delta r$ can be written as follows:

$$\left(4\pi r^2 e_r\right)\big|_r - \left(4\pi r^2 e_r\right)\big|_{r+\Delta r} = 0 \tag{10.50}$$

Here, e_r is the energy flux within the pellet in the r direction. The following differential equation is obtained by dividing each term of this equation with $4\pi r^2\Delta r$ and taking the limit as Δr going to zero:

$$\frac{1}{r^2}\frac{d(r^2 e_r)}{dr} = 0 \tag{10.51}$$

Energy flux through the catalyst pellet can be expressed as a summation of the heat conduction term and the energy flux term due to the molecular transport of species. While the species diffuse in the radial direction within a spherical catalyst pellet, they also carry some enthalpy.

$$e_r = -\lambda_e \frac{dT_p}{dr} + \sum_i (N_{i_r}\hat{H}_i) \tag{10.52}$$

Here, λ_e is the effective thermal conductivity of the pellet and $\hat{H}_i$ is the molar enthalpy of species i. Substitution of Eq. (10.52) into Eq. (10.51) gives the following equation:

$$-\frac{\lambda_e}{r^2}\frac{d}{dr}\left(r^2\frac{dT_p}{dr}\right) + \frac{1}{r^2}\left[\sum_i \hat{H}_i \frac{d}{dr}(r^2 N_{i_r}) + \sum_i r^2 N_{i_r}\frac{d\hat{H}_i}{dr}\right] = 0 \tag{10.53}$$

Combining this equation with the species conservation equation (Eq. (10.14)) and expressing $\frac{d\hat{H}_i}{dr}$ as $C_{p_i}\frac{dT_p}{dr}$, Eq. (10.53) reduces to

$$-\frac{\lambda_e}{r^2}\frac{d}{dr}\left(r^2\frac{dT_p}{dr}\right) + \sum_i R_{i,p}\hat{H}_i + \sum_i \left(N_{i_r} C_{p_i}\frac{dT_p}{dr}\right) = 0 \tag{10.54}$$

In this equation, the last term corresponds to the transport of sensible enthalpy due to diffusing molecules within the pellet. For many cases, this last term is negligibly small as compared to the heat conduction term through the pellet (first term of Eq. (10.54)). Neglecting the last term of Eq. (10.54) and considering the relations between the rates of different species in terms of their stoichiometric coefficients

$$\frac{R_{i,p}}{\nu_i} = \frac{R_{j,p}}{\nu_j} = \cdots = R_p \tag{10.55}$$

and noting that $\sum_i (\nu_i \hat{H}_i)$ is the heat of reaction (ΔH^o), energy balance for the pellet becomes

$$\frac{\lambda_e}{r^2}\frac{d}{dr}\left(r^2\frac{dT_p}{dr}\right) - R_p \Delta H^o = 0 \tag{10.56}$$

Here, R_p is the intrinsic rate of the reaction based on catalyst pellet volume. The form of Eq. (10.56) is quite similar to the form of the corresponding species conservation equation.

$$\frac{D_e}{r^2}\frac{d}{dr}\left(r^2\frac{dC_{A_p}}{dr}\right) - R_p = 0 \tag{10.57}$$

Since the intrinsic reaction rate R_p is a function of both temperature and concentration within the catalyst pellet, evaluation of concentration and temperature profiles within the catalyst pellet requires the simultaneous solution of Eqs. (10.56) and (10.57). The coupling term in these equations is R_p. Elimination of the rate term from these equations gives

$$\frac{\lambda_e}{\Delta H^o r^2}\frac{d}{dr}\left(r^2\frac{dT_p}{dr}\right) = \frac{D_e}{r^2}\frac{d}{dr}\left(r^2\frac{dC_{A_p}}{dr}\right) \tag{10.58}$$

Considering the symmetry condition at the center of the pellet

$$\frac{dC_{A_p}}{dr} = \frac{dT_p}{dr} = 0 \quad \text{at } r = 0 \tag{10.59}$$

and taking the temperature and the concentration at the external surface of the pellet as

$$T_p = T_s \quad \text{and} \quad C_{A_p} = C_{A_s} \quad \text{at } r = r_p \tag{10.60}$$

Equation (10.58) can be integrated. A relation between the temperature and the concentration at any point within the catalyst pellet is then obtained by the integration of this equation with the boundary conditions given by Eqs. (10.59) and (10.60).

$$\left(T_p - T_s\right) = \frac{D_e(-\Delta H^o)}{\lambda_e}\left(C_{A_s} - C_{A_p}\right) \tag{10.61}$$

In the absence of the external film mass and heat transfer resistances, T_s and C_{A_s} are equal to the bulk fluid-phase temperature (T) and bulk concentration of limiting reactant A (C_A) in the reactor.

An expression for the maximum possible temperature difference across the pellet can be obtained by taking the concentration at the center point of the catalyst pellet zero. This case corresponds to the complete conversion of reactant A. Hence, the maximum temperature rise within a catalyst pellet can be predicted by taking $C_{A_p} = 0$ in Eq. (10.61).

$$\frac{\Delta T_{\max}}{T_s} = \frac{\left(-\Delta H_R^o\right) D_e C_{A_s}}{\lambda_e T_s} = \beta \tag{10.62}$$

The order of magnitude of the dimensionless Prater group β, which appears on the right-hand side of Eq. (10.62), gives an idea about the relative importance of heat generation rate within the pellet compared to the heat removal rate by conduction through the pellet [2, 9, 10].

Effective thermal conductivity λ_e, which appears in Eq. (10.62), depends strongly upon the pore structure of the catalyst pellet, as well as its chemical nature. The effective thermal conductivity values for many of the catalysts are in the range of 0.15–0.35 J/s m K [2, 5]. For instance, its values were reported as 0.31 and 0.18 J/s m K for silica-alumina and alumina (boehmite) pellets, respectively [11]. Experimental values and prediction methods of the effective diffusion coefficient D_e are discussed in Chapter 12.

For a non-isothermal catalyst pellet, the effectiveness factor depends upon both concentration and temperature variations across the pellet. For a first-order reaction, the energy balance equation (Eq. (10.56)) and the species conservation equation (Eq. (10.57)) can be written in dimensionless form, as follows [2]:

$$\frac{1}{\zeta^2}\frac{d}{d\zeta}\left(\zeta^2\frac{d\psi}{d\zeta}\right) - \varphi^2\left(\frac{(k)_{T_p}}{(k)_{T_s}}\right)\psi = 0 \tag{10.63}$$

$$\frac{1}{\zeta^2}\frac{d}{d\zeta}\left(\zeta^2\frac{d\Theta}{d\xi}\right) + \beta\varphi^2\left(\frac{(k)_{T_p}}{(k)_{T_s}}\right)\psi = 0 \tag{10.64}$$

Here, the dimensionless parameter $\varphi^2 = \frac{r_p}{3}\left(\frac{k}{D_e}\right)$ should be evaluated at the external surface temperature of T_s. The ratio of reaction rate constants, which are evaluated at a temperature of T_p within the pellet, and the surface temperature T_s, can be expressed as follows:

$$\frac{(k)_{T_p}}{(k)_{T_s}} = \exp\left(-\frac{E_a}{R_g}\left(\frac{1}{T_p} - \frac{1}{T_s}\right)\right) \tag{10.65}$$

Hence, Eqs. (10.63) and (10.64) can be expressed in terms of the three dimensionless parameters φ, β, and γ, as follows:

$$\frac{1}{\zeta^2}\frac{d}{d\zeta}\left(\zeta^2\frac{d\psi}{d\zeta}\right) - \varphi^2\psi\exp\left[\gamma\left(\frac{\Theta-1}{\Theta}\right)\right] = 0 \tag{10.66}$$

$$\frac{1}{\zeta^2}\frac{d}{d\zeta}\left(\zeta^2\frac{d\Theta}{d\xi}\right) + \beta\varphi^2\psi\exp\left[\gamma\left(\frac{\Theta-1}{\Theta}\right)\right] = 0 \tag{10.67}$$

Here, ζ is defined by Eq. (10.4). In these equations, Θ is the dimensionless temperature

$$\Theta = T_p/T_s \tag{10.68}$$

The dimensionless Arrhenius parameter γ is defined as follows:

$$\gamma = \frac{E_a}{R_g T_s} \tag{10.69}$$

Physically speaking, this dimensionless group gives us information about the sensitivity of the reaction rate to temperature change within the catalyst pellet. As the magnitude of this group increases, the effect of temperature rise within the catalyst pellet on the observed reaction rate becomes more significant.

Prediction of the effectiveness factor in a nonisothermal pellet requires the simultaneous solution of Eqs. (10.66) and (10.67). By combining Eq. (10.66) with Eq. (10.61), the temperature in the species conservation equation can be eliminated, and the species conservation equation can be expressed in dimensionless form as follows:

$$\frac{1}{\zeta^2}\frac{d}{d\zeta}\left(\zeta^2\frac{d\psi}{d\zeta}\right) - \varphi^2\psi\exp\left[\gamma\frac{\beta(1-\psi)}{1+\beta(1-\psi)}\right] = 0 \tag{10.70}$$

In this equation, the only dependent variable is dimensionless concentration, and the solution of this equation is expected to be a function of three dimensionless parameters, namely Thiele modulus φ, Prater number β, and Arrhenius number γ. The effectiveness factor for a non-isothermal catalytic system can then be expressed in terms of these three dimensionless groups. The numerical solution of Eq. (10.70) can be used for the prediction of the effectiveness factor, using its definition given in Eq. (10.5).

As it would be expected, for exothermic reactions (β being positive), effectiveness factor values being higher than 1 are possible. As the value of the parameter β becomes large, temperature gradients within the catalyst pellet become more significant. The increase of temperature within the catalyst pellet may also increase the reaction rate constant, resulting in higher effectiveness factor values than 1. On the other hand, in the case of endothermic reactions, β values become negative, and an increase in the absolute value of this parameter causes a decrease in the value of the effectiveness factor.

Another significant factor for the effect of temperature rise within a catalyst pellet on the effectiveness factor is the Arrhenius number γ. The value of the Arrhenius number indicates the temperature sensitivity of the reaction rate. If the value of γ is very small, the temperature rise within the pellet will not have a significant influence on the observed reaction rate.

Prediction of effectiveness factor for a nonisothermal pellet is developed in the early publication of Weisz and Hicks [12]. For endothermic reactions, both diffusion and heat effects cause a decrease in the value of the effectiveness factor. However, for exothermic reactions, heat effects may cause higher reaction rate values within the catalyst pellet than the rate values evaluated at the bulk fluid temperature surrounding the pellet. Due to the non-linearity of the temperature dependence of the rate constant in the energy equation, multiple steady-state effectiveness factor values are also possible for large values of β ($\beta > 0.2$). For many of the industrially important reactions, the value of the parameter β is less than 0.1. Some typical values of β and γ are listed in Table 10.1. For small values of β (for $\beta(1-\psi) \ll 1$), Eq. (10.70) reduces to the following form:

$$\frac{1}{\zeta^2}\frac{d}{d\zeta}\left(\zeta^2\frac{d\psi}{d\zeta}\right) - \varphi^2\psi \exp\left[\gamma\beta(1-\psi)\right] = 0 \tag{10.71}$$

There are two dimensionless groups in this equation. These dimensionless groups are the Thiele modulus φ and $\gamma\beta$ (Eq. (10.72)). Hence, it is possible to predict the effectiveness factor, knowing the values of these two parameters. We can plot the effectiveness factor as a function of the observable modulus Φ (Eq. (10.38)) by taking $\gamma\beta$ as a parameter. Note that the values of observable modulus can be estimated from the observed reaction rate data.

Table 10.1 Dimensionless parameters for some exothermic reactions

Reaction	β	γ	$\beta\gamma$	φ_s, Φ_s	References
The reaction of hydrogen with oxygen to yield water	0.14	7.0	0.98	$\Phi_s = 4.4$	[2, 5, 13]
Oxidation of ethylene to ethylene oxide	0.13	13.4	1.76	$\varphi_s = 0.08$	[2, 14, 15]
Oxidative coupling of methane	6.5×10^{-3}	17.1	0.111	$\Phi_s = 0.12$	[16]
Dimethyl ether synthesis from synthesis gas	0.12	8.7	1.04	$\Phi_s = 2.5 \times 10^{-4}$	

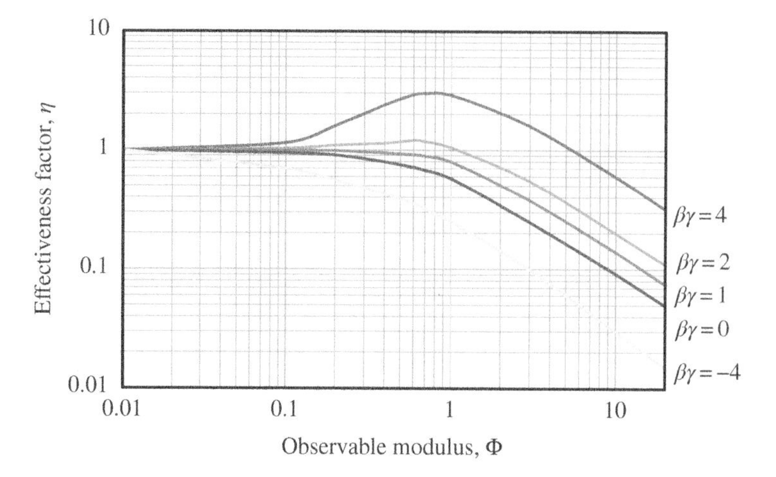

Figure 10.8 Non-isothermal effectiveness factor as a function of observable modulus, for a first-order reaction, for small values of β. *Source:* Dogu [2], Reproduced with permission of Taylor & Francis.

$$\gamma\beta = \frac{(-\Delta H_R^o)D_e C_{A_s} E_a}{\lambda_e R_g T_s^2} \tag{10.72}$$

The variation of effectiveness factor as a function of observable modulus ($\gamma\beta$ being a parameter) is illustrated in Figure 10.8, for a first-order reaction in a nonisothermal spherical catalyst pellet. As illustrated in Figure 10.8, the effect of temperature rise within a catalyst pellet on the observed rate of the reaction becomes relatively small for $|\beta\gamma| \ll 1.0$.

The product of β and γ gives a good indication of heat effects in catalytic reactions [1, 2]. Effectiveness factor values higher than 1 are possible for the $\beta\gamma$ values being higher than unity, especially for the observable modulus values higher than 1 (in the diffusion-control region). For the examples listed in Table 10.1, oxidation of ethylene to ethylene oxide has a $\beta\gamma$ value of about 1.76, which implies that the heat effect within the catalysts of this reaction is notable. However, the value of the Thiele modulus is given as 0.08 for this reaction [15]. This low value of Thiele modulus indicates negligible diffusion resistance on the observed rate and the effectiveness factor approaches 1 (Figure 10.8).

In the case of oxidative coupling of methane reported by Vandewalle et al. [16], the value of $\beta\gamma$ was about 0.111.

$$2CH_4 + O_2 \rightarrow C_2H_4 + 2H_2O$$

This result indicated that the effect of heat transfer resistance within an Sn/Li/MgO type catalyst having a pellet radius of 125 μm was small. Hence, the isothermal pellet assumption is acceptable in that study. The value of the observable modulus is about 0.12 for this reaction, indicating that the effect of pore-diffusion resistance on the observed rate of oxidative coupling of methane was also negligibly small for this catalyst. As discussed later in this chapter, the following criterion can be used for neglecting any heat effects on the observed rate within the catalyst pellets.

$$\Phi_s\gamma\beta = \left(\frac{r_p^2 R_{p_{obs}}}{D_e C_{A_s}}\right)\left(\frac{(-\Delta H_R^o)D_e C_{A_s} E_a}{\lambda_e R_g T_s^2}\right) < 0.75 \tag{10.73}$$

This criterion was first proposed by Andersen [17]. It was also called the Mears criterion [18]. For the oxidative coupling example, the left-hand side of Eq. (10.73) is about 0.013, supporting the conclusion of the negligible effect of temperature rise within the catalyst pellet on the observed rate.

In the case of oxidation of hydrogen to yield water with a Pt-Al_2O_3 catalyst having a diameter of 1.86 cm, the value of β was reported as 0.14 [5, 13]. According to Eq. (10.62), the value of β is equal to the ratio of maximum possible temperature rise in the catalyst pellet to the surface temperature. Since the surface temperature was 363 K for this reaction, the maximum possible temperature rise within this catalyst pellet was estimated as about 50 K. Observable modulus for this reaction is given as about 4.4, indicating strong effect of pore-diffusion resistance on the observed rate. As indicated in Figure 10.8, the effectiveness factor values for this non-isothermal pellet are expected to be somewhat higher than the corresponding values for an isothermal pellet.

Direct synthesis of dimethyl ether from synthesis gas is a highly exothermic reaction ($-\Delta H^o$ = 246 000 J/mol). This reaction was studied at 50 bar and 548 K with a reactor feed stream containing an equimolar mixture of carbon monoxide and hydrogen [19]. For a heteropolyacid impregnated Cu–Zn–Alumina catalyst of 15 μm in diameter, the values of the parameters β, γ, and the observable modulus Φ_s were estimated as 0.12, 8.7, and 2.5×10^{-4}, respectively. The very small value of the observable modulus is due to the very small size of the catalyst pellet and shows that the pore-diffusion effect on the observed rate is negligible. Even though the value of $\beta\gamma$ is about 1.0, the product of observable modulus with $\beta\gamma$ (Andersen criterion) is about 2.6×10^{-4}, indicating negligible heat effects on the observed rate.

10.5 Interphase and Intrapellet Temperature Gradients for Catalyst Pellets

The effectiveness factor is expected to depend upon the heat and mass transfer resistances at the external surface of a catalyst pellet, in addition to intrapellet diffusion and heat conduction limitations within the catalyst [1]. For many of the catalytic reactions, external heat transfer resistance may be more significant than the heat conduction resistance within the catalyst pellet. Hence, isothermal catalyst pellet assumption with a temperature gradient in the boundary layer on the external surface of the pellet is made in some cases. On the other hand, pore-diffusion resistance is usually more significant than the external mass transfer resistance.

For a system with significant temperature and concentration gradients across the boundary layer at the external surface of the catalyst pellet, the boundary conditions for the energy balance (Eq. (10.56)) and species conservation equations (Eq. (10.57)) should be written as follows:

$$h\left(T - T_p\big|_{r=r_p}\right) = \lambda_e \frac{dT_p}{dr}\bigg|_{r=r_p} \tag{10.74}$$

$$k_m\left(C_A - C_{A_p}\big|_{r=r_p}\right) = D_e \frac{dC_{A_p}}{dr}\bigg|_{r=r_p} \tag{10.75}$$

For this case, the relation between the temperature (T_p) and concentration $\left(C_{A_p}\right)$ at any point within the catalyst pellet can then be obtained in terms of bulk temperature (T) and bulk concentration (C_A) by combining Eqs. (10.74) and (10.75) with Eq. (10.61) [2]. Note that $T_p\big|_{r=r_p} = T_s$ and $C_{A_p}\big|_{r=r_p} = C_{A_s}$.

$$\frac{(T_p - T)}{T} = \left(\frac{D_e(-\Delta H_R^o)C_A}{\lambda_e T}\right)\left[\left(\frac{C_A - C_{A_p}}{C_A}\right) - \left(\frac{D_e}{k_m} - \frac{\lambda_e}{h}\right)\left(\frac{1}{C_A}\right)\frac{dC_{A_p}}{dr}\bigg|_{r=r_p}\right] \quad (10.76)$$

Noting that

$$V_p R_{p_{obs}} = A_e D_e \frac{dC_{A_p}}{dr}\bigg|_{r=r_p} \quad (10.77)$$

the following dimensionless relation can be written between the temperature and concentration at any point within the pellet:

$$\frac{(T_p - T)}{T} = \beta\left[\left(\frac{C_A - C_{A_p}}{C_A}\right) + \Phi\left(\frac{Bi_m - Bi_h}{Bi_m Bi_h}\right)\right] \quad (10.78)$$

Here, dimensionless parameters for heat transfer Biot number (Bi_h) and the observable modulus Φ are defined by Eqs. (10.79) and (10.80), respectively.

$$Bi_h = \frac{h\left(\frac{V_p}{A_e}\right)}{\lambda_e} \quad (10.79)$$

$$\Phi = \frac{\left(\frac{V_p}{A_e}\right)^2 R_{p_{obs}}}{D_e C_A} \quad (10.80)$$

The maximum temperature difference between the center of the pellet and the surface temperature can then be expressed for this case, as follows [2]:

$$\frac{\Delta T_{max}}{T} = \beta\left[1 + \Phi\left(\frac{Bi_m - Bi_h}{Bi_m Bi_h}\right)\right] \quad (10.81)$$

In the case of negligible film heat transfer resistance, Eq. (10.81) reduces to

$$\frac{\Delta T_{max}}{T} = \beta\left[1 - \Phi\left(\frac{1}{Bi_m}\right)\right] \quad (10.82)$$

However, if film mass transfer resistance is negligible, but film heat transfer resistance is significant, Eq. (10.81) reduces to

$$\frac{\Delta T_{max}}{T} = \beta\left[1 + \Phi\left(\frac{1}{Bi_h}\right)\right] = \beta + \beta\Phi\left(\frac{1}{Bi_h}\right) \quad (10.83)$$

Noting that β corresponds to the ratio of maximum temperature rise to the bulk temperature in the absence of any film transport resistances, the second term on the right-hand side of Eq. (10.83) corresponds to the temperature gradient within the external heat transfer film. This term can be expressed as follows:

$$\beta\Phi\left(\frac{1}{Bi_h}\right) = \frac{(-\Delta H^o)R_{p_{obs}}\left(\frac{V_p}{A_e}\right)}{hT_s} \quad (10.84)$$

Physically speaking, this dimensionless group is proportional to the ratio of heat generation rate due to reaction within the catalyst pellet to the heat removal rate from the external surface of the pellet by convection.

10.6 Pore Structure Optimization and Effectiveness Factor Analysis for Catalysts with Bi-modal Pore-Size Distributions

The pore structure of a catalyst is one of the most important factors affecting its catalytic performance. Catalysts can be categorized as microporous, mesoporous, or macroporous (Table 10.2). The order of magnitude of effective diffusion coefficient values is also reported in Table 10.2.

Microporous materials, like molecular sieves, have quite high surface area values, which promote activity. However, pore diameters of these materials are very small, causing very high diffusion resistance for the transport of reactant molecules to the active sites on the surfaces of the pores within the catalyst. On the other hand, macroporous materials have quite low surface area values. Pore-diffusion resistance for the transport of molecules to the active sites is quite low in macroporous materials. Mesoporous materials with ordered pore structures, like MCM-41 and SBA-15, also have the advantage of not having high diffusion resistance with a reasonably high surface area. Optimization of the meso–macropore network of hierarchically structured zeolite catalysts and transport effects in such catalytic materials are critically discussed by Coppens and coworkers [20, 21].

Solid catalysts having bidisperse pore-size distributions are used in many of the industrial processes. Catalysts with bi-modal pore-size distributions contain both macropores/mesopores and micropores. Most supported alumina catalyst pellets, ion-exchange resin catalysts, pelletized zeolite catalysts are some examples of catalysts having bidisperse pore-size distributions. Such catalyst pellets can be visualized as composed of agglomerated microporous particles, while the pores between these agglomerated microporous particles are called macropores/mesopores (Figure 10.9). In the case of macroreticular ion-exchange resin catalysts, gel-like micrograins contain most of the acidic sites.

Table 10.2 Pore structures of catalysts.

	Microporous	Mesoporous	Macroporous
Pore diameter (nm)	$d_{pore} < 2$	$2 < d_{pore} < 50$	$d_{pore} > 50$
Surface area (m^2/g)	$S_g \geq 10^3$	$10^2 \leq S_g \leq 10^3$	$S_g \leq 10^2$
Effective diffusivity (m^2/s)	10^{-8}–10^{-14}	10^{-7}–10^{-8}	10^{-5}–10^{-6}

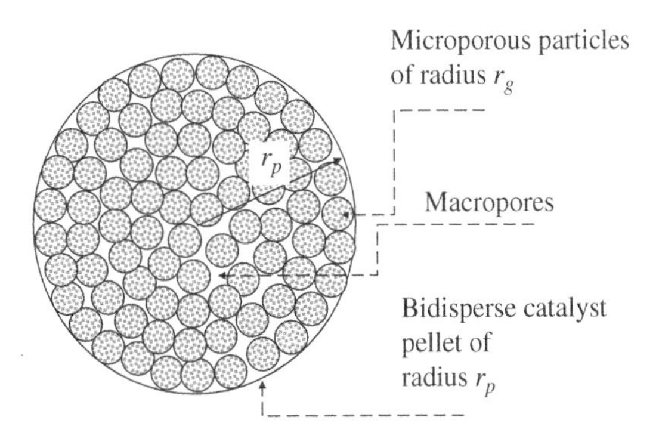

Figure 10.9 Schematic diagram of a catalyst pellet with a bidisperse pore-size distribution.

Catalysts with bidisperse pore structures have the advantage of better utilization of the active sites. Most of the active sites lie within the microporous particles, while macropores contribute to the transport of reactants into the catalyst pellet. Diffusion resistance for the transport of reactants into the bidisperse catalyst pellet is much less than the catalyst pellets containing only micropores.

One approach for analyzing diffusion and reaction processes in a catalyst with a bidisperse pore-size distribution is to consider the structure described in Figure 10.9. In this model, microporous particles or gel-like micrograins are considered agglomerated to form the catalyst pellet. The ratio of catalyst pellet radius to particle radius is higher than 10^2 in most cases [22]. This pellet–particle model can be used to derive the effectiveness factor expression for a bidisperse catalyst pellet [23]. This model assumes that the catalytic reaction takes place mainly within the microporous particles. Pseudo-homogeneous species conservation equation for an nth -order catalytic reaction, in spherical microporous particles (micrograins), can be written as follows:

$$\frac{D_i}{r_i^2}\frac{d}{dr_i}\left(r_i^2\frac{dC_{A_i}}{dr_i}\right) - kC_{A_i}^n = 0 \qquad (10.85)$$

Here, D_i, C_{A_i}, and r_i are the effective diffusivity, reactant concentration, and the radial coordinate within the microporous particles, respectively.

Species conservation equation for the catalyst pellet should also be expressed to model reaction and diffusion in a catalyst with a bi-modal pore-size distribution.

$$\frac{D_a}{r^2}\frac{d}{dr}\left(r^2\frac{dC_{A_a}}{dr}\right) - \frac{3}{r_g}(1-\varepsilon_a)D_i\frac{dC_{A_i}}{dr_i}\bigg|_{r_i = r_g} = 0 \qquad (10.86)$$

In this equation, D_a, C_{A_a} are the effective diffusion coefficient and reactant concentration in the macropores, respectively; ε_a is the macro-porosity of the pellet; r is the radial direction in the catalyst pellet; and r_g is the radius of the microporous grains. The second term of Eq. (10.86) corresponds to the input rate of reactant A from the macropores into the microporous particles per unit volume of the pellet. Simultaneous solution of Eqs. (10.85) and (10.86) can be obtained using the following boundary conditions.

$$\text{Symmetry conditions: } \frac{dC_{A_i}}{dr_i}\bigg|_{r_i = 0} = 0; \quad \frac{dC_{A_a}}{dr}\bigg|_{r = 0} = 0 \qquad (10.87)$$

$$\text{At } r_i = r_g; \quad C_{A_i} = C_{A_a} \qquad (10.88)$$

$$\text{At } r = r_p; \quad C_{A_a} = C_A \qquad (10.89)$$

Here, C_A is the concentration of reactant A at the external surface of the catalyst pellet, which is assumed to be the same as its bulk concentration within the reactor. The possible contribution of the mass transfer resistance over the catalyst external surface is neglected in this derivation. The following dimensionless variables can be defined to dimensionalize Eqs. (10.85) and (10.86):

$$\psi_i = (C_{A_i}/C_A) \qquad (10.90a)$$

$$\psi_a = (C_{A_a}/C_A) \qquad (10.90b)$$

$$\zeta = (r/r_p) \qquad (10.90c)$$

$$\zeta_i = (r_i/r_g) \qquad (10.90d)$$

Dimensionless forms of the species conservation equations for the micrograins and the macroporous pellet can be expressed for a first-order reaction, as follows:

$$\frac{1}{\zeta_i^2}\frac{d}{d\zeta_i}\left(\frac{\zeta_i^2 d\psi_i}{d\zeta_i}\right) - \varphi_i^2\psi_i = 0 \tag{10.91a}$$

$$\frac{1}{\zeta^2}\frac{d}{d\zeta}\left(\zeta^2\frac{d\psi_a}{d\zeta}\right) - \alpha\left.\frac{d\psi_i}{d\varsigma}\right|_{\varsigma=1} = 0 \tag{10.91b}$$

Here, particle Thiele modulus φ_i, and the dimensionless parameter α, which characterizes the ratio of diffusion times in the macro- and micropore regions, are defined as follows:

$$\varphi_i = r_g\left(\frac{k}{D_i}\right)^{1/2} \tag{10.92}$$

$$\alpha = 3(1-\varepsilon_a)\frac{D_i}{D_a}\left(\frac{r_p}{r_g}\right)^2 \tag{10.93}$$

The solution of this set of equations for a first-order reaction gives the following relations for the dimensionless concentrations within the micrograins and the catalyst pellet [23]:

$$\psi_i = \frac{1}{\zeta_i}\left(\frac{\psi_a}{\sinh(\varphi_i)}\right)\sinh(\varphi_i\zeta_i) \tag{10.94}$$

$$\psi_a = \frac{1}{\zeta\sinh\left[\alpha\left(\frac{\varphi_i}{\tanh\varphi_i}-1\right)\right]^{1/2}}\sinh\left[\left(\alpha\left(\frac{\varphi_i}{\tanh\varphi_i}-1\right)\right)^{1/2}\zeta\right] \tag{10.95}$$

The effectiveness factor of a spherical bidisperse catalyst can then be evaluated using its definition, as follows:

$$\eta = \left(\frac{3}{r_p}\right)\frac{D_a\left.\frac{dC_{A_a}}{dr}\right|_{r=r_p}}{k(1-\varepsilon_a)C_A} = \frac{9}{\varphi_i^2\alpha}\left(\frac{d\psi_a}{d\zeta}\right)_{\zeta=1} \tag{10.96}$$

$$\eta = \frac{9}{\varphi_i^2\alpha}\left\{\frac{\left[\alpha\left(\frac{\varphi_i}{\tanh\varphi_i}-1\right)\right]^{1/2}}{\tanh\left[\alpha\left(\frac{\varphi_i}{\tanh\varphi_i}-1\right)\right]^{1/2}}-1\right\} \tag{10.97}$$

The effectiveness factor for a bidisperse catalyst is found to be a function of two dimensionless parameters, namely particle Thiele modulus and the parameter α. Variation of effectiveness factor with these two parameters is given by Ors and Dogu (Figure 10.10) for a first-order reaction [23].

By the rearrangement of these relations, the effectiveness factor can be expressed in terms of an observable modulus Φ_{bi}, defined for a bidisperse catalyst pellet (Figure 10.11).

$$\Phi_{bi} = \eta\varphi_i^2\alpha = 3(1-\varepsilon_a)\left(\frac{(r_p)^2 R_{p_{obs}}}{D_a C_A}\right) \tag{10.98}$$

The physical significance of the parameter α can be described as

$$\alpha = \frac{\text{Diffusion resistance in the macropores of the pellet}}{\text{Diffusion resistance in the microporous particle}} \tag{10.99}$$

A wide range of values is possible for this parameter. Effective diffusivity values of different molecules were reported in the literature [2]. Effective macropore-diffusivity values were reported to be in the range of 7×10^{-7} to 7×10^{-6} m^2/s. However, a much wider range of effective

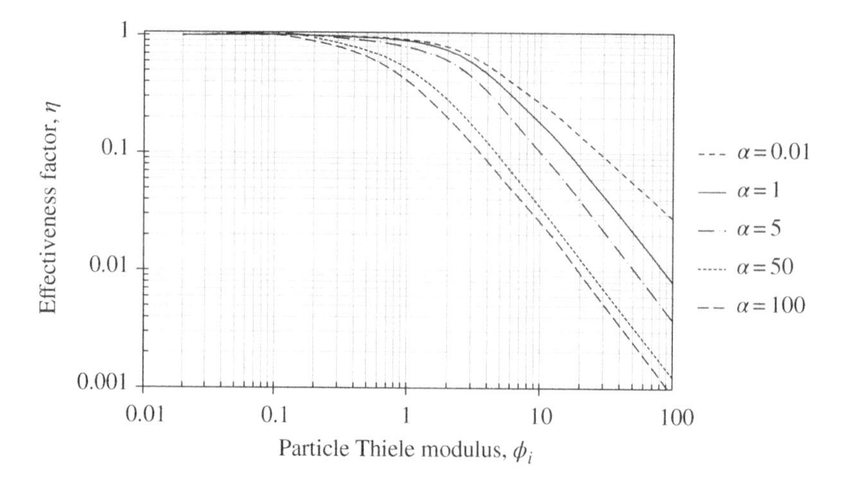

Figure 10.10 Variation of effectiveness factor for a first-order reaction, with respect to particle Thiele modulus and the parameter α, in a bidisperse pellet. *Source:* Örs and Dogu [23], Reproduced with permission of John Wiley & Sons.

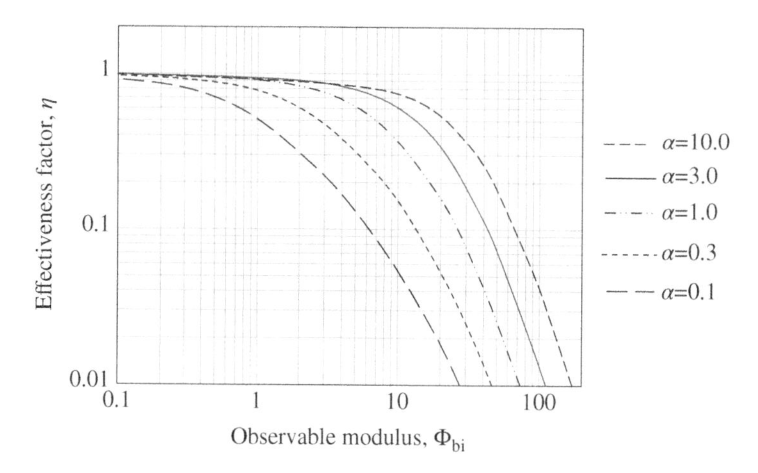

Figure 10.11 Effectiveness factor of a bidisperse catalyst in terms of observable modulus Φ_{bi} and α. *Source:* Dogu [2], Reproduced with permission of Taylor & Francis.

micropore-diffusivity values is reported in the literature. Most micropore-diffusivity values are in the range of 10^{-8}–10^{-10} m^2/s. On the other hand, intercrystalline diffusivity values for zeolites and diffusivities within the gel-like micrograins of macro-reticular resins (such as Amberlyst 15) were reported in the range of 10^{-14}–10^{-19} m^2/s.

Diffusion mechanisms in capillaries and porous catalysts are discussed in Chapter 12 in detail. As discussed in Chapter 12, the value of the effective diffusion coefficient depends on the pore structure of the pellet. In the case of having pore sizes smaller than the mean free path of the diffusing molecules, Knudsen diffusion controls the transport of molecules in the pores. The effective micropore-diffusion coefficient can be evaluated using the Knudsen diffusion coefficient in the micropores and porosity and tortuosity factor (see Chapter 12) values of the microporous particles.

$$D_i = \frac{\varepsilon_i D_{Ki}}{\tau_i} \tag{10.100}$$

The ratio of micro- and macropore effective diffusion coefficients is less than 10^{-2} in most bidisperse catalysts [22]. However, the ratio of pellet to particle radius is higher than 10^2 for most cases. Some typical values are as follows:

$$\text{Alumina pellets:}\ \frac{D_i}{D_a} \approx 10^{-2} - 10^{-4} \tag{10.101}$$

$$\text{Molecular sieve pellets:}\ \frac{D_i}{D_a} \approx 10^{-6} - 10^{-7} \tag{10.102}$$

If the value of α is very high (approaching infinity), diffusion resistance in the micropores becomes negligible. In this case, the effectiveness factor expression approaches the effectiveness factor expression of a monodisperse catalyst pellet containing only macropore-diffusion resistance. The effectiveness factor expressions for this limit are the same as given by Eqs. (10.23), (10.28), and (10.29) for the spherical, slab, and cylindrical geometries, respectively.

10.7 Criteria for Negligible Transport Effects in Catalytic Reactions

Concentration and temperature gradients formed within the catalyst pellets and within the boundary layer at the external surface of these materials may cause deviation of observed reaction rates from the reaction rates that would be evaluated at the bulk fluid concentration and temperature. Intrapellet diffusion and heat conduction, as well as external heat and mass transfer resistances over the catalyst pellets, may have significant effects on the observed reaction rate. In interpreting the experimental kinetic data and for the design of solid-catalyzed reactors, it is crucial to know the importance of such transport processes on the observed rates of reactions. Due to the nonlinear dependence of the reaction rate on temperature and the complex dependence on the reactant concentrations, for most cases, it is quite difficult to obtain analytical solutions of the species concentration and the energy balance equations for many of the reaction systems.

10.7.1 Criteria for Negligible Diffusion and Heat Effects on the Observed Rate of Solid-Catalyzed Reactions

From the practical point of view, it is highly useful to have criteria to test the importance of heat and mass transfer resistances on the observed rate. Weisz and Prater proposed a criterion to test the importance of intraparticle diffusion on the observed rate for an isothermal catalyst pellet of slab shape [8]. This criterion contains only observable variables (Eq. (10.103)), such as observed reaction rate, pellet dimension, the concentration of the reactant at the external surface of the catalyst pellet, and effective diffusion coefficient within the porous pellet.

$$R_{p_{\text{obs}}} \frac{L_s^2}{D_e C_{A_s}} \ll 1.0 \tag{10.103}$$

Here, L_s corresponds to the half-thickness of a slab-shaped catalyst pellet. For the general shape of a catalyst, this criterion can be expressed as follows:

$$\frac{\left(\frac{V_p}{A_e}\right)^2 R_{p_{\text{obs}}}}{D_e C_{A_s}} \ll 1.0 \tag{10.104}$$

Later, Hudgins developed a more general criterion to test the pore-diffusion effect on the observed rate in an isothermal spherical catalyst pellet for reaction rate expressions of any form [24].

$$\Phi_s = \frac{(r_p)^2 R_{p_{obs}}}{D_e C_{A_s}} \ll \frac{1}{C_{A_s}} \left(\frac{R_p(C_{A_s})}{R'_p(C_{A_s})} \right) \tag{10.105}$$

Here, Φ_s is the observable modulus for a spherical catalyst pellet, evaluated using the reactant concentration at the external surface of the pellet, and R'_p is the derivative of the rate expression evaluated at the external surface concentration of the pellet.

In the case of a power-law type nth -order reaction taking place in a spherical catalyst pellet, this criterion can be written as follows:

$$\frac{(r_p)^2 R_{p_{obs}}}{D_e C_{A_s}} \ll \frac{1}{n} \tag{10.106}$$

Both Weisz–Prater and Hudgins criteria discussed above assume an isothermal pellet. However, in some cases, thermal effects may also have quite an important role on the observed rate of a reaction. As discussed in Section 10.4, in addition to concentration gradients, temperature gradients may also form within a catalyst pellet. This may cause changes in the reaction rate along the radial coordinate of a pellet. For exothermic reactions, the heat generated within the pellet may cause temperature rise within the catalyst. On the other hand, in the case of endothermic reactions, temperature decrease from the surface to the inner parts of the pellet may cause a decrease in the effectiveness factor. Andersen derived a criterion for testing the isothermal behavior of a catalyst pellet [17]. According to Andersen criterion, the effect of temperature gradients within a spherical catalyst pellet on the observed rate can be neglected, and isothermal catalyst pellet assumption can be made if the following criterion is satisfied:

$$|\Phi_s \beta \gamma| \ll 0.75 \tag{10.107}$$

Here, observable modulus Φ_s, Prater group β, and Arrhenius group γ are defined by Eqs. (10.108), (10.62), and (10.69), respectively.

$$\Phi_s = \frac{(r_p)^2 R_{p_{obs}}}{D_e C_{A_s}} \tag{10.108}$$

10.7.2 Relative Importance of Concentration and Temperature Gradients in Catalyst Pellets

Intrapellet concentration and temperature gradients within a catalyst pellet cause variations of reaction rate with respect to position. For an exothermic reaction, the increase of reaction rate due to an increase in temperature within the catalyst pellet may offset the decrease of the rate due to a decrease of reactant concentration, which may be caused by pore-diffusion resistance. On the other hand, some temperature decrease may be expected within the pellet, compared to the surface temperature for an endothermic reaction. A decrease in both temperature and concentration from the surface to the center of a catalyst pellet is expected to decrease in reaction rate. Hence, effectiveness factor values being less than 1 are expected due to both of these effects. Dogu and Dogu derived a criterion for testing the relative importance of diffusion and thermal effects on the observed reaction rate [25]. The criterion is developed for a multiple reaction system containing

n independent reactions involving m species. This criterion is developed by allowing a maximum 5% deviation of effectiveness factor from unity, that is:

$$1.05 > \eta > 0.95 \tag{10.109}$$

The selection of a 5% deviation of effectiveness factor from unity is arbitrary and can be taken as a different value if preferred.

The simplified version of the derivation of the criterion for testing the relative importance of diffusion and the thermal effects for a single reaction is presented here. In this derivation, the intrinsic reaction rate is expanded in a Taylor series about the external surface concentrations and the temperature. The second- and higher-order terms of the Taylor expansion can be neglected for small deviations of concentration and temperature from the corresponding external surface values. With this assumption, the rate of a reaction at any point within the catalyst pellet can be approximated as follows:

$$R_p = R_p(C_{i_s}, T_s) + \left(\frac{\partial R_p}{\partial C_i}\right)_s (C_i - C_{i_s}) + \left(\frac{\partial R_p}{\partial\left(\frac{1}{T}\right)}\right)_s \left(\frac{1}{T} - \frac{1}{T_s}\right) + \cdots \tag{10.110}$$

Here, C_{i_s} and T_s are the concentration of reactant i and the temperature at the external surface of the catalyst pellet, respectively. In the case of small deviations from the surface values, concentration and temperature profiles within the spherical pellet can be approximated by parabolic equations.

$$(C_i - C_{i_s}) = -p_1\left(1 - \frac{r^2}{r_p^2}\right) \tag{10.111}$$

$$\left(\frac{1}{T} - \frac{1}{T_s}\right) = p_2\left(1 - \frac{r^2}{r_p^2}\right) \tag{10.112}$$

Here, the parameters p_1 and p_2 can be evaluated by equating the diffusion and heat conduction rates at the surface of the catalyst pellet to the observed reaction rate and the heat generation rate, respectively.

$$D_e \left.\frac{dC_i}{dr}\right|_{r=r_p} 4\pi r_p^2 = R_{p_{obs}}\left(\frac{4}{3}\pi r_p^3\right) \tag{10.113}$$

$$\lambda_e \left.\frac{dT}{dr}\right|_{r=r_p} 4\pi r_p^2 = (\Delta H^o) R_{p_{obs}}\left(\frac{4}{3}\pi r_p^3\right) \tag{10.114}$$

The parameters p_1 and p_2 are then evaluated and expressed as

$$p_1 = \left(\frac{r_p^2}{6D_e}\right) R_{p_{obs}} \tag{10.115}$$

$$p_2 = \left(\frac{r_p^2}{6\lambda_e T_o^2}\right) (\Delta H^o) R_{p_{obs}} \tag{10.116}$$

Substitution of Eqs. (10.111)–(10.112) and (10.115)–(10.116) into Eq. (10.110) gives an expression for the reaction rate at any point within the catalyst pellet. This rate expression is then substituted into the integral form of the effectiveness factor definition (Eq. (10.2)). For a spherical catalyst pellet, this expression can be expressed as follows:

$$\eta = \frac{\text{Observed reaction rate}}{\text{Ideal reaction rate}} = \left(\frac{3}{r_p}\right)\frac{\int_0^{r_p} R_p(C_i, T)r^2 dr}{R_p(C_{i_s}, T_s)} \tag{10.117}$$

The integration in Eq. (10.117) is performed by substituting the rate relation given in Eq. (10.110). This gives the following approximate expression for the effectiveness factor:

$$\eta \cong 1 - \frac{2}{5R_p(C_{i_s}, T_s)}\left(\frac{r_p^2}{6D_e}\right)R_{p_{\text{obs}}}\left(\frac{\partial R_p}{\partial C_i}\right)_s + \frac{2}{5R_p(C_{i_s}, T_s)}\left(\frac{r_p^2}{6\lambda_e T_s^2}\right)(\Delta H^o)R_{p_{\text{obs}}}\left(\frac{\partial R_p}{\partial\left(\frac{1}{T}\right)}\right)_s \tag{10.118}$$

Hence, the criterion for having an effectiveness factor being close to unity can be expressed by using the condition set in Eq. (10.109). In the case of an nth -order reaction rate expression, this criterion can be expressed as follows:

$$E = \left|\frac{(r_p)^2 R_{p_{\text{obs}}}}{D_e C_{A_s}} n\left(1 - \frac{\beta\gamma}{n}\right)\right| \ll \frac{3}{4} \tag{10.119}$$

Here, β and γ are the Prater and Arrhenius groups defined by Eqs. (10.62) and (10.69), respectively.

In this criterion, the term in front of the parentheses is the observable modulus for the spherical catalyst pellet, and it gives information about diffusion limitations. This term is the same as the term in the Weisz criterion. On the other hand, the term within the parentheses gives us information about the heat effects. For an endothermic reaction, the Prater parameter β is negative; hence, the value of parentheses becomes positive. In that case, heat effects cause a decrease in temperature from the surface to the inner positions of the pellet. Both diffusional resistance and heat effects work in the same direction to decrease the observed rate from the ideal surface reaction rate. Only if the condition expressed in Eq. (10.119) is satisfied, we can have effectiveness factor values close to unity. On the other hand, in the case of an exothermic reaction, diffusion limitations and heat effects are expected to have opposite effects on the observed rate. The decrease of reaction rate due to diffusion limitations may be compensated by the increase of the rate due to the increase of temperature within the pellet.

For $\frac{\beta\gamma}{n} < 1$, heat effects are expected to be insignificant. On the other hand, for $\frac{\beta\gamma}{n} > 1$, the effect of temperature rise within the pellet on the reaction rate is expected to be quite high so that it may offset the decreasing effect of diffusional limitations.

To illustrate this, we can consider the following two examples:

Case 1: For an arbitrary selection of $\frac{(r_p)^2 R_{p_{\text{obs}}}}{D_e C_{A_s}} n = 2.0$, $\frac{\beta\gamma}{n} = 0.1$, $\left(1 - \frac{\beta\gamma}{n}\right) = 0.9$, the value of the dimensionless group in Eq. (10.119) becomes $E = 1.8$.

In this case, diffusion resistance is quite high. It is a mildly exothermic reaction with quite low heat effects, such that the compensation effect of temperature rise on the reaction rate is not very high. The value of E is greater than 3/4. This result indicates that the effectiveness factor is less than unity due to the diffusion resistance of reactant into the pellet.

Case 2: If $\frac{(r_p)^2 R_{p_{\text{obs}}}}{D_e C_{A_s}} n = 2.0$, $\frac{\beta\gamma}{n} = 4.0$, $\left(1 - \frac{\beta\gamma}{n}\right) = -3$, the value of the dimensionless group in Eq. (10.119) becomes $E = 6.0$.

In this case, diffusion resistance is again quite high. However, heat effects are also very high. The reaction is highly exothermic. The heat effect is so high that it more than offsets the decreasing effect of diffusion resistance on the reaction rate. The value of E is much higher than 3/4, indicating that the effectiveness factor is higher than 1. This analysis was further extended in the literature by considering the second- and third-order terms in the Taylor expansion of the reaction rate expression given by Eq. (10.110) [26].

10.7.3 Intrapellet and External Film Transport Limitations

Considering a reaction system with m species and n independent reactions, a criterion was developed to test the relative significance of intrapellet and external transport limitations for a non-isothermal catalyst pellet [26]. The simplified form of this criterion for an isothermal single reaction system in a spherical catalyst pellet is as follows:

$$\frac{(r_p)^2 R_{p_{obs}}}{D_e C_{A_s}}\left(1 + \frac{5}{3}Bi_m^{-1}\right) \ll \frac{3}{4}\frac{1}{C_{A_s}}\left(\frac{R_p(C_{A_s})}{R_p'(C_{A_s})}\right) \tag{10.120}$$

Here, Biot number for mass transfer is defined as

$$Bi_m = \frac{k_m \dfrac{V_p}{A_e}}{D_e} = \frac{k_m r_p}{3D_e}$$

For an nth- order power-law type rate expression, this criterion reduces to

$$\frac{(r_p)^2 R_{p_{obs}}}{D_e C_{A_s}}\left(1 + \frac{5}{3}Bi_m^{-1}\right) \ll \frac{1}{n} \tag{10.121}$$

This expression is the general form of the Hudgins criterion. In the case of very high Biot numbers, film mass transfer resistance becomes negligible, and the criterion reduces to Weisz criterion. This criterion can be used to test the effect of diffusion and film mass transfer resistances on the observed rate for an isothermal pellet. If this criterion is satisfied, we can safely assume that the effectiveness factor is very close to unity in an isothermal catalyst pellet.

10.7.4 A Criterion for Negligible Diffusion Resistance in Bidisperse Catalyst Pellets

Diffusion effects in catalysts with bi-modal pore-size distributions are discussed in Section 10.6. The following criterion is proposed [27] for the negligible macro- and micropore-diffusion resistances on the observed rate in a catalyst pellet with a bi-modal pore-size distribution.

$$\Phi_{obs}(1 + g)\frac{R_p'(C_{A_b})}{R_p(C_{A_b})}C_{A_b} \ll \frac{3}{4} \tag{10.122}$$

For a first-order reaction, this criterion reduces to

$$\Phi_{obs}(1 + g) \ll \frac{3}{4} \tag{10.123}$$

Here, the parameter g is inversely proportional to α, which is defined by Eq. (10.93).

$$g = \left(\frac{r_g}{r_p}\right)^2 \frac{D_a}{D_i(1-\varepsilon_a)} \tag{10.124}$$

10.8 Transport Effects on Product Selectivities in Catalytic Reactions

10.8.1 Film Mass Transfer Effect

For reactions catalyzed by solid catalysts, both external mass transfer and pore-diffusion resistances may influence the point selectivity of the desired product [1]. Let us consider a multiple reaction system catalyzed by a non-porous solid catalyst, as illustrated in Figure 10.12. For the parallel reaction system described in Chapter 6

$$\mathrm{A + D \rightarrow B}\ \text{(desired reaction)};\quad R_{B_s} = k_{1_s} C_A^n\ (\text{mol/m}^2\,\text{s})$$

$$\mathrm{A + E \rightarrow C};\quad R_{C_s} = k_{2_s} C_A^m\ (\text{mol/m}^2\,\text{s})$$

if the mass transfer resistance over the external surface of the catalyst is not negligible, point selectivity of the desired product B should be expressed in terms of surface concentrations of the reacting species instead of their bulk concentrations.

$$s_{BC_s} = \frac{k_{1_s} C_{A_s}^n}{k_{2_s} C_{A_s}^m} = \left(\frac{k_{1_s}}{k_{2_s}}\right) C_{A_s}^{(n-m)} \tag{10.125}$$

The surface concentration of the reactant A should then be estimated by equating the steady-state molar flux of A to the surface reaction rate.

$$k_m(C_A - C_{A_s}) = k_{1_s} C_{A_s}^n + k_{2_s} C_{A_s}^m \tag{10.126}$$

For this case, the ratio of the point selectivity expressions for the mass transfer affected and non-affected cases can be expressed as a function of the ratio of surface and bulk concentrations of reactant A, as follows:

$$\frac{s_{BC_s}}{s_{BC}} = \left(\frac{C_{A_S}}{C_A}\right)^{(n-m)} \tag{10.127}$$

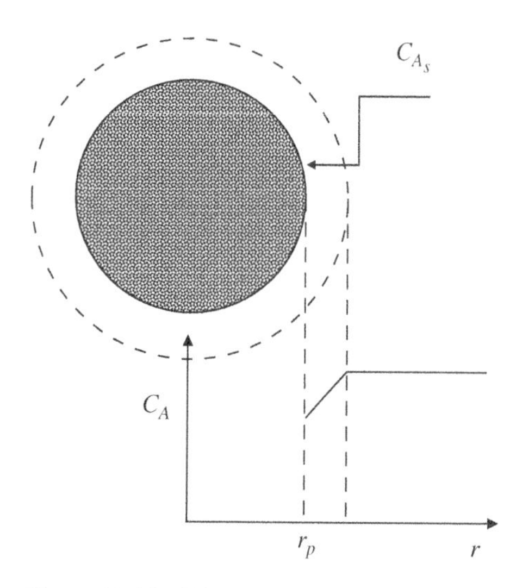

Figure 10.12 Schematic representation of external mass transfer resistance over a catalyst pellet.

In the case of significant mass transfer resistance, the surface concentration of A will be less than the bulk concentration. Hence, the ratio of surface and bulk concentrations will be less than 1.0 for the case of significant mass transfer resistance. If the order of the desired reaction is higher than the reaction order of the undesired reaction, $(n - m) > 0$, the selectivity ratio expressed in Eq. (10.127) is negatively affected. Film mass transfer resistance causes a decrease of concentration across the boundary layer. The effect of the decrease of concentration on the rates of the surface reactions will be more significant for reactions having higher orders. If the reaction rate orders of the desired and the undesired reactions are the same $(n = m)$, the selectivity of the desired product will not be affected by the film mass transfer resistance. However, in the case of having $(n - m) < 0$, an increase in film mass transfer resistance affects the selectivity ratio in the positive direction.

10.8.2 Pore-Diffusion Effect

Pore-diffusion resistance within a catalyst pellet is also expected to influence the catalyst selectivity. Here, this effect is discussed for parallel and consecutive reaction cases. For the two parallel reactions taking place within a catalyst pellet

$$\mathrm{A} \rightarrow \mathrm{B}\ (\mathrm{B\ is\ the\ desired\ product}); \quad R_{B,p} = k_1 C_{A_p}^n$$

$$\mathrm{A} \rightarrow \mathrm{C}; \quad R_{C,p} = k_2 C_{A_p}^m$$

point selectivity of B with respect to C at any point within the catalyst pellet is expressed as follows:

$$s_{BC} = \frac{R_{B,p}}{R_{C,p}} = \frac{k_1 C_{A_p}^n}{k_2 C_{A_p}^m} = \left(\frac{k_1}{k_2}\right) C_{A_p}^{(n-m)} \tag{10.128}$$

Here, C_{A_p} is the concentration of reactant A within the pores, and it is expected to be a function of radial position within the catalyst pellet. For this case, the observed selectivity of B with respect to C can be evaluated as follows:

$$s_{BC_p} = \frac{\int_{V_p} R_{B,p} dV}{\int_{V_p} R_{C,p} dV} = \frac{\int_0^{r_0} k_1 C_{A_p}^n 4\pi r^2 dr}{\int_0^{r_0} k_2 C_{A_p}^m 4\pi r^2 dr} \tag{10.129}$$

Evaluation of the observed selectivity for a catalyst pellet requires the solution of the species conservation equation for reactant A within the catalyst pellet, which will give us an expression for the variation of C_{A_p} with respect to the radial position within the pellet.

In the case of significant pore-diffusion resistance, the concentration of A within the catalyst pellet is less than the bulk concentration of A. If the order of the rate of the desired reaction is higher than the order of the rate of the undesired reaction $(n - m) > 0$, higher concentrations favor the desired reaction. Hence, in the case of significant pore-diffusion resistance, the selectivity of the desired reaction is negatively affected. On the other hand, if $(n - m) < 0$, a higher concentration of reactant A favors the occurrence of the undesired reaction. In this case, we expect an increase in the selectivity of the desired product with an increase in pore-diffusion resistance.

For a consecutive reaction system

$$\mathrm{A} \xrightarrow{k_1} \mathrm{B} \xrightarrow{k_2} \mathrm{C}$$

point selectivity of the desired product B with respect to the limiting reactant A can be expressed within the pores of a catalyst pellet as follows:

$$s_{BA} = \frac{R_{B,p}}{-R_{A,p}} = \frac{k_1 C_{A_p} - k_2 C_{B_p}}{k_1 C_{A_p}} = 1 - \frac{k_2}{k_1}\left(\frac{C_{B_p}}{C_{A_p}}\right) \tag{10.130}$$

For the consecutive reaction system described above, the observed selectivity of B with respect to A can be evaluated from the following relations:

$$s_{BA_p} = \frac{\int_{V_p} R_{B,p} dV}{\int_{V_p} (-R_{A,p}) dV} = \frac{\int_0^{r_0} (k_1 C_{A_p} - k_2 C_{B_p}) 4\pi r^2 dr}{\int_0^{r_0} k_1 C_{A_p} 4\pi r^2 dr} \tag{10.131}$$

The integration of the expressions in Eq. (10.131) requires the simultaneous solution of the species conservation equations of A and B within the catalyst pellet.

$$\frac{D_{A_e}}{r^2}\left(\frac{d}{dr}\left(r^2\frac{dC_{A_p}}{dr}\right)\right) - k_1 C_{A_p} = 0 \quad (10.132)$$

$$\frac{D_{B_e}}{r^2}\left(\frac{d}{dr}\left(r^2\frac{dC_{B_p}}{dr}\right)\right) + \left(k_1 C_{A_p} - k_2 C_{B_p}\right) = 0 \quad (10.133)$$

A similar analysis can be performed for other reaction orders in consecutive reactions. The solution of the model equations may require numerical integration procedures.

Problems and Questions

10.1 A porous solid catalyst catalyzes a first-order reaction. It was shown that, if the pellet diameter of this porous catalyst was doubled, the observed reaction rate dropped to half of its original value. Explain this observation.

It was also reported that the value of the observed reaction rate increased 64 times when the reaction temperature was increased from 500 to 600 °C. Evaluate the actual activation energy of this reaction.

10.2 Observed rate data obtained for the solid-catalyzed reaction (A + B → C), using two different sizes of the same catalysts, are reported in the table. Data reported in this table were obtained in a system where the concentration of B was much higher than the concentration of A. Hence, the reaction can be assumed as pseudo-first order.

T, °C	420	440	460	475	500
k, s^{-1} (Catalyst 1)	6.5	18.0	50.0	100.0	295.0
k, s^{-1} (Catalyst 2)	2.9	5.0	10.0	15.0	27.0

Catalyst 1: Powder ~0.05 mm in diameter
Catalyst 2: 8 mm in diameter

(a) Evaluate the activation energy values obtained with Catalysts 1 and 2. Discuss the possible reason for the differences in activation energy values.
(b) Estimate the effectiveness factor and the corresponding Thiele modulus values for Catalyst 2 at 460 °C. Discuss the assumptions involved in this estimation.
(c) If we use another dimension of the same catalyst (5 mm in diameter), estimate the possible value of the observed rate constant at 475 °C.

10.3 Observed rate data reported in the table were obtained for a first-order gas-phase reaction, catalyzed by a porous Ni impregnated alumina catalyst having a pellet diameter of 0.4 cm.

(a) Obtain the observed and the actual activation energy values of the reaction.
(b) Estimate the value of the effective diffusion coefficient of reactant A.

T, K	500	599	649	752	826	901	1000	1250
k_{obs}, s^{-1}	5.06×10^{-8}	1.25×10^{-6}	3.33×10^{-6}	9.17×10^{-6}	1.59×10^{-5}	2.37×10^{-5}	4.11×10^{-5}	9.61×10^{-5}
$1/T$, K^{-1}	2×10^{-3}	1.67×10^{-3}	1.54×10^{-3}	1.33×10^{-3}	1.21×10^{-3}	1.11×10^{-3}	1.0×10^{-3}	0.8×10^{-3}
ln (k_{obs})	−16.80	−13.59	−12.61	−11.60	−11.05	−10.65	−10.10	−9.25

Exercises

10.1 Derive expressions for the effectiveness factors in a slab-shaped porous catalyst pellet for the cases in which the reaction order is 0 and 1. Also, find the asymptotic limits of the effectiveness factor expressions for the cases of Thiele modulus approaching zero and infinity.

10.2 Derive the effectiveness factor expression for a first-order irreversible reaction in a spherical catalyst pellet, for which both pore-diffusion and external film mass transfer resistances are significant. Plot effectiveness factor as a function of Thiele modulus, for a set of Biot numbers in the log–log scale. Discuss the physical significance of your results.

10.3 Consider a first-order catalytic reaction taking place in a CSTR.

$$A \rightarrow \text{Products}$$

The space-time in the reactor is 20 seconds. Experiments were performed using a spinning-basket reactor to achieve perfect mixing. The following set of experimental results were obtained using the same porous catalyst having different pellet sizes (Table 10.3).

For the same reaction, the following data were obtained at different temperatures using catalyst pellets with a diameter of 1.2 cm (Table 10.4).

(a) Discuss the possible reasons for the variations observed in fractional conversion values reported in Table 10.3.

(b) Evaluate the observed and the actual activation energy values of the reaction using the data reported in Table 10.4.

(c) Estimate the fractional conversion, which could be obtained at 100 °C with the catalyst pellets having diameters of 0.01 cm in the same reactor.

Table 10.3 Fractional conversion values obtained with porous catalysts having different pellet sizes (T = 50 °C).

d_p (cm)	0.01	0.05	0.2	0.5	0.8	1.2
x_{A_f}	0.7	0.7	0.65	0.38	0.2	0.12

Table 10.4 Fractional conversion values obtained at different temperatures (catalyst pellet diameter: 1.2 cm).

T (°C)	50	80	100	150
x_{A_f}	0.12	0.33	0.52	0.85

10.4 The following observations were made in the experiments performed in a fixed-bed catalytic reactor. Explain the possible reasons for each of these observations.

(a) By doubling the catalyst amount in the reactor (reactor volume) and at the same time by doubling the volumetric flow rate of the feed stream, observed fractional conversion of reactant A showed some increase. Note that the reactor operates isothermally.

(b) In the experiments performed with catalyst particles having diameters of 2.0 and 0.1 cm, observed reaction order was found to increase from 3/2 to 2, respectively.

(c) Experiments were performed in two fixed-bed reactors having the same volume (and the same amount of the same catalyst). However, different diameter to length ratios and different fractional conversion values were observed. The volumetric flow rates of the feed streams were also the same in these reactors. However, in the reactor with a higher length to diameter ratio, observed fractional conversion was also higher.

10.5 Consider a slab-shaped catalyst pellet for which external mass transfer resistance might also significantly affect the observed rate. Write down the observed rate expression and discuss the effect of the Biot number on the observed rate for a first-order reaction.

(a) Plot the effectiveness factor as a function of Thiele modulus, Biot number being the parameter.

(b) Discuss the effect of the Biot number on the observed activation energy.

10.6 An exothermic reaction is taking place in a porous spherical catalyst, with negligible external film mass and heat transfer resistances. Intraparticle mass transfer limitations on the observed rate are negligible. However, intraparticle heat transfer limitations alone affect the observed rate.

(a) Write down a differential equation for the energy balance for this catalyst pellet in terms of dimensionless temperature (T_p/T_s) and dimensionless position $[r/(V_p/A_e)]$. Consider an nth- order reaction and Arrhenius-type temperature dependence of the rate constant.

(b) Obtain an integral equation (which could be evaluated numerically) for the effectiveness factor and discuss the physical significances of the dimensionless groups, which appear in this equation.

10.7 Consider a partially poisoned (deactivated) spherical catalyst pellet as given in Figure 10.13. To predict the effectiveness factor for this system, it is necessary to find the concentration profile of the reactant within the pellet.

(a) Set up necessary equations and boundary conditions for this isothermal pellet and derive the effectiveness factor expression for a first-order reaction. The radius of the active core section of the catalyst "r_c" may appear as a parameter in your equation.

(b) Set up the necessary equations and boundary conditions for a non-isothermal pellet for a first-order reaction. Reduce these equations to a single equation for concentration (eliminate temperature).

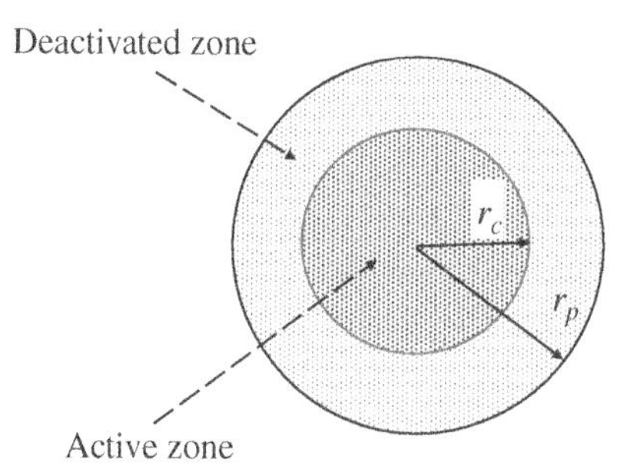

Figure 10.13 Schematic representation of deactivated catalyst pellet for Problem 10.7.

10.8 Hegedus and Petersen investigated the kinetics of poisoning phenomena of porous catalysts in a single-pellet reactor [28]. Consider the single-pellet system shown in Figure 10.14, which is designed to investigate the poisoning of a catalyst pellet. The reactant mixture with a concentration of C_{A_s} is passed over one face of the flat catalyst pellet. The other face of the pellet is exposed to a small chamber of volume V. The reactant mixture contains a poisoning compound at a very small concentration, and this poison also diffuses into the pellet and poisons the active sites of the catalyst. Assume that the poisoning of the catalyst pellet takes place progressively and very slowly. The portion of the catalyst to a depth of L_p ($L_p = pL$) is poisoned and inactive, while the rest of the catalyst pellet is active. Here, p is the fraction of the pellet. The rate of change of p with respect to time can be taken as constant $\left(\frac{dp}{dt} = c\right)$.

(a) Set up all of the necessary equations and the necessary boundary conditions for the active and inactive zones of the catalyst pellet and predict the variation of concentration of the reactant A in the small chamber with respect to time, for an isothermal system, by assuming a first-order reaction taking place in the active zone of the pellet.

(b) Set up the necessary equations of the system described above if the pellet is thermally insulated, and the variation of temperature throughout the pellet is not to be ignored.

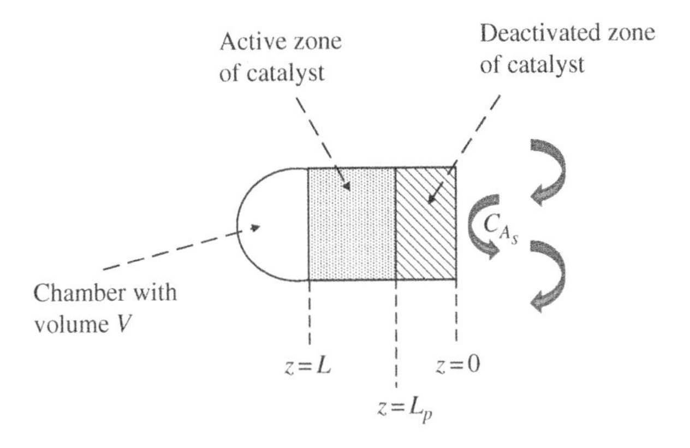

Figure 10.14 Single-pellet system for Problem 10.8.

10.9 Consider an isothermal annular flow reactor with a porous catalytic layer covering the annulus surface between r_2 and r_3 (see Figure 10.15). Reactants flow through the annular region. There is no homogeneous reaction in the fluid phase within the annular region. The catalytic reaction taking place within the catalyst layer is first order. Pore diffusion in the radial direction into the catalyst layer should be taken into account. However, negligible diffusion flux in the z-direction within the porous catalyst layer can be assumed.

Set up all of the necessary equations and the boundary conditions to model this reactor for the following cases:

(a) Plug flow in the annular region, no film mass transfer resistance on the surface of the catalyst layer.

(b) Plug flow in the annular region, significant film mass transfer resistance on the surface of the catalyst layer at $r = r_2$.

(c) The concentration of reactant changes both in the radial and axial directions in the annular region.

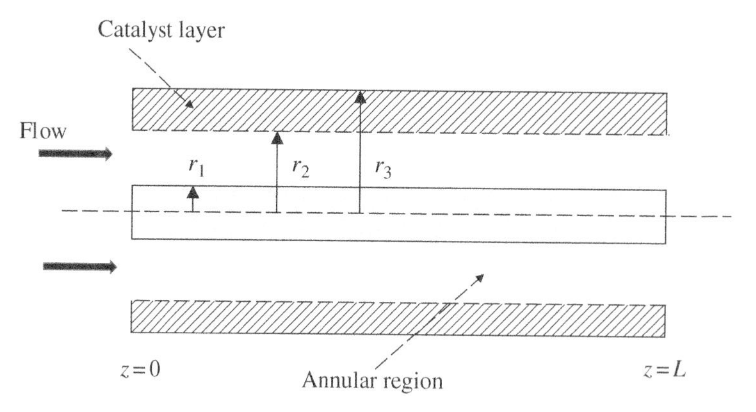

Figure 10.15 Schematic representation of the reactor for Problem 10.9.

10.10 Consider a consecutive catalytic reaction system taking place in a CSTR.

$$A \rightarrow B \rightarrow C \text{ (both reactions are elementary)}$$

The concentration of reactant A is C_{A_0}, and there is no B and C in the inlet stream of the reactor. Pore-diffusion resistance within the catalyst pellets is not negligible. Answer the following questions:

(a) Assuming slab geometry for the catalyst pellets obtain expressions for the variation of C_A and C_B within the catalyst pellets.

(b) Obtain expressions for the observed rates for A and B in the catalyst pellet.

(c) For the given values of reactor volume, volumetric flow rate, catalyst properties, dimensions, and the amount of catalyst within the reactor, set up expressions for the reactor outlet concentrations of A and B.

10.11 Spherical porous alumina catalyst pellets, having radii of 1.2 cm, were impregnated by Ni to increase its activity in a dry reforming reaction. Characterization results of the catalyst pellets indicated that nickel was not uniformly distributed but mainly deposited on the pore surfaces at the outer layer of the catalyst pellet with a thickness of 0.3 cm. Assuming that only the Ni-impregnated zone is active for the reaction under investigation, and assuming that the effective diffusion coefficient of the reactant is the same in the active and inactive zones of the catalyst pellet, answer the following questions.

(a) Derive an effectiveness factor expression for this catalyst pellet.

(b) Plot and discuss the effect of Thiele modulus on the effectiveness factor.

10.12 An exothermic catalytic reaction with an arbitrary order is taking place in a cylindrical catalyst pellet with a high length to diameter ratio so that the temperature and concentration gradients are significant only in the radial direction. Considering that the external film heat transfer resistance is also significant, but the external film mass transfer resistance is negligible, derive an equation for the maximum possible temperature rise within the pellet as compared to the bulk fluid temperature. Also, obtain an expression for the ratio of temperature gradients in the external film and the total maximum temperature gradient.

10.13 Weisz proposed a criterion for testing experimental data for intraparticle diffusion effects on the observed rate. Derive a criterion for a zero-order reaction taking place in a spherical porous catalyst. Note that diffusion resistance may cause a sharp decrease of reactant concentration from the surface to the inner positions of the catalyst pellet. The concentration of the reactant may reach a value of zero at a particular radial position (may be called r^*) for a zero-order reaction.

10.14 During catalytic oxidation of ethylene to ethylene oxide, complete oxidation of ethylene may also take place.

1) $C_2H_4 + 1/2O_2 \rightarrow C_2H_4O$
2) $C_2H_4 + 3O_2 \rightarrow 2CO_2 + 2H_2O$

The observed rate of reactions (1) and (2) are $R_{1,p_{obs}}$ and $R_{2,p_{obs}}$.

(a) Derive a criterion for testing the significance of intra-pellet diffusion for this multiple reaction system. The rate expressions of both of these reactions are functions of concentrations of both ethylene and oxygen. This derivation should hold for any type of rate expressions.

(b) If the ideal reaction rate expressions for these two reactions are given as:

$$R_{1,p} = k_1 C_{C_2H_4}^{1/2} C_{O_2}; \quad R_{2,p} = k_2 C_{C_2H_4}^{1/2} C_{O_2}$$

What is the final form of the criterion that is derived in part (a)?

References

1 Carberry, J.J. (1976). *Chemical and Catalytic Reaction Engineering*. New York: McGraw Hill.
2 Dogu, G. (1986). Diffusion limitations for reactions in porous catalysts. In: *Handbook of Heat and Mass Transfer*, vol. 2 (ed. N.P. Chremisinoff), 433–496. Houston, TX: Gulf Publ. Co.
3 Thiele, E.W. (1939). *Ind. Eng. Chem.* 31: 916–920.
4 Aris, R. (1975). *The Mathematical Theory of Diffusion and Reaction in Permeable Catalysts*. Oxford: Clarendon Press.
5 Petersen, E.E. (1965). *Chemical Reaction Analysis*. New Jersey: Prentice Hall.
6 Satterfield, C.N. (1970). *Mass Transfer in Heterogeneous Catalysis*. Cambridge: MIT Press.
7 Froment, G.F. and Bischoff, K.B. (1979). *Chemical Reactor Analysis and Design*. New York: Wiley.
8 Weisz, P.B. and Prater, C.D. (1954). *Adv. Catal.* 6: 143.
9 Prater, C.D. (1958). *Chem. Eng. Sci.* 8: 284–286.
10 Cordoso, S.S. and Rodrigues, A.E. (2006). *AIChE J.* 52: 3924.
11 Mischke, R.A. and Smith, J.M. (1962). *Ind. Eng. Chem. Fundam.* 1: 288–292.
12 Weisz, P.B. and Hicks, J.S. (1962). *Chem. Eng. Sci.* 17: 265–275.
13 Maymo, J.A. and Smith, J.M. (1966). *AIChE J.* 12: 845–854.
14 Hlavacek, V., Kubicek, M., and Marec, M. (1969). *J. Catal.* 1: 17–30.
15 Slinko, M.G., Malinovskaja, O.A., and Beskov, V.S. (1967). *Chim. Prom.* 42: 641.
16 Vandewalle, L.A., De Vijuer, R.V., Van Greem, K.M., and Marin, G.B. (2019). *Chem. Eng. Sci.* 198: 268–289.
17 Andersen, J.B. (1963). *Chem. Eng. Sci.* 18: 147–148.
18 Mears, D.E. (1971). *J. Catal.* 20: 127–131.
19 Karaman, B.P., Oktar, N., Dogu, G. et al. (2020). *Catal. Lett.* 150: 2744–2761.

20 Rao, S.M., Saraçi, E., Gkaser, R., and Coppens, M.O. (2017). *Chem. Eng. J.* 329: 45–55.

21 Coppens, M.O. and Wang, G. (2009). Optimal design of hierarchically structured porous catalysts. In: *Design of Heterogeneous Catalysts* (ed. U. Ozkan), 25–58. Weinheim: Wiley-VCH.

22 Dogu, T. (1998). *Ind. Eng. Chem. Res.* 37: 2158–2171.

23 Ors, N. and Dogu, T. (1979). *AIChE J.* 25: 723–725.

24 Hudgins, R.R. (1968). *Chem. Eng. Sci.* 23: 93–94.

25 Dogu, T. and Dogu, G. (1984). *AIChE J.* 30: 1002–1004.

26 Dogu, T. (1985). *Can. J. Chem. Eng.* 63: 37–42.

27 Dogu, T. and Dogu, G. (1980). *AIChE J.* 26: 287–288.

28 Hegedus, L.L. and Petersen, E.E. (1973). *Chem. Eng. Sci.* 28: 69–82.

11

Introduction to Catalysis and Catalytic Reaction Mechanisms

In solid-catalyzed reactions, contact of the catalyst surface with the reactant involves both transport resistances and surface reaction steps. In many cases, supported catalysts are used, mainly to increase the surface area and hence to increase the activity of the catalytic material. Catalyst support materials are usually porous materials, like alumina, silica, aluminum silicate, porous carbon, etc. Active metals, like Pt, Pd, Ni, Co, Cu, Ag, are the catalytic agents that are incorporated into the support material to enhance their catalytic performances [1, 2]. For instance, hydrogenation catalysts used in many industrial processes contain Ni, Cu, Co, Pt, Pd, Rh, or Ru as the active metals. Copper modified by ZnO is used in the methanol synthesis processes. In some cases, oxides of metals are used as catalytic agents. A typical example of a dehydrogenation reaction is the synthesis of styrene from ethylbenzene. This reaction is catalyzed by iron oxide and potassium carbonate. Iron oxide (magnetite) is also used as a catalyst in ammonia synthesis. Vanadium pentoxide catalyst is used to convert sulfur dioxide to sulfur trioxide in a sulfuric acid plant. Silver is a well-known catalyst for the oxidation of ethylene to ethylene oxide.

11.1 Basic Concepts in Heterogeneous Catalysis

The dispersion of active metals on the surface of the catalyst supports is of utmost importance for the performance of the solid catalysts. The dispersion of active metal on the catalyst surface is generally defined as the percentage of the number of active surface atoms out of the total number of metal atoms. The dispersion of active metal increases as its particle size decreases.

The primary function of a catalyst is to decrease the overall activation energy of the conversion process, hence to increase the reaction rate. The activity of a catalyst for a specific reaction is the main factor considered in its selection. Another crucial factor is the selectivity toward the desired product. The catalyst used for a particular process should increase the rate of the desired product, but not the rates of formation of the undesired products, which may form through parallel or consecutive parasitic reactions.

Support materials are usually catalytically inactive, and their primary role is to supply the high surface area for the dispersion of active metals. However, sometimes, they may also have some activity for the catalytic conversion or may act as a reservoir for the reacting molecules. Species within the reaction medium may also adsorb on the support material, and these adsorbed species may diffuse to the active surface sites by surface diffusion. Hence, support materials may also act as a reservoir that gives and/or receives species to/from the active metal sites. Surface acidity of the support material is also an important property for the performance of the catalysts. For instance,

Fundamentals of Chemical Reactor Engineering: A Multi-Scale Approach, First Edition. Timur Doğu and Gülşen Doğu.

Companion website: www.wiley.com/go/dogu/chemreacengin

dehydration of methanol to dimethyl ether may be achieved on the Bronsted acid sites of alumina. Catalyst support materials may also contribute to coke formation, which is a significant problem in some of the catalytic reactions. The high acidity of the support material is considered as one of the factors enhancing coke formation.

Besides the support material and active metals, some promoters are also incorporated into the solid catalysts, in many cases. Promoters may improve dispersion of the active metal, increase the catalytic activity and selectivity toward the desired reaction, help to reduce coke formation, decrease catalyst deactivation, and may prevent sintering of the active metals, etc. For instance, ZnO incorporated into the methanol synthesis catalyst improves the dispersion and stability of the active metal Cu.

When a molecule collides with a catalytic surface, it may adsorb on the surface if it has sufficient energy. There are two types of adsorption phenomena, namely physical adsorption (physisorption) and chemical adsorption (chemisorption). A typical potential energy diagram for an adsorption process is shown in Figure 11.1.

The physically adsorbed state is a precursor state to the chemically adsorbed state, and activation energy of physisorption is generally small (non-activated). Molecules are held on the surface by van der Waals forces, and usually, dissociation of molecules does not take place during physisorption. The heat of adsorption is also quite low, in the order of magnitude of 5–20 kJ/mol, in physisorption. Physical adsorption is usually reversible, and multi-layer physisorption is also possible.

In the case of chemisorption, molecules are more strongly bound to the surface, with the typical heat of adsorption values of 100 kJ/mol or higher. Chemisorption may be activated in many cases and may be irreversible. Molecules may also dissociate during chemisorption, and chemisorbed molecules do not form multi-layer on the surface.

Various adsorption isotherms are used for the adsorption of molecules on solid surfaces. These isotherms give relations between the fluid-phase concentration (partial pressure) and the adsorbed state concentration of the adsorbing molecules. The most commonly used adsorption isotherm is Langmuir isotherm, which is based on the assumption that the surface is energetically uniform, and adsorption takes place as a monolayer. Possible variations of activity of the surface sites and interaction forces between the adsorbed species are neglected in the derivation of Langmuir isotherm. In this derivation, adsorption energies and heat of adsorption of adsorbing molecules are assumed to

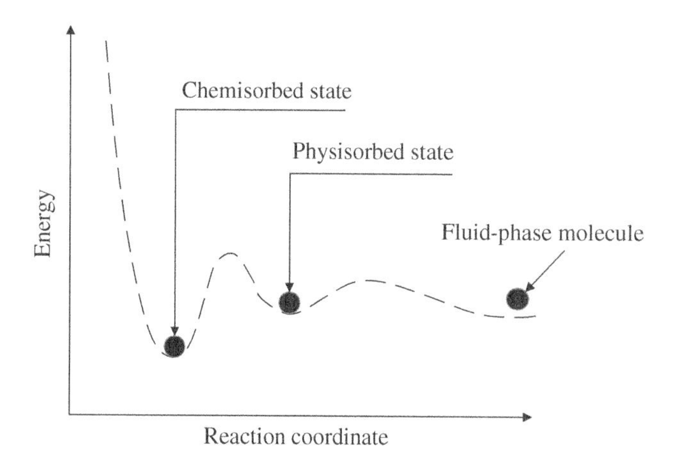

Figure 11.1 Schematic diagram of physisorbed and chemisorbed states.

be independent of surface coverage. Actually, in many cases, the assumption of uniform energetics of the surface sites may not be correct. Variation of the heat of adsorption with surface coverage is reported in many cases. Hence, several adsorption isotherms were proposed in the literature to take into account the heterogeneity in surface energetics. Further details of adsorption isotherms are discussed in Section 11.3.

11.2 Surface Reaction Mechanisms

Let us consider a gas-phase reaction catalyzed by a solid catalyst.

$$A + B \rightleftarrows C + D$$

In the mechanism illustrated in Figure 11.2, it assumed that both species A and B are adsorbed on the catalyst surface, and products C and D are formed as a result of a surface reaction between the adsorbed species. Product molecules may or may not be adsorbed on the catalyst surface. In this example, all of the surface sites are assumed to have the same energetics (uniform activity), and possible adsorption of product molecules is not considered. As illustrated in Figure 11.2, a fraction of surface sites are covered by adsorbed A (θ_A), and another fraction of surface sites are covered by reactant B (θ_B). Considering that the total number of sites available on the surface (S_o) is equal to the summation of the number of sites occupied by the adsorbed molecules and unoccupied sites available for adsorption, the following relation can be written:

$$S_o = [\underline{s}] + [\underline{As}] + [\underline{Bs}] \tag{11.1}$$

Here, $[\underline{As}]$, $[\underline{Bs}]$, and $[\underline{s}]$ correspond to the adsorbed concentrations of A, B, and the surface concentration of available sites (vacant sites), respectively.

Assuming a reaction mechanism, which is composed of reversible elementary adsorption steps of reactants A and B (Eqs. (11.2) and (11.3)) and a surface reaction step (Eq. (11.4)) for the production of C and D, we can derive a reaction rate expression for this surface-catalyzed reaction.

$$A + \underline{s} \rightleftarrows \underline{As} \tag{11.2}$$

$$B + \underline{s} \rightleftarrows \underline{Bs} \tag{11.3}$$

$$\underline{As} + \underline{Bs} \xrightarrow{k_r} C + D + 2\underline{s} \quad \text{(rate-determining step)} \tag{11.4}$$

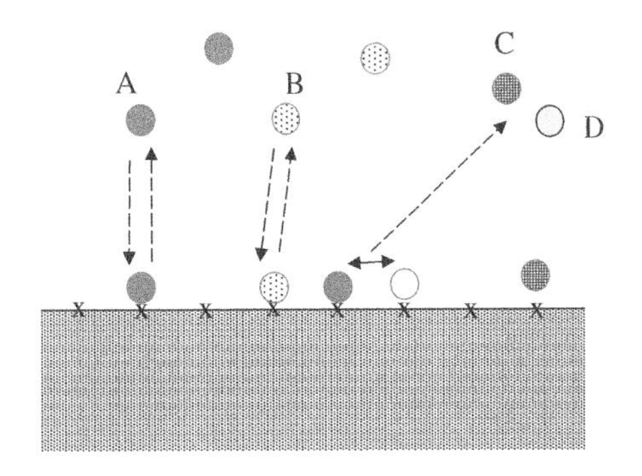

Figure 11.2 Schematic representation of a surface-catalyzed reaction.

Assuming that the surface reaction between the adsorbed molecules $\underline{As}$ and $\underline{Bs}$ is the rate-determining step (slowest step of the mechanism), the rate of formation of C and D can be expressed in terms of adsorbed concentrations of A and B, as follows:

$$R_C = -R_A = k_r[\underline{As}][\underline{Bs}] \tag{11.5}$$

In a catalytic reaction mechanism, reversible elementary steps, which occur before the rate-determining step, can be assumed in pseudo-equilibrium. Hence, the following adsorption equilibrium relations can be written for the adsorption steps of Eqs. (11.2) and (11.3).

$$K_A = \frac{[\underline{As}]}{[\underline{s}]C_A}; \quad [\underline{As}] = K_A[\underline{s}]C_A \tag{11.6}$$

$$K_B = \frac{[\underline{Bs}]}{[\underline{s}]C_B}; \quad [\underline{Bs}] = K_B[\underline{s}]C_B \tag{11.7}$$

Elimination of $[\underline{As}]$ and $[\underline{Bs}]$ from Eqs. (11.6) and (11.7) and substitution into Eqs. (11.5) and (11.1) yield the following relations:

$$-R_A = k_r K_A K_B [\underline{s}]^2 C_A C_B \tag{11.8}$$

$$S_o = [\underline{s}] + K_A C_A [\underline{s}] + K_B C_B [\underline{s}] \tag{11.9}$$

Elimination of the surface concentration of available (vacant) sites for adsorption from Eq. (11.9) gives the following relation for $[\underline{s}]$.

$$[\underline{s}] = \frac{S_o}{1 + K_A C_A + K_B C_B} \tag{11.10}$$

The substitution of this relation to Eq. (11.8) gives the typical Langmuir–Hinshelwood-type rate expression for a surface-catalyzed reaction.

$$-R_A = \frac{k_r K_A K_B S_o^2 C_A C_B}{(1 + K_A C_A + K_B C_B)^2} = \frac{k^* C_A C_B}{(1 + K_A C_A + K_B C_B)^2} \tag{11.11}$$

Here, K_A and K_B are the adsorption equilibrium constants of species A and B. k^* is the apparent rate constant of the surface-catalyzed reaction, which is defined as follows:

$$k^* = k_r K_A K_B S_o^2 \tag{11.12}$$

Since physical adsorption is an exothermic process, adsorption equilibrium constants K_A and K_B are expected to decrease with an increase in temperature. As for the temperature dependence of the apparent rate constant k^* is concerned, it depends upon the activation energy of the surface reaction rate constant k_r, as well as the heat of adsorption values for species A and B.

If adsorption equilibrium constants of both of the species are very small, Eq. (11.11) reduces to a second-order rate expression.

$$\text{If } K_A C_A \ll 1.0 \text{ and } K_B C_B \ll 1.0; \quad -R_A = k^* C_A C_B \tag{11.13}$$

On the other hand, if the adsorption equilibrium constant of one of the adsorbing species is very high, that species covers most of the surface and does not leave many available sites for the adsorption of the other reactant. In that case, the rate expression reduces to Eq. (11.14).

$$\text{If } K_A C_A \gg K_B C_B \text{ and } K_A C_A \gg 1.0; \quad -R_A = \frac{k_r K_B S_o^2}{K_A}\left(\frac{C_B}{C_A}\right) \tag{11.14}$$

In this case, the increase of concentration of reactant A causes a decrease in the reaction rate.

In the derivation of Eq. (11.11), only A and B are considered as the adsorbed species. However, in many cases, one or more of the products may also be adsorbed on the catalyst surface. For instance, if product C is also reversibly adsorbed on the catalyst surface, an adsorption equilibrium relation can also be written for this species.

$$K_C = \frac{[\underline{Cs}]}{[\underline{s}]C_C}; \quad [\underline{Cs}] = K_C C_C[\underline{s}] \tag{11.15}$$

In this case, the expressions for the concentration of free sites on the catalyst surface and the reaction rate expression become as follows:

$$[\underline{s}] = \frac{S_o}{1 + K_A C_A + K_B C_B + K_C C_C} \tag{11.16}$$

$$-R_A = \frac{k_r K_A K_B S_o^2 C_A C_B}{(1 + K_A C_A + K_B C_B + K_C C_C)^2} \tag{11.17}$$

Note that adsorption equilibrium terms ($K_i C_i$) of all the adsorbed species appear in the denominator of the Langmuir–Hinshelwood-type rate expressions.

The power of the parentheses in the denominator of the rate expression (Eq. (11.17)) is equal to the number of active sites involved in the rate-determining step of the reaction mechanism. The rate-determining step expressed in Eq. (11.4) involves two active sites. Hence, the power of the denominator is 2.0 in this expression. In some cases, an empty site may also contribute to the surface reaction step. For instance, if the rate-determining step is in the following form

$$\underline{As} + \underline{Bs} + \underline{s} \xrightarrow{k_r} C + D + 3\underline{s} \text{ (rate-determining step)} \tag{11.18}$$

the rate expression should contain the third power of $\underline{s}$

$$-R_A = k_r K_A K_B [\underline{s}]^3 C_A C_B \tag{11.19}$$

and hence the rate law becomes

$$-R_A = \frac{k_r K_A K_B S_o^3 C_A C_B}{(1 + K_A C_A + K_B C_B)^3} \tag{11.20}$$

In the reaction schemes illustrated above, all of the sites are assumed as identical. However, it is possible to have different types of active sites on the catalyst surface [3]. If there are two different types of sites on the catalyst surface and if reactants A and B are adsorbed on different types of sites ($\underline{s_1}$ and $\underline{s_2}$), the reaction mechanism may be written as follows:

$$A + \underline{s_1} \rightleftharpoons \underline{As_1} \tag{11.21}$$

$$B + \underline{s_2} \rightleftharpoons \underline{Bs_2} \tag{11.22}$$

$$\underline{As_1} + \underline{Bs_2} \xrightarrow{k_r} C + D + \underline{s_1} + \underline{s_2} \text{ (rate-determining step)} \tag{11.23}$$

For this case, the rate expression becomes

$$-R_A = \frac{k_r K_A K_B S_{1o} S_{2o} C_A C_B}{(1 + K_A C_A)(1 + K_B C_B)} \tag{11.24}$$

where S_{1o} and S_{2o} are the total number of Type 1 and Type 2 sites available on the catalyst surface per unit mass (or per unit area) of the catalytic material.

Another relation derived for the surface-catalyzed reactions is Rideal–Eley-type rate expression. In this case, one of the reactants is assumed to be adsorbed on the catalyst surface, while the second reactant is in the fluid phase, and the reaction takes place between the adsorbed A and fluid phase B.

$$A + \underline{s} \rightleftharpoons \underline{As} \tag{11.25}$$

$$\underline{As} + B \xrightarrow{k_r} C + D + \underline{s} \text{ (rate-determinig step)} \tag{11.26}$$

The following relations can be written in this case:

$$S_o = [\underline{s}] + K_A C_A [\underline{s}]; \quad [\underline{s}] = \frac{S_o}{1 + K_A C_A} \tag{11.27}$$

The Rideal–Eley-type rate expression then becomes

$$-R_A = \frac{k_r K_A S_o C_A C_B}{(1 + K_A C_A)} \tag{11.28}$$

In some other cases, dissociative adsorption of molecules should be considered in the derivation of the rate expression. Suppose that the reactant A_2 is a diatomic molecule, and it dissociates upon adsorption. The other reactant B is also adsorbed on the surface. The overall reaction stoichiometry is

$$A_2 + B \rightarrow C \tag{11.29}$$

The reaction mechanism is given as

$$A_2 + 2\underline{s} \rightleftharpoons 2\underline{As} \text{ (fast, equilibrium)} \tag{11.30}$$

$$B + \underline{s} \rightleftharpoons \underline{Bs} \text{ (fast, equilibrium)} \tag{11.31}$$

$$\underline{As} + \underline{Bs} \xrightarrow{k_r} \underline{Es} + \underline{s} \text{ (slow, rate-determining step)} \tag{11.32}$$

$$\underline{As} + \underline{Es} \rightarrow \underline{Cs} + \underline{s} \text{ (fast)} \tag{11.33}$$

$$\underline{Cs} \rightleftharpoons C + \underline{s} \text{ (fast, equilibrium)} \tag{11.34}$$

Here, E is an adsorbed intermediate, which immediately reacts with an adsorbed atom of A, giving adsorbed product molecule C. The rate of this reaction can be written from the rate-determining step as follows:

$$R = -R_A = k_r [\underline{As}][\underline{Bs}] \tag{11.35}$$

Note that the fast elementary steps after the rate-determining step are not kinetically significant. The surface concentrations of adsorbed A and B can be estimated from the equilibrium relations of Eqs. (11.30) and (11.31), respectively.

$$K_A = \frac{[(\underline{As})]^2}{[\underline{s}]^2 C_{A_2}} \tag{11.36}$$

$$K_B = \frac{\underline{Bs}}{C_B [\underline{s}]} \tag{11.37}$$

Then, the rate expression can be written as

$$R = k_r K_A^{1/2} K_B [\underline{s}]^2 (C_{A_2})^{1/2} C_B \tag{11.38}$$

The total number of sites on the catalyst surface is expressed as follows:

$$S_o = [\underline{s}] + K_A^{1/2} C_{A_2}^{1/2} [\underline{s}] + K_B C_B [\underline{s}] + K_C C_C [\underline{s}] \tag{11.39}$$

The concentration of available sites for adsorption $[\underline{s}]$ is then eliminated from Eq. (11.39) and substituted into Eq. (11.38) to obtain the following rate expression:

$$-R_A = \frac{k_r K_A^{1/2} K_B S_o^2 C_{A_2}^{1/2} C_B}{\left(1 + K_A^{1/2} C_{A_2}^{1/2} + K_B C_B + K_C C_C\right)^2} \tag{11.40}$$

As indicated in this expression, the square root of the concentration of this molecule appears in the denominator of the rate expression in the case of dissociative adsorption of a diatomic molecule A_2 (like hydrogen).

In all of the reaction schemes illustrated above, the overall reaction is considered as being irreversible, and the reverse reaction terms are neglected. In the case of reversible reactions, the net rate of reaction should become zero when the equilibrium condition is reached. The rate expressions for a reversible reaction can then be predicted, knowing the forward rate law. For instance, for the Langmuir–Hinshelwood-type rate expression derived above for the reaction mechanism described by Eqs. (11.2)–(11.4), if the rate-determining step is taken as reversible, the following rate law can be obtained:

$$-R_A = \frac{k_r K_A K_B S_o^2 \left(C_A C_B - \frac{1}{K} C_C C_D\right)}{(1 + K_A C_A + K_B C_B)^2} \tag{11.41}$$

Here, K is the equilibrium constant of the overall reaction ($A + B \rightleftarrows C + D$), and the reaction rate approaches zero at equilibrium.

11.3 Adsorption Isotherms

Any surface-catalyzed reaction involves the adsorption of reacting species on the catalyst surface. The most commonly used approach for the estimation of surface coverages of reacting species and the products is the use of Langmuir isotherms. According to the Langmuir approach, the surface of the adsorbent is considered as an array of equivalent sites available for the adsorption of molecules. In the ideal case, any lateral interactions between the adsorbed molecules are negligible, and the energetics of the adsorption/desorption process is independent of surface coverage. According to the Langmuir formulation, adsorption is monolayer, and equilibrium is reached between the adsorbed and the fluid-phase concentrations. When adsorption equilibrium is reached, the adsorption rate becomes the same as the desorption rate.

$$k_a P_A (1 - \theta) = k_d \theta \tag{11.42}$$

Here, θ is the fraction of sites covered by the adsorbing molecules, and, k_a and k_d are the adsorption and desorption rate constants, respectively. Noting that the ratio of adsorption and desorption rate constants gives the adsorption equilibrium constant ($k_A/k_d = K_A$) and expressing the adsorbed concentration of the adsorbent A as, $[\underline{As}] = n_{A,\text{ads}} = S_o\theta$, the following equilibrium relation can be written:

$$n_{A,\text{ads}} = \theta S_o = \frac{K_A S_o P_A}{1 + K_A P_A} \tag{11.43}$$

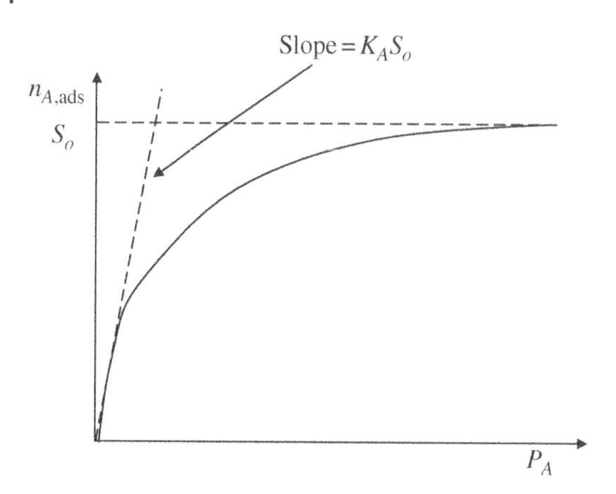

Figure 11.3 Langmuir adsorption isotherm for species A.

Variation of $n_{A,\text{ads}}$ with respect to the pressure of the adsorbing gas is shown in Figure 11.3. Rearrangement of Eq. (11.43) yields a linear relationship between $P_A/n_{A,\text{ads}}$ and P_A.

$$\frac{P_A}{n_{A,\text{ads}}} = \frac{1}{K_A S_o} + \left(\frac{1}{S_o}\right) P_A \tag{11.44}$$

Using the relation given in Eq. (11.44), the surface area of the adsorbent and the adsorption equilibrium constant of adsorbing species A can be estimated from the slope and the intercept of the $P_A/n_{A,\text{ads}}$ vs. P_A plot. Such a plot can be drawn using the experimental adsorption equilibrium data obtained by volumetric or gravimetric techniques. In such experiments, adsorbed quantities of adsorbing gas can be determined at different gas-phase concentrations (pressures). Note that the surface area of the adsorbent can be estimated from the product of S_o and the surface area occupied by a single adsorbing molecule (estimated from its molecular dimensions). In many cases, nitrogen gas is used in the surface area measuring instruments, and the surface area occupied by a single nitrogen molecule is used as 16.2×10^{-16} cm^2/molecule.

Langmuir adsorption isotherm can be used for systems of uniform surface energetics and also for monolayer adsorption cases. Monolayer adsorption assumption fails in the case of high pressures of the adsorbing gas or at low temperatures. Brunauer, Emmett, and Teller (BET) extended the Langmuir approach to multi-layer adsorption and obtained the following linear BET equation [4].

$$\frac{P_A}{n_{A,\text{ads}}\left(P_A^o - P_A\right)} = \frac{1}{S_o c} + \frac{(c-1)}{c S_o}\left(\frac{P_A}{P_A^o}\right) \tag{11.45}$$

Here, P_A^o is the saturation vapor pressure of adsorbing species and c is a temperature-dependent constant. The constant c corresponds to the adsorption equilibrium constant, which appears in the Langmuir isotherm. It can be exponentially related to the heat of adsorption (or molar energy of adsorption). BET isotherms are commonly used for the measurement of the surface area of the catalytic materials. The linear relation predicted by the BET isotherm is shown in Figure 11.4. The values of S_o and the parameter c can be evaluated from the slope and the intercept of the linear relation given in Figure 11.4.

For a quick evaluation of the surface area, a single point of adsorption isotherm can be used. For nitrogen, the value of the parameter c is quite high, so that the intercept of the BET equation can be neglected. In this case, the single-point BET method can be used for the evaluation of the surface area. The following expression can be used for the evaluation of S_o.

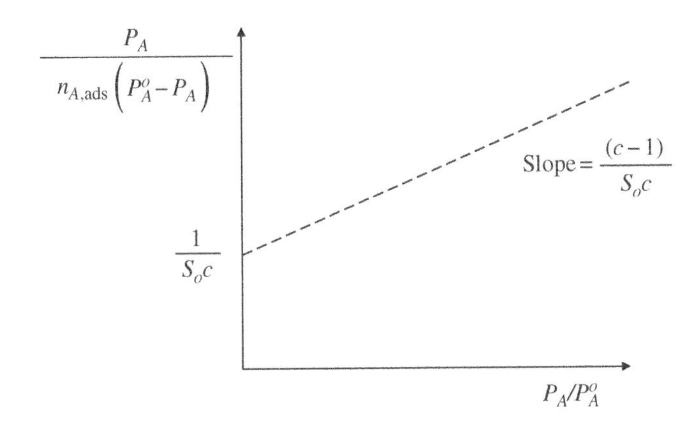

Figure 11.4 Schematic representation of BET isotherm.

$$S_o = n_{A,\text{ads}}\left(1 - \frac{P_A}{P_A^o}\right) \tag{11.46}$$

In many cases, the assumptions of Langmuir adsorption isotherm do not apply for gas–solid catalytic systems. Measured heat of adsorption values indicates a decline as a function of surface coverage θ of the adsorbing molecules. The activation energy of the adsorption rate constant may also increase with an increase in surface coverage [4]. Such a behavior can be explained either by the nonuniformity of the energetics of the catalyst surface or by recognizing the interaction forces between the adsorbing molecules.

Several different models were proposed as adsorption isotherms for nonuniform surface energetics. One of the most commonly used adsorption isotherms for nonideal adsorption is Freundlich isotherm (Eq. (11.47)).

$$n_{A,\text{ads}} = KP_A^{1/m} \quad [K \text{ and } m \text{ are constants } (m > 1)] \tag{11.47}$$

Nitrogen adsorption/desorption data can also be used to evaluate the pore volume and the mean pore radius of the porous catalyst. The total pore volume is determined from the amount of nitrogen gas adsorbed at a pressure close to the saturation pressure P_A^o. For pressure values close to the saturation, capillary condensation occurs, and the pores can be assumed to be filled with the liquid adsorbate. The moles of nitrogen adsorbed at the saturation pressure can then be used to predict the pore volume. The evaluation of pore volume by the nitrogen adsorption/desorption isotherms is generally used to evaluate the microporous and mesoporous materials. Mercury porosimeter should be used for the evaluation of the macropore volume.

The radii of the pores can be estimated by assuming that all of the pores are cylindrical. For instance, the average value of the pore radius can be estimated using the following expression:

$$a = \frac{2V_{\text{pore}}}{S_g} \tag{11.48}$$

Here, V_{pore} and S_g are pore volume and pore surface area per unit mass of the porous material, respectively.

11.4 Deactivation of Solid Catalysts

In many cases, the activity of the catalysts decreases with reaction time. Several factors cause catalyst deactivation. Most of the deactivation processes may be due to the following factors:

1) *Coke formation during reaction:* In many of the reactions in the petroleum and organic industry, carbon deposition may take place on the active sites of the catalysts. For instance, in fluidized-bed catalytic cracking (FCC) units, a significant amount of coke forms on the catalyst particles in a quite short time, and regeneration of the catalyst is needed by burning the carbon deposit. Coke deposition on the catalyst surface may also plug the pores of the catalyst. This type of catalyst deactivation is reversible in most cases, and regeneration of the deactivated catalytic material can be achieved.
2) *Deactivation by catalyst poisons:* Several catalyst poisons may be chemisorbed on the active sites of the catalysts and cause deactivation. For instance, sulfur compounds act as poisons for Ni, Cu, Pt-based catalysts. Another gaseous compound, which may be chemisorbed on the active sites of Ni and Pt, is carbon monoxide. This is the main reason for the use of almost CO-free hydrogen in PEM fuel-cells, which contain Pt catalysts. These poisons may be strongly or weakly chemisorbed on the catalyst surface, causing deactivation. If the adsorption is not very strong and reversible, regeneration may be possible. However, in some cases, chemisorption of the poison on the active site may be very strong and irreversible. In such cases, regeneration may not be possible.
3) *Sintering:* The third group of deactivation processes is due to the change of the structure of the catalyst during the reaction. For instance, if the temperature of the reactor is higher than the maximum stable temperature of the catalyst, active metal clusters distributed over the supported catalyst may coagulate, forming larger clusters as a result of sintering. Another example is the presence of water in the reaction mixture, which may interact with the catalyst support material, causing changes in the pore structure. For instance, for the MCM-41-type silicate-structured catalyst support material having an ordered pore structure, the presence of water vapor may cause pore structure changes, especially at elevated temperatures. This type of catalyst deactivation is generally irreversible.

There are several models proposed in the literature for the deactivation of the catalysts. One of the earlier and most common approaches is the inclusion of an activity term to the rate expression. For instance, for an nth- order catalytic reaction, the rate expression can be expressed as follows:

$$-R_A = ka_cC_A^n \tag{11.49}$$

Here, a_c corresponds to the activity factor of the catalyst, which is unity at the initial time. This factor can also be related to the fraction of sites that are active at a reaction time of t.

$$a_c = \frac{S_t}{S_o} \tag{11.50}$$

Here, S_o and S_t correspond to the initial number of active sites and the number of active sites at reaction time t, respectively.

In the case of parallel poisoning by a poison present in the reactor, the deactivation rate may be proportional to the concentration of the poison (C_p).

$$-\frac{da_c}{dt} = k_da_cC_p^m \tag{11.51}$$

Here, k_d is the deactivation rate constant. In the case of a constant concentration of the poison in the reactor, the integration of Eq. (11.51) gives the following relation for activity.

$$a_c = \exp\left(-k_d C_p^m t\right) = \exp\left(-k_d^* t\right) \quad (11.52)$$

The rate equation for an nth- order reaction can then be written as follows:

$$-R_A = k\left(\exp\left(-k_d^* t\right)\right)C_A^n \quad (11.53)$$

For slow deactivation processes, the pseudo-steady-state assumption can be made for the species conservation equation around the reactor. The relation, which is given in Eq. (11.53), can then be used in setting up the design equation for the reactor. For instance, for an nth -order reaction taking place in a CSTR or a plug-flow reactor, species conservation equation can be expressed, for the constant volumetric flow rate case, as follows:

$$\text{CSTR:} \quad Q\left(C_{A_o} - C_{A_f}\right) - Vk\left(\exp\left(-k_d^* t\right)\right)C_A^n = 0 \quad (11.54)$$

$$\text{Plug Flow:} \quad -Q\frac{dC_A}{dV} - k\left(\exp\left(-k_d^* t\right)\right)C_A^n = 0 \quad (11.55)$$

In some cases, the reactant A may also act as a poison. For instance, in the case of carbon deposition on the catalyst surface, the increase in reactant concentration may also increase the coke deposition rate. In this case, the deactivation rate can be expressed as follows:

$$-\frac{da_c}{dt} = k_d a_c C_A^m \quad (11.56)$$

In the case of concentration-independent deactivation ($m = 0$), activity expression becomes

$$a_c = \exp\left(-k_d t\right) \quad (11.57)$$

For a first-order reaction with concentration-independent deactivation process plug-flow equation becomes

$$-Q\frac{dC_A}{dV} - k\exp\left(-k_d t\right)C_A = 0 \quad (11.58)$$

The solution of Eq. (11.58) gives the following expression:

$$C_A = C_{A_o}\exp\left[-\frac{kV}{Q}\exp\left(-k_d t\right)\right] \quad (11.59)$$

As discussed in Chapter 15, concerning non-catalytic gas–solid reactions, in the case of concentration-dependent deactivation rate ($m = 1$ in Eq. (11.56)), iterative solution of the species conservation equation gives the following relation.

$$C_A = C_{A_o}\exp\left\{\frac{\left[1 - \exp\left(\frac{kV}{Q}\left(1 - \exp\left(-k_d t\right)\right)\right)\right]}{\left(1 - \exp\left(-k_d t\right)\right)}\exp\left(-k_d t\right)\right\} \quad (11.60)$$

More sophisticated models, considering changes in the structure, progressive poisoning in the pores of a catalyst, etc. are available in the literature.

Exercises

11.1 Consider a catalyst deactivation process, such that the rate of decrease of activity at any point in a spherical pellet can be expressed as follows:

$$-\frac{da_c}{dt} = k_d a_c$$

Obtain an expression for the effectiveness factor considering a first-order surface-catalyzed reaction, and plot effectiveness factor with respect to the initial Thiele modulus, time being the parameter. Discuss the results.

11.2 Epitaxial growth of crystalline films on a substrate is performed by chemical vapor deposition (CVD) techniques. Epitaxial growth process involves adsorption of gas-phase species on the substrate surface, followed by a surface reaction to incorporate adsorbed species into the crystal structure.

$$A(g) + \underline{s} \rightleftharpoons \underline{As} \quad \text{(adsorption equilibrium)}$$

$$\underline{As} \rightarrow C(c) + B(g) \quad \text{(rate-determining step)}$$

Another mechanism proposed for epitaxial growth involves the formation of adsorbed surface atoms as an intermediate step in the reaction mechanism.

$$A(g) + \underline{s} \rightleftharpoons \underline{As}$$

$$\underline{As} \rightleftharpoons \underline{Cs} + B(g)$$

$$\underline{Cs} \rightarrow C(c)$$

Derive rate expressions for both of the proposed mechanisms mentioned above.

References

1 Rase, F.F. (2000). *Handbook of Commercial Catalysts*. New York: CRC Press.

2 Somorjai, G.A. (1994). *Introduction to Surface Chemistry and Catalysis*. New York: Wiley.

3 Carberry, J.J. (1977). Catalysis and catalytic kinetics: an introduction. In: *Chemical Reactor Theory: A Review* (eds. L. Lapidus and N.R. Amundson), 156–189. Englewood Cliffs: Prentice-Hall.

4 Rouquerol, F., Rouquerol, J., and Sing, K. (1999). *Adsorption by Powders and Porous Solids, Principles Methodology and Applications*. London: Academic Press.

12

Diffusion in Porous Catalysts

As discussed in Chapter 9, most industrially important reactions are catalytic, and in many cases, porous solid catalysts are used to accelerate the reaction rate. In these heterogeneously catalyzed reaction systems, transport of species through porous catalysts should take place before the adsorption-reaction steps on the active sites (Chapters 9 and 10). In a packed-bed reactor filled with porous solid catalysts, the reactants should diffuse into the pores of the catalysts to reach active sites, which are on the pore surfaces. After the surface reaction, back diffusion of products to reach the external catalyst surface and then to the bulk of the fluid within the reactor takes place. Therefore, it is essential to understand and analyze the diffusion of species through porous solids.

12.1 Diffusion in a Capillary

Transport of a species through a porous solid may occur by diffusion and convective flow within the pores and also by surface diffusion. A detailed review of diffusion of gases in porous media was reported in the early work of Youngquist [1], Scott and Dullien [2], Dogu [3]. Since the pore structure in a porous catalyst is very complex, the diffusion process within these materials is not simple. The pores are not generally regularly arranged, and they are highly interconnected. It is crucial to analyze and understand the diffusion mechanism inside a capillary.

The diffusion process in a capillary takes place by two mechanisms:

- Molecular diffusion
- Knudsen diffusion.

Molecular diffusive transport of a gaseous species A in a mixture of A and B takes place through the collision of molecules. It is generally represented by Fick's law of diffusion as follows:

$$J_A = -D_{AB}C\frac{dy_A}{dz} \tag{12.1}$$

In this equation J_A represents the flux of species A in direction z. D_{AB} is the molecular diffusion coefficient of A in B, and it is a function of the molecular properties of the two substances, temperature and total pressure. It has the unit of $(\text{length})^2/\text{time}$. Total concentration and the mole fraction of A in the mixture are represented by C and y_A, respectively. In the molecular diffusion region, the transport of species A in a binary mixture can be written considering both diffusive and convective fluxes, as follows:

Fundamentals of Chemical Reactor Engineering: A Multi-Scale Approach, First Edition. Timur Doğu and Gülşen Doğu.

Companion website: www.wiley.com/go/dogu/chemreacengin

$$N_A = (N_A + N_B)y_A - CD_{AB}\frac{dy_A}{dz} \tag{12.2}$$

Here, N_A and N_B represent the fluxes of A and B in the fixed coordinates. The first term on the right-hand side of Eq. (12.2) represents convective flux, while the second term represents molecular diffusive flux.

If the radius of the pore is much greater than the mean free path of the molecules, the diffusive transport is governed by the collision of molecules. In this case, diffusion is said to be in the molecular region. However, if the radius of the pore is much less than the mean free path of the molecules, transport of molecules takes place by their collision with the pore walls. This type of diffusion is called Knudsen diffusion. If the diffusive transport in a capillary is in the Knudsen region, the molecules of A are transported through the collision of molecules with the pore walls.

Consider the schematic diagram of a porous solid given in Figure 12.1a. Before modeling the transport of molecules through porous media, we should model the transport in a single pore (Figure 12.1b). The flux of A in the Knudsen region is written in terms of a Knudsen diffusion coefficient D_{K_A} as follows:

$$N_A = -D_{K_A}\frac{dC_A}{dz} \tag{12.3}$$

Pore radius to mean free path ratio (a/λ) determines whether the diffusion is in molecular or in Knudsen region:

If $\frac{a}{\lambda} \gg 10$; molecular diffusion dominates

If $\frac{a}{\lambda} \ll 0.1$; Knudsen diffusion dominates

If $0.1 < \frac{a}{\lambda} < 10$; transition region

Both molecular and Knudsen diffusion regimes contribute to the diffusive transport of molecules in the transition region.

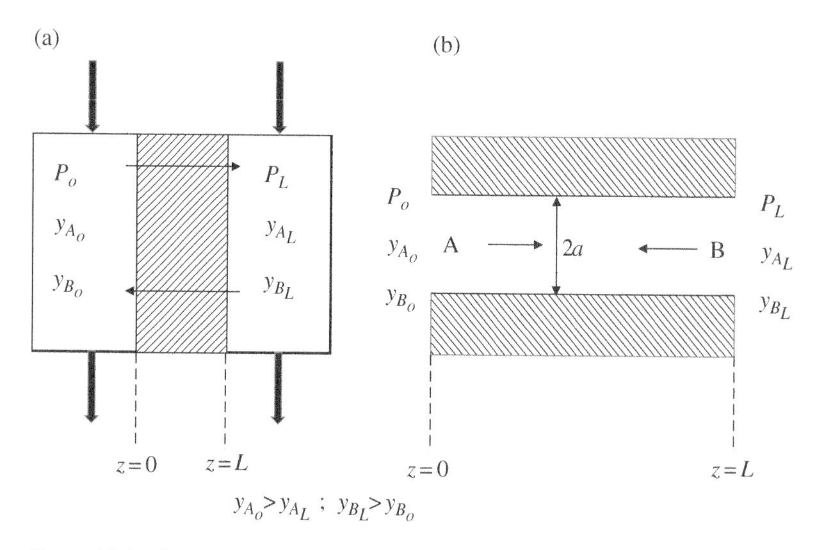

Figure 12.1 Schematic representation of diffusion of A in a binary mixture: (a) through a single porous pellet and (b) through a single capillary.

Molecular diffusion coefficients can be predicted using Eq. (12.4), which was derived based on a simplified model in gas kinetic theory [3, 4].

$$D_{AB} = \frac{1}{3}\bar{u}_A\lambda \tag{12.4}$$

Here, $\bar{u}_A$ is the average velocity of molecule A. Equations for the mean free path (λ) and $\bar{u}_A$ were also derived using gas kinetic theory.

$$\bar{u}_A = \left(\frac{8R_gT}{\pi M_A}\right)^{½} \tag{12.5}$$

$$\lambda = \frac{1}{2^{1/2}\pi d^2 N} \tag{12.6}$$

Here, M_A is the molecular weight of A, R_g is the gas constant, and N is the total number density of molecules (total number of molecules per unit volume), and d is the molecular diameter.

Knudsen diffusion coefficient can be predicted by replacing λ with the pore diameter in Eq. (12.4).

$$D_{K_A} = \frac{2}{3}\bar{u}_A a = \frac{2}{3}\left(\frac{8R_gT}{\pi M_A}\right)^{½} a \tag{12.7a}$$

$$D_{K_A} = 9700a\left(\frac{T}{M_A}\right)^{½} \quad (\text{cm}^2/\text{s}) \tag{12.7b}$$

In this equation, "a" should be in the unit of cm, and T is in K. As for the molecular diffusion coefficient is considered, it can be predicted using the Chapman and Enskog expression [4].

$$D_{AB} \propto \frac{\sqrt{T^3\left(\frac{1}{M_A} + \frac{1}{M_B}\right)}}{P\sigma_{AB}^2\Omega_{D,AB}} \tag{12.8}$$

The values of the Lennard–Jones parameter $\left(\sigma_{AB} = \frac{\sigma_A + \sigma_B}{2}\right)$ and the collision integral $\Omega_{D,AB}$, which appear in this expression, are available for different molecules in the literature [4].

As seen from Eqs. (12.7a) and (12.7b), when diffusion is in the Knudsen region, the diffusion coefficient is proportional to $T^{1/2}$. However, in the molecular diffusion region, D_{AB} is proportional to $T^{3/2}$. The Knudsen flux of A through the capillary is given by Eq. (12.3), even if the pressures on both sides of the capillary are not the same. The number of collisions between molecules increases with an increase in pressure. This is essential in the molecular diffusion region. However, in a purely Knudsen diffusion region, transport takes place by the collision of molecules with the walls of the pores. Hence, the convective transport contribution can be neglected. Considering the model system shown in Figure 12.1, when the concentration of A at position $z = 0$ (which is C_{A_0}) is higher than the concentration at position $z = L$ (namely C_{A_L}), molecules of A are transported in the positive z direction, while there is a back diffusion of molecules of B in the opposite direction. If the diffusion is in the Knudsen region, the fluxes for an ideal gas mixture can be related to partial pressures by the following equations:

$$N_A = -D_{K_A}\frac{dC_A}{dz} = -\frac{D_{K_A}}{R_gT}\frac{dP_A}{dz} \tag{12.9}$$

$$N_B = -D_{K_B}\frac{dC_A}{dz} = -\frac{D_{K_B}}{R_gT}\frac{dP_B}{dz} \tag{12.10}$$

Integral forms of these equations are

$$N_A = -D_{K_A}\left(\frac{P_{A_L} - P_{A_o}}{R_g TL}\right) \tag{12.11}$$

$$N_B = -D_{K_B}\left(\frac{P_{B_L} - P_{B_o}}{R_g TL}\right) \tag{12.12}$$

$$\frac{N_A}{D_{K_A}} + \frac{N_B}{D_{K_B}} = -\left(\frac{1}{R_g TL}\right)[(P_{A_L} + P_{B_L}) - (P_{A_o} + P_{B_o})] = -\left(\frac{1}{R_g TL}\right)(P_L - P_o) \tag{12.13}$$

For a constant pressure system $(P_L - P_o) = 0$, the following relation can be written between the fluxes of A and B.

$$\frac{N_A}{N_B} = -\frac{D_{K_A}}{D_{K_B}} \tag{12.14}$$

Since the Knudsen diffusion coefficient is inversely proportional to the molecular weight of the diffusing species, Eq. (12.14) can be expressed in the following form:

$$\frac{N_A}{N_B} = -\left(\frac{M_B}{M_A}\right)^{1/2} \tag{12.15}$$

This relation is shown to apply also for the molecular diffusion region in a capillary. The molar flux of species A in a binary mixture of A and B is expressed in the molecular diffusion range as

$$N_A = -D_{AB}C\frac{dy_A}{dz} + y_A\left(1 + \frac{N_B}{N_A}\right)N_A = -D_{AB}C\frac{dy_A}{dz} + y_A\overline{\alpha}N_A \tag{12.16}$$

The rearrangement of this expression yields

$$N_A = -\left(\frac{D_{AB}C}{1 - \overline{\alpha}y_A}\right)\frac{dy_A}{dz} \tag{12.17}$$

The parameter $\overline{\alpha}$ is expressed as

$$\overline{\alpha} = 1 - \left(\frac{M_A}{M_B}\right)^{1/2} \tag{12.18}$$

If the diffusion is in the transition region, the flux of A in the capillary can then be expressed as

$$N_A = -\frac{C}{\frac{1 - \overline{\alpha}y_A}{D_{AB}} + \frac{1}{D_{K_A}}}\frac{dy_A}{dz} \tag{12.19a}$$

This expression can also be written in terms of the total pressure of the system:

$$N_A = -\frac{\frac{P}{R_g T}}{\frac{1 - \overline{\alpha}y_A}{D_{AB}} + \frac{1}{D_{K_A}}}\frac{dy_A}{dz} \tag{12.19b}$$

In the case of constant total concentration, Eq. (12.19a) can be expressed as follows:

$$N_A = -\frac{1}{\frac{1 - \overline{\alpha}y_A}{D_{AB}} + \frac{1}{D_{K_A}}}\frac{dC_A}{dz} \tag{12.19c}$$

The following expression can be used in the transition region, where molecular and Knudsen diffusivities contribute to the transport in a capillary:

$$D_T = \frac{1}{\dfrac{1-\bar{\alpha}y_A}{D_{AB}} + \dfrac{1}{D_{K_A}}} \tag{12.20}$$

This composite diffusivity is a function of both the capillary radius and molecular properties of the diffusing species. At low concentrations or for equimolar countercurrent diffusion, Eq. (12.20) reduces to the well-known Bosanquet formula.

$$D_T = \frac{1}{\dfrac{1}{D_{AB}} + \dfrac{1}{D_{K_A}}} \tag{12.21}$$

12.2 Effective Diffusivities in Porous Solids

The equations derived above give flux relations for species A within the pores of a porous solid. In analyzing the transport of A through a porous solid, it is customary to express a pseudo-homogeneous diffusion expression. Diffusive flux for a pseudo-homogeneous catalyst is based on the total cross-sectional area (open pore area plus solid area) of the pellet, which is normal to the direction of diffusion (Figure 12.2). This diffusive flux is called effective diffusion (N_{A_e}) in the porous catalyst.

If there are m number of pores per unit cross-sectional area of the pellet, the porosity of the pellet can be expressed as follows:

$$\varepsilon = \frac{\pi a^2 L_e m}{L} \tag{12.22}$$

Here, L_e is the equivalent length of a pore, which is the actual distance traveled by a molecule in a pore (Figure 12.2).

The effective flux N_{A_e} is expressed as the moles of A transported per total cross-sectional area of the porous solid per unit time. Therefore, based on the simple parallel-pore model illustrated in Figure 12.2, the relation between the effective flux and flux within the pores can be expressed as follows:

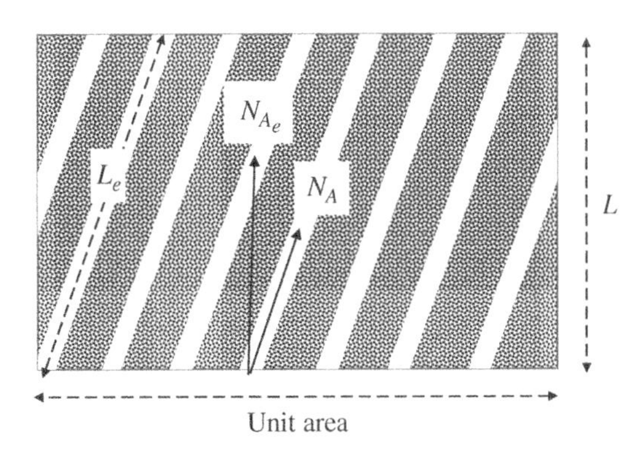

Figure 12.2 Schematic representation of a porous solid.

$$N_{A_e} = N_A(\pi a^2)m\ \cos\theta = N_A\varepsilon\left(\frac{L}{L_e}\right)^2 \tag{12.23}$$

In this relation, the $\left(\frac{L}{L_e}\right)^2$ term accounts for the fact that the actual diffusion path in the capillaries is not the same as the effective diffusion path in the direction normal to the external surface of the porous pellet. Thus, the relation between N_{A_e} and the flux within a capillary can be expressed as follows:

$$N_{A_e} = N_A\frac{\varepsilon}{\tau} \tag{12.24}$$

Here, τ is the tortuosity factor of the pellet. Effective diffusivity within a porous catalyst pellet can then be expressed by using the relations given in Eqs. (12.24) and (12.21), as follows:

$$D_e = D_T\frac{\varepsilon}{\tau} = \left(\frac{1}{\frac{1}{D_{AB}} + \frac{1}{D_{K_A}}}\right)\frac{\varepsilon}{\tau} \tag{12.25}$$

Since Knudsen diffusivity is a function of pore radius, composite diffusivity in the transition region is also expected to be a function of pore radius. Knudsen diffusivity is commonly evaluated using the average pore radius, in Eq. (12.7b). However, if the pore size distribution of the porous solid is not narrow, this assumption may not give satisfactory results. In that case, Knudsen and molecular contributions should be summed over the complete range of pore sizes. Thus, the effective diffusivity can be evaluated by using the following integral expression:

$$D_e = \frac{1}{\tau}\int_0^\infty D_T(a)f(a)da \tag{12.26}$$

Here, $f(a)da$ is the pore volume in the pores of radius between "a" and "$a + da$" per unit pellet volume.

12.3 Surface Diffusion

The transport of adsorbed species on the catalyst surface is called surface diffusion. To have an appreciable contribution of surface diffusion to the total flux, the species should be adsorbed on the surface, and the adsorbed molecules should be mobile. When a molecule strikes the pore surface, it may reflect back to the gas phase or remain on the surface for a while. If the vibrational energy of the molecule is higher than the bond energy of the adsorbed molecule and the surface, the molecule desorbs after a short period. As also discussed in Chapter 11, this kind of adsorption is called reversible adsorption. Suppose the energy of the adsorbed molecule at a particular surface site is sufficiently high as compared to the energy barrier between the adjacent sites. In that case, the adsorbed molecules may jump from one site to another site and hence move over the surface. If the energy barrier of the adjacent sites is very low, the mobility of the adsorbed molecules on the surface may be considered as non-localized surface diffusion. Lateral interaction of the adsorbing molecules should also be considered at high concentrations.

Surface diffusion flux through a porous solid may be expressed as follows:

$$N_{A_s} = -\frac{D_s \rho_p K}{\tau_s}\frac{dC_A}{dz} = -D_s^*\frac{dC_A}{dz} \tag{12.27}$$

Here, $\rho_p K$ is the adsorption equilibrium constant, giving the relation between the gas phase and adsorbed concentrations of A, per unit pellet volume ($n_{A,\text{ads}} = \rho_p K C_A$). Here, τ_s corresponds to the tortuosity factor for surface diffusion. Surface diffusion is an activated process, and hence temperature dependence of the surface diffusion coefficient is exponential, as in Arrhenius expression.

The contribution of surface diffusion on the effective diffusion flux may become significant for strongly adsorbed species and if the energy barrier between the adjacent sites is quite low [7]. For this case, effective diffusivity can be expressed as follows:

$$D_e \cong \frac{\varepsilon}{\tau}\left(\frac{1}{\frac{1}{D_{K_A}} + \frac{1}{D_{AB}}}\right) + \frac{D_s \rho_p K}{\tau_s} \tag{12.28}$$

12.4 Models for the Prediction of Effective Diffusivities

In chemical reaction processes dealing with gas-porous solid catalytic systems, the prediction of diffusive transport rate in the porous solid is often required. For the prediction of the effective diffusion coefficient, information on the pore size distribution and the tortuosity of the pellet are required. The less the amount of experimental data required for the prediction of diffusivities, the more useful the model is. Several diffusion models used to determine effective diffusivities in porous solid catalysts were reviewed in an early publication [3]. Some of the models available in the literature for the prediction of effective diffusivities are summarized below.

12.4.1 Random Pore Model

This early model, proposed by Wakao and Smith, requires porosity and pore size distribution data for the estimation of effective diffusivity [5]. The random pore model is one of the earliest but well-accepted models, which is proposed to estimate the diffusive flux through a bidisperse catalyst pellet, containing both macro- and micropores. Predictions of this model can also be reduced to catalysts with a monodisperse pore structure. A bidisperse porous catalyst is formed as an agglomeration of microporous particles with macropores in between these particles.

The volume fraction of macropores is equal to the void area fraction of the surface of the catalyst. Hence, diffusion through the bidisperse catalyst is assumed to take place by three parallel paths:

- Diffusion through the macropore–macropore region in series, proportional to ε_a^2
- Diffusion through the micropore–micropore region in series, proportional to $(1 - \varepsilon_a)^2$
- Diffusion through the macropore–micropore region in series, proportional to $2\varepsilon_a(1 - \varepsilon_a)$.

Effective flux expression was then derived as given in Eq. (12.28).

$$N_{A_e} = \left[-\varepsilon_a^2 D_{T_a} - (1-\varepsilon_a)^2 D_{T_i} - 2\varepsilon_a(1-\varepsilon_a)\frac{1}{\frac{1}{D_{T_a}} + \frac{1}{D_{T_i}}}\right]\frac{P}{R_g T}\frac{dy_A}{dz} \tag{12.29}$$

Here, ε_a is the macroporosity of the pellet. D_{T_a} and D_{T_i} are the combined diffusivity values in the macropore and micropore regions, respectively.

$$D_{T_a} = \frac{1}{\dfrac{1}{D_{AB}} + \dfrac{1}{D_{K_a}}} \tag{12.30}$$

$$D_{T_i} = \frac{(\varepsilon_i/(1-\varepsilon_a))^2}{\dfrac{1}{D_{AB}} + \dfrac{1}{D_{K_i}}} \tag{12.31}$$

For a monodisperse pellet, ε_i (microporosity) is zero, and Eq. (12.29) reduces to

$$N_{A_e} = -\varepsilon_a^2 D_{T_a} \frac{P}{R_g T} \frac{dy_A}{dz} \tag{12.32}$$

Here, $\varepsilon_a^2 D_{T_a}$ corresponds to the effective diffusivity. Hence, the tortuosity factor of a monodisperse catalyst pellet can be estimated by the Wakao and Smith model using the following relation:

$$\tau = \frac{1}{\varepsilon_a} \tag{12.33}$$

12.4.2 Grain Model

There are several other models, like Johnson and Stewart model, Foster and Butt model, etc. in the early literature, to estimate the tortuosity factor. Another practical model proposed in the literature is the grain model [6]. This model applies to both mono- and bidisperse structured porous solids. In this model development, a unit cell of the catalyst pellet was defined as the smallest element representing the pore structure. The unit cell was considered to contain a single grain. If the grain is microporous, the pellet has a bidisperse pore structure. If the ratio of macro- and micropore diffusivity values is higher than 10^2, the contribution of micropores to the net flux through the pellet is considered to be negligible. For this case, the tortuosity factor expression proposed by the grain model was reported as follows:

$$\tau = \frac{\varepsilon_a}{1 - \pi\left[(1-\varepsilon_a)\dfrac{3}{4\pi}\right]^{2/3}} \tag{12.34}$$

12.5 Diffusion and Flow in Porous Solids

For a non-isobaric system, transport of molecules through a porous solid is a quite complex phenomenon, involving viscous flow, Knudsen flow, surface flow, as well as effective pore diffusion. If we consider a slab-shaped porous pellet, the effective flux of an adsorbing gas through this non-isobaric pellet can be expressed as follows [7–9]:

$$N_{A_e} = -D_e \frac{dC_{A_p}}{dz} - P_D C_{A_p} \tag{12.35}$$

Here, the parameter P_D is proportional to the pressure gradient across the pellet, and it contains the convective transport terms.

For catalyst pellets containing mainly macropores, viscous flow contribution may become significant, and the parameter P_D, which appears in Eq. (12.35), can be expressed as follows:

$$P_D = \left(\frac{-\Delta P}{L}\right)\frac{a^2\varepsilon}{8\mu}\left(\frac{1}{\tau}\right) = \left(\frac{-\Delta P}{L}\right)\frac{C_d}{\mu} \tag{12.36}$$

Here, C_d is the Darcy coefficient defined for convective flux in a porous solid.

12.6 Experimental Methods for the Evaluation of Effective Diffusion Coefficients

Effective diffusion coefficients in porous solids can be experimentally evaluated following steady-state or unsteady-state (dynamic) methods.

12.6.1 Steady-State Methods

Effective diffusivities can be evaluated either directly from diffusion experiments with inert molecules or indirectly from the adsorption or reaction data. One of the most commonly used steady-state methods for evaluating the effective diffusion coefficient involves the use of a Wicke–Kallenbach-type diffusion cell (Figure 12.3).

In this diffusion cell, gases A and B flow through the upper and lower chambers of the cell, sweeping the upper and the lower faces of the porous catalyst pellet. The system is kept at constant pressure to achieve counter-diffusion of species A and B through the porous solid. Compositions of the gases in the upper and the lower chambers can be determined by the detectors connected to the upper- and lower-exit gas streams. Diffusion flux through the pellet can be experimentally evaluated, knowing the compositions and flow rates of these streams. Then, the effective diffusion coefficient of species A can be determined using Eq. (12.37).

$$N_{A_e} = -D_e\left(\frac{P}{R_gT}\right)\left(\frac{y_{A_2}-y_{A_1}}{L}\right) \tag{12.37}$$

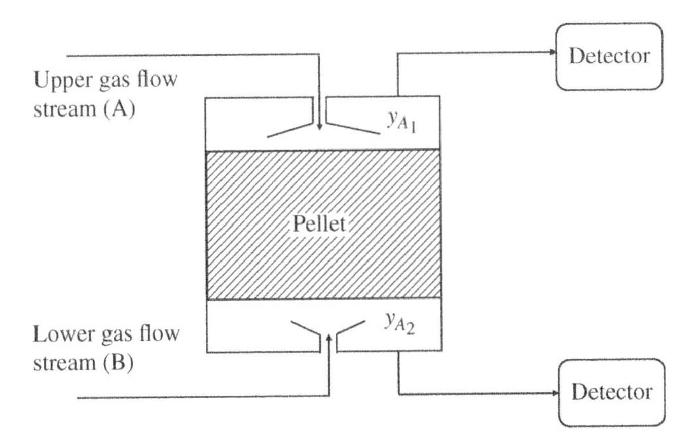

Figure 12.3 Wicke–Kallenbach-type diffusion cell for steady-state experiments.

12.6.2 Dynamic Methods

Details of dynamic analysis of the axial dispersion coefficient in a tubular reactor by the moment technique are described in Section 8.6. As described in Section 8.6, this technique is based on injecting a tracer into the inlet stream of a vessel and analyzing the moments of the response peaks to evaluate the system parameters. If we consider injection of a tracer A into the inlet stream of a vessel as an impulse, distribution of the residence times of the tracer molecules within this vessel will depend upon the hydrodynamics of this system, as well as the diffusion, adsorption, and reaction processes taking place within the reactor/adsorber (Figure 12.4).

Chromatographic processes involving sorption and diffusion of gases and liquids are extensively investigated in the literature. One of the earlier studies for the dynamic analysis of a fixed-bed adsorber for the evaluation of effective diffusion coefficient and adsorption parameters of an adsorbing tracer is by Schneider and Smith [10]. Pseudo-homogeneous species conservation equation for an adsorbing tracer A injected as a pulse into the inlet stream of a fixed-bed adsorber can be expressed as follows:

$$\varepsilon_b \frac{\partial C_A}{\partial t} = -U_0 \frac{\partial C_A}{\partial z} + D_z \frac{\partial^2 C_A}{\partial z^2} - \frac{3}{r_p} D_e (1-\varepsilon_b) \left(\frac{\partial C_{A_p}}{\partial r} \right)_{r=r_p} \tag{12.38}$$

Here, U_0 is the superficial velocity in the fixed bed, ε_b is the bed void fraction, D_z is the axial dispersion coefficient, and D_e is the effective diffusivity of tracer within the spherical porous pellets having a radius of r_p. Corresponding equations for the concentration of tracer C_{A_p} within the porous spherical adsorbent and for the adsorbed concentration of tracer $n_{A,\text{ads}}$ are given in the following equations:

$$\varepsilon \frac{\partial C_{A_p}}{\partial t} = \frac{D_e}{r^2} \frac{\partial}{\partial r} \left(r^2 \frac{\partial C_{A_p}}{\partial r} \right) - \rho_p \frac{\partial n_{A,\text{ads}}}{\partial t} \tag{12.39}$$

$$\frac{\partial n_{A,\text{ads}}}{\partial t} = k_{\text{ads}} \left(C_{A_p} - \frac{n_{A,\text{ads}}}{K_A} \right) \tag{12.40}$$

In these equations, ε is the porosity of the pellet, and reversible adsorption of the tracer is considered on the pore surfaces of the pellet. The primary assumption of these equations is a linear adsorption process on the surface of the adsorbent. Mass transfer resistance at the external surface of the adsorbent pellets is also considered in the derivation of the moment expressions.

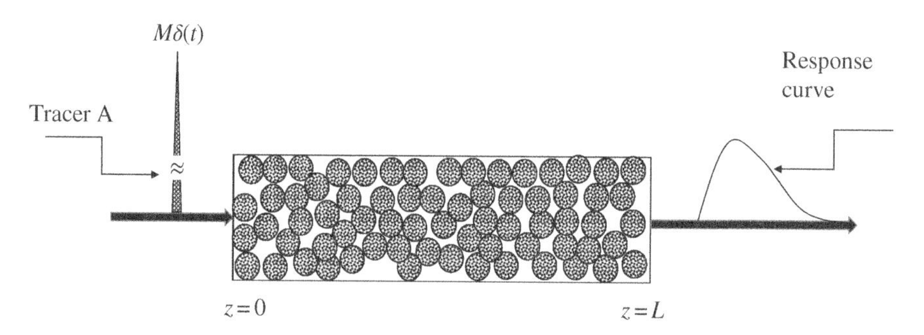

Figure 12.4 Schematic diagram of the pulse-response analysis for a fixed-bed reactor.

$$D_e\left(\frac{\partial C_{A_p}}{\partial r}\right)_{r = r_p} = k_m\left(C_A - C_{A_p}\big|_{r = r_p}\right) \tag{12.41}$$

Details of the moment analysis and the definitions of the first absolute and the second central moments are given in Section 8.6. First absolute and the second central moment expressions were then derived and reported by Schneider and Smith [10].

$$\mu_1 = \frac{L}{U_0}\varepsilon_b(1 + \delta_0) \tag{12.42}$$

$$\mu_2' = \frac{2L\varepsilon_b}{U_0}\left[\delta_1 + D_z\varepsilon_b(1 + \delta_0)^2\frac{1}{U_0^2}\right] \tag{12.43}$$

where

$$\delta_0 = \left[(1-\varepsilon_b)\frac{\varepsilon}{\varepsilon_b}\right]\left(1 + \frac{\rho_p}{\varepsilon_b}K_A\right) \tag{12.44}$$

$$\delta_1 = (1-\varepsilon_b)\frac{\varepsilon}{\varepsilon_b}\left[\frac{\rho_p K_A^2}{\varepsilon k_{\text{ads}}} + \frac{r_p^2\varepsilon}{15}\left(1 + \frac{\rho_p}{\varepsilon}K_A\right)^2\left(\frac{1}{D_e} + \frac{5}{k_m r_p}\right)\right] \tag{12.45}$$

In the case of negligible film mass transfer resistance on the surface of the adsorbent pellets and also for a fast adsorption rate process, the expression for the parameter δ_1 becomes as follows:

$$\delta_1 = (1-\varepsilon_b)\frac{\varepsilon}{\varepsilon_b}\left[\frac{r_p^2\varepsilon}{15D_e}\left(1 + \frac{\rho_p K_A}{\varepsilon}\right)\right] \tag{12.46}$$

In the case of an inert tracer, the only parameters that appear in the second central moment expression are the axial dispersion coefficient and the effective diffusivity in the pellet. All of the transport rate parameters appear in the second central moment expression in fixed-bed moment analysis. Hence, independent evaluation of axial dispersion coefficient, external mass transfer rate, or effective pore diffusion coefficient from the second central moment data requires careful experimentation. Moment analysis of diffusion and adsorption in packed columns was also investigated in detail in the recent literature [11, 12].

12.6.3 Single-Pellet Moment Method

Evaluation of the intrapellet rate and equilibrium parameters, from the response peaks to concentration pulses injected into the inlet stream of fixed beds, requires the elimination of axial dispersion and fluid–pellet mass transfer processes. A single-pellet moment technique was developed by Dogu and Smith [13, 14] considering these disadvantages of dynamic analysis of fixed-bed systems. Investigation of effective diffusion coefficients, and intrapellet rate and equilibrium processes were achieved by eliminating the contributions of axial dispersion and external transport parameters to the moments of the response peaks. In this method, a dynamic version of the Wicke–Kallenbach diffusion cell was used (Figure 12.5).

In dynamic experiments performed with the single-pellet system, a pulse of inert or adsorptive tracer gas is introduced into the carrier gas stream flowing over the upper face of a cylindrical porous pellet. Then, the response is measured with a detector, which is connected to the outlet stream, leaving the lower chamber. The differential species conservation equation for an inert tracer A within the pellet is written as follows:

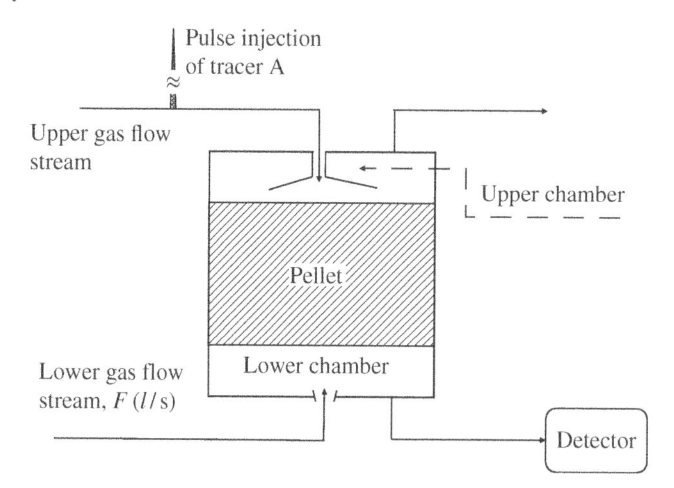

Figure 12.5 Schematic representation of the dynamic single-pellet technique.

$$\varepsilon\frac{\partial C_{A_p}}{\partial t} = D_e\frac{\partial^2 C_{A_p}}{\partial x^2} \tag{12.47}$$

The solution of this equation with negligible mass transfer resistance in the upper chamber of the cell gives the following expression for the first absolute moment.

$$\mu_1 = \frac{L^2\varepsilon}{6D_e}\left(\frac{3\frac{A}{L}D_e + F}{\frac{A}{L}D_e + F}\right) \tag{12.48}$$

The effective diffusion coefficient is the only rate parameter in this first-moment expression. It can be evaluated from the experimental first-moment data obtained at different flow rates (F) of the lower stream carrier gas. The limit of Eq. (12.48) for very high values of F can also be used for the evaluation of D_e.

$$\lim_{F\to\infty}\mu_1 = \frac{L^2\varepsilon}{6D_e} \tag{12.49}$$

In the case of an adsorbing tracer, the species conservation equation can be written in the single pellet, as follows:

$$\varepsilon\frac{\partial C_{A_p}}{\partial t} = D_e\frac{\partial^2 C_{A_p}}{\partial x^2} - \rho_p K_A\frac{\partial C_{A_p}}{\partial t} \tag{12.50}$$

In writing this equation, a linear adsorption equilibrium relation is considered. For this case, the first absolute and the second central moment expressions become as follows:

$$\mu_1 = \frac{L^2\varepsilon}{6D_e}\left(1 + \frac{\rho_p K_A}{\varepsilon}\right)\left(\frac{3\frac{A}{L}D_e + F}{\frac{A}{L}D_e + F}\right) \tag{12.51}$$

$$\mu_2' = \left(\frac{L^2\varepsilon + L^2\rho_p K_A}{D_e}\right)\left[\frac{\left(\frac{A}{L}D_e\right)^2 + \frac{2}{5}\left(\frac{A}{L}D_e\right)F + \frac{1}{15}F^2}{\left(\frac{A}{L}D_e + F\right)^2}\right] \tag{12.52}$$

The application of the single-pellet moment technique to the evaluation of diffusion and adsorption parameters in different systems is available in the literature [13–16]. This method was further developed to evaluate the thermal conductivity of catalyst pellets [17].

Exercises

12.1 The effective diffusion coefficient of He in a macroporous catalyst was experimentally evaluated as 0.046 cm^2/s at 25 °C by the single-pellet diffusion cell method, in which the carrier gas was nitrogen. The mean pore diameter of the porous catalyst was given as 270 nm, and its porosity was 0.41.

- **(a)** Which diffusion mechanism (molecular of Knudsen) controls the diffusion process in the pellet?
- **(b)** Estimate the tortuosity factor from the experimental data.
- **(c)** Estimate the tortuosity factor using the Wakao–Smith and the grain models. Compare these predictions with the experimental value of tortuosity.

12.2 For a Cu–ZnO–alumina-based methanol synthesis catalyst, surface area and pore-volume values are reported as 65 m^2/g and 0.25 cm^3/g, respectively. Apparent density and porosity of this material are given as 2.0 g/cm^3 and 0.50, respectively.

- **(a)** Estimate the value of pore diameter, assuming parallel cylindrical pores having the same diameter.
- **(b)** Estimate the Knudsen diffusion coefficient of methanol in this catalyst at 350 °C and 200 atm. Compare this with the molecular diffusion coefficient of methanol in hydrogen.
- **(c)** Estimate the tortuosity factor of this catalyst and also the effective diffusion coefficient of methanol.

12.3 The following data are obtained in a diffusion experiment in monodispersed porous catalysts having different porosities. Diffusion data are obtained for the diffusion of He in nitrogen at 45 °C. At this temperature, the molecular diffusion coefficient of He in N_2 is 0.77 cm^2/s. Obtain the tortuosity factors of the pellets using the following experimental data, and compare these values with the tortuosity factors estimated from the Wakao–Smith and the grain models.

Pellet no.	Porosity, ε	Average pore diameter, "a" (nm)	D_e (cm^2/s)
1	0.41	200	0.040
2	0.20	20	0.005

References

1 Youngquist, G.R. (1970). *Ind. Eng. Chem.* 62: 52–63.
2 Scott, D.S. and Dullien, F.A.L. (1962). *AIChE J.* 8: 113–117.

3 Dogu, G. (1986). Diffusion limitations for reactions in porous catalysts. In: *Handbook of Heat and Mass Transfer*, vol. 2 (ed. N.P. Cheremisinoff), 433–483. Houston: Gulf Publishing Company.
4 Bird, R.B., Stewart, W.E., and Lightfoot, E.N. (2002). *Transport Phenomena*. New York: Wiley.
5 Wakao, N. and Smith, J.M. (1962). *Chem. Eng. Sci.* 17: 825–834.
6 Dogu, G. and Dogu, T. (1991). *Chem. Eng. Commun.* 103: 1–9.
7 Dogan, M. and Dogu, G. (2003). *AIChE J.* 49: 3188.
8 Satterfield, C.N. (1970). *Mass Transfer in Heterogeneous Catalysis*. Cambridge: MIT Press.
9 Dogu, G., Pekediz, A., and Dogu, T. (1989). *AIChE J.* 35: 1370–1375.
10 Schneider, P. and Smith, J.M. (1968). *AIChE J.* 14: 762.
11 Qamar, S. and Seidel-Morgenstern, A. (2016). *TrAC, Trends Anal. Chem.* 81: 87–101.
12 Qamar, S., Abbasi, J.N., Mehwish, A., and Seidel-Morgenstern, A. (2015). *Chem. Eng. Sci.* 137: 352–363.
13 Dogu, G. and Smith, J.M. (1975). *AIChE J.* 21: 58–61.
14 Dogu, G. and Smith, J.M. (1976). *Chem. Eng. Sci.* 31: 123–135.
15 Paek, S., Ahn, D.H., Kim, K.R. et al. (2007). *J. Ind. Eng. Chem.* 13: 121–126.
16 Guo, J.H., Shah, D.B., and Talu, O. (2007). *Ind. Eng. Chem. Res.* 46: 600–607.
17 Dogu, G., Murtezaoglu, K., Dogu, T. et al. (1988). *AIChE J.* 35: 683–686.

13

Process Intensification and Multifunctional Reactors

Inefficiencies involved in many of the conventional chemical processes led to the development of new approaches. The concept of process intensification involves the development of more efficient, smaller, and safer processing techniques than conventional processes used in the chemical industry. Process intensification requires the design and development of new equipment for chemical processing. The aim of process intensification is to produce useful products using less raw materials, less energy, and less space [1, 2]. Significant benefits of process intensification can be summarized as follows:

- Smaller and more efficient processes
- Less energy consumption per unit mass of product
- Waste minimization
- Improved yields of the desired product in multifunctional units
- Safer and sustainable processes.

From the point of view of chemical reaction engineering, the idea of integration of reaction and separation processes into a single unit initiated the development of multifunctional reactors. The use of alternative energy sources, like microwave, solar energy, ultrasound, etc. in chemical processing and design of novel reactors, such as micro-reactors, supercritical reactors, monolithic reactors, and fuel cells, opened new avenues in reaction engineering applications. A partial list of multifunctional reactors involving reaction and separation within the same unit is

- Membrane reactors
- Reactive distillation
- Sorption-enhanced reactors
- Reactive absorption
- Chromatographic reactors.

As a result of the integration of reaction and separation processes within the same unit, we expect the following improvements to the conventional systems:

- Lower investment and land cost due to the use of compact equipment
- Smaller inventory
- Safer operation
- Improved energy utilization, higher energy efficiency
- Improved yield of the desired product in equilibrium-limited reactions
- Elimination of some separation problems, like in azeotropic distillation.

Fundamentals of Chemical Reactor Engineering: A Multi-Scale Approach, First Edition. Timur Doğu and Gülşen Doğu.

Companion website: www.wiley.com/go/dogu/chemreacengin

Some of the multifunctional reactor examples, namely membrane reactors, equilibrium-stage, and continuous reactive distillation units, sorption-enhanced reaction processes, and chromatic reactors, are illustrated with examples in Sections 13.1–13.3 and 13.5. Modeling of monolithic reactors and some information about the microchannel reactors are also given in this chapter, with examples.

13.1 Membrane Reactors

Membrane reactors are typical examples of multifunctional reactors, in which improved yield of the desired product can be achieved [3]. Typically, membrane reactors can be used to increase the product yields in dehydrogenation reactions, like the production of ethylene from ethane, which is equilibrium limited.

$$C_2H_6 \leftrightarrow C_2H_4 + H_2$$

This is an endothermic reaction, and the equilibrium conversion of ethane to ethylene is quite low at moderate temperatures. Hence, high temperatures (over 800 °C) are needed to increase the conversion of ethane to ethylene (Figure 13.1). However, at high temperatures, further dehydrogenation of ethylene to acetylene and coke formation may also take place.

$$C_2H_4 \leftrightarrow C_2H_2 + H_2$$

$$C_2H_2 \rightarrow 2C + H_2$$

Hence, to achieve high conversions (over the equilibrium limit) at low temperatures, one of the products may be continuously removed from the reaction zone of the reactor. This can be achieved using a membrane reactor by continuous removal of hydrogen molecules from the reaction zone of a fixed-bed reactor, as illustrated in Figure 13.2. Hydrogen produced as a result of dehydrogenation reaction can be continuously transferred from the outer reaction zone to the inner tube through the membrane wall of this tube. Pd-based materials are known to be permeable to hydrogen but not to hydrocarbons. Hence, the inner tube wall is composed of a Pd membrane layer deposited on a porous support. As a result of the continuous removal of hydrogen from the reaction zone, equilibrium limitation can be eliminated, and very high ethane conversions can be achieved.

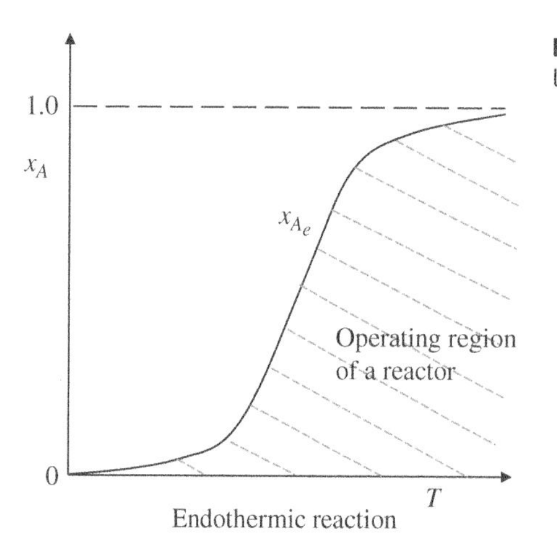

Figure 13.1 Schematic representation of equilibrium limitation for an endothermic reaction.

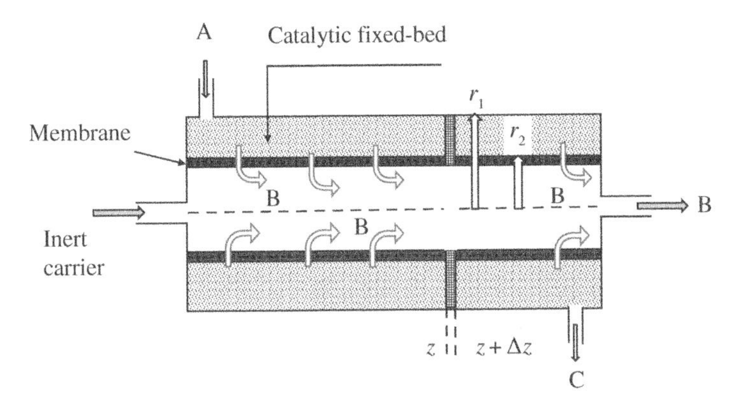

Figure 13.2 Schematic diagram of a membrane reactor for a dehydrogenation reaction.

Improved yields and production of high-purity hydrogen can also be obtained in steam reforming of methane and water–gas shift reaction in a fixed-bed or a fluidized-bed membrane reactor. Equilibrium limitations are also quite important in these reactions.

$$CH_4 + H_2O \rightleftarrows CO + 3H_2 \text{ (steam reforming of methane)}$$

$$CO + H_2O \rightleftarrows CO_2 + H_2 \text{ (water–gas shift reaction)}$$

Membrane reactors may also be used for the selective oxidation and oxidative dehydrogenation reactions. For instance, in oxidative dehydrogenation of ethane to ethylene, oxygen concentration should be kept low within the reaction zone to minimize the undesired total oxidation reactions.

$$C_2H_6 + \frac{1}{2}O_2 \rightarrow C_2H_4 + H_2O \text{ (oxidative dehydrogenation)}$$

$$C_2H_6 + \frac{7}{2}O_2 \rightarrow 2CO_2 + 3H_2O \text{ (combustion)}$$

A membrane reactor can be designed for this purpose. Oxygen may be continuously introduced to the reaction zone through the membrane walls of the inner tube in this reactor (Figure 13.3).

13.1.1 Modeling of a Membrane Reactor

Let us consider an elementary gas-phase reversible catalytic reaction taking place in a membrane reactor.

$$A \rightleftarrows B + C; \quad -R_A = R_B = R_C = k_f\left(P_A - \frac{1}{K_c}P_BP_C\right)$$

As illustrated in Figure 13.2, product B is continuously removed from the reaction zone through the membrane wall of the inner tube. To achieve transport of B through the membrane, there should be a concentration gradient of this species between the reaction zone and the inner tube. This can be achieved by applying a suction to the inner tube or increasing the pressure of the reaction zone. Species conservation equations for the reactant A and the products B and C, over the differential volume element between z and $z + \Delta z$ in the reaction section can be expressed as follows:

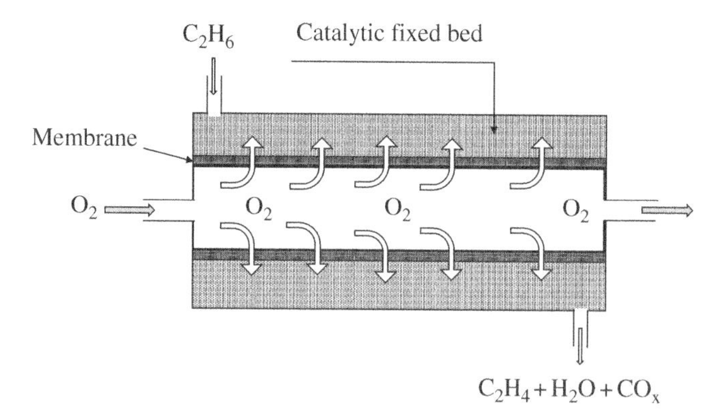

Figure 13.3 Schematic representation of an oxidative dehydrogenation reaction in a membrane reactor.

Reactant A:

$$\left(\pi r_1^2 - \pi r_2^2\right)\left(N_{A_z}\big|_z - N_{A_z}\big|_{z+\Delta z}\right) + \left(\pi r_1^2 - \pi r_2^2\right)\Delta z(R_A) = 0 \quad (13.1)$$

$$-\frac{dF_A}{dV} - k_f\left(P_A - \frac{1}{K_c}P_B P_C\right) = 0 \quad (13.2)$$

Product B:

$$\left(\pi r_1^2 - \pi r_2^2\right)\left(N_{B_z}\big|_z - N_{B_z}\big|_{z+\Delta z}\right) - 2\pi r_2 \Delta z\left(N_{B_r}\big|_{r=r_2}\right) + \left(\pi r_1^2 - \pi r_2^2\right)\Delta z(R_B) = 0 \quad (13.3)$$

$$-\frac{dF_B}{dV} - \frac{2r_2}{\left(r_1^2 - r_2^2\right)}\left(N_{B_r}\big|_{r=r_2}\right) + R_B = 0 \quad (13.4)$$

$$-\frac{dF_B}{dV} - \frac{2r_2}{\left(r_1^2 - r_2^2\right)}\prod\nolimits_B\left(P_B^{1/2} - P_{B,\mathrm{pm}}^{1/2}\right) + k_f\left(P_A - \frac{1}{K_c}P_B P_C\right) = 0 \quad (13.5)$$

Here, $a_{\mathrm{mem}} = \frac{2r_2}{\left(r_1^2 - r_2^2\right)}$ and $\prod_B$ are the membrane surface area per unit volume of the reaction zone and the permeability constant of product B in the membrane.
Product C:

$$\left(\pi r_1^2 - \pi r_2^2\right)\left(N_{C_z}\big|_z - N_{C_z}\big|_{z+\Delta z}\right) + \left(\pi r_1^2 - \pi r_2^2\right)\Delta z(R_C) = 0 \quad (13.6)$$

$$-\frac{dF_C}{dV} + k_f\left(P_A - \frac{1}{K_c}P_B P_C\right) = 0 \quad (13.7a)$$

Instead of Eq. (13.7a), we may also use the following stoichiometric relation for the molar flow rate of product C.

$$F_C = F_{A_o} - F_A \quad (13.7b)$$

These equations are written with the assumption of plug flow.

Flux through a membrane is generally considered proportional to the differences of nth powers of the partial pressures (or concentrations) of the penetrating molecule at the two sides of the membrane. In many cases, n is close to 1/2. In writing Eq. (13.5), the flux of product B through the membrane is expressed as

$$N_{Br}\big|_{r=r_2} = \prod_B \left(P_B^{1/2} - P_{B,\text{pm}}^{1/2}\right) \tag{13.8}$$

Here, $P_{B,\text{pm}}$ is partial pressure in the permeate side of the membrane reactor. Temperature dependence of the permeability constant is generally considered in the Arrhenius form. The design of this membrane reactor requires the simultaneous solution of these species conservation equations (Eqs. (13.2), (13.5), and (13.7a)). The partial pressures of the species involved in these equations may be expressed in terms of their molar flow rates as follows:

$$P_A = P\frac{F_A}{F_T}, \quad P_B = P\frac{F_B}{F_T}, \quad P_C = P\frac{F_C}{F_T} \tag{13.9}$$

Here P and F_T are the total pressure and the total molar flow rate in the reaction section, respectively.

$$F_T = F_A + F_B + F_C \tag{13.10}$$

If the partial pressure of the permeating product B in the permeate side ($P_{B,\text{pm}}$) is much less than its value in the reaction side, it may be taken as zero in Eq. (13.5). If not, we should also write a species conservation equation in the permeate section.

If the rate expression is given in terms of concentrations of species, instead of their partial pressures, these species conservation equations may also be expressed in terms of their concentrations. We may express the molar flow rates as the products of the species concentrations and the total volumetric flow rate. Constant total volumetric flow rate assumption may also be made, especially if the concentration of the limiting reactant is not high. For instance, if the steam to hydrocarbon ratio is quite high in a steam-reforming reaction, this assumption may be justified.

13.1.2 General Conservation Equations and Heat Effects in a Membrane Reactor

For a single reaction taking place in a non-isothermal fixed-bed membrane reactor, the conservation equation for the permeating product i through the membrane and the energy balance equations in the reaction and the sweeping gas (permeate) sections can be expressed in terms of the molar flow rates of the species as follows:

Reaction section:

$$-\frac{1}{A_c}\frac{dF_i}{dz} + (1-\varepsilon_b)\eta\,\nu_i R_p - a_{\text{mem}}\prod_i \left(P_i^{1/2} - P_{i,\text{pm}}^{1/2}\right) = 0 \tag{13.11}$$

$$-\frac{1}{A_c}\sum_i F_i C_{p_i}\frac{dT}{dz} + (1-\varepsilon_b)\eta R_p(-\Delta H^o) - a_{\text{mem}}U\left(T - T_{\text{pm}}\right) = 0 \tag{13.12}$$

Permeate section:

$$-\frac{1}{A_{\text{pm}}}\frac{dF_{i,\text{pm}}}{dz} + a_{\text{mem}}\prod_i \left(P_i^{1/2} - P_{i,\text{pm}}^{1/2}\right)\frac{A_c}{A_{\text{pm}}} = 0 \tag{13.13}$$

$$-\frac{1}{A_{\text{pm}}}\sum_i F_{i,\text{pm}} C_{p_i}\frac{dT_{\text{pm}}}{dz} + a_{\text{mem}}U\left(T - T_{\text{pm}}\right)\frac{A_c}{A_{\text{pm}}} = 0 \tag{13.14}$$

Here, R_p is the rate of the reaction based on catalyst volume ($R_{i,p} = \nu_i R_p$). A_c and A_{pm} are the cross-sectional areas of the reaction and the permeate sections of the membrane reactor, and a_{mem} is the surface area of the membrane per unit volume of the reaction zone. $F_{i,\text{pm}}$ and T_{pm} correspond to the molar flow rate of species i and the temperature in the permeate section, respectively. The contribution of convective heat flux through the membrane is neglected in writing these equations. Equation (13.11) may also be applied to the non-permeating species involved in this reaction. In the case of non-permeating species, the last term of Eq. (13.11) (permeation rate term) should be taken as zero.

In the case of a multiple reaction system, the reaction rate and the heat generation terms of Eqs. (13.11) and (13.12) should be modified accordingly. The reaction rate expressions may also be expressed as a function of molar flow rates of the species. An example of a membrane reactor is illustrated in Online Appendix O6.4.

13.2 Reactive Distillation

Reactive distillation is another example of a multifunctional reactor system. In a reactive distillation system, chemical conversion and the separation processes occur simultaneously within the same vessel, namely in a distillation column. Reactive distillation may be used for liquid-phase chemical conversions, which are limited by chemical equilibrium. Immediate removal of the products from the reaction zone of a reactive distillation column enhances the conversion of reactants. Hence, it causes an increase in the yield of the desired product by minimizing the equilibrium limitation. Significant advantages of a reactive distillation process can be listed as follows:

- Increase in the yield of the desired product due to overcoming of equilibrium limitations
- Decrease of energy consumption as a result of heat integration of distillation and exothermic reaction processes
- Increase of the selectivity of the desired product
- Avoidance of azeotropes
- Achievement of better temperature control
- Reduction of capital and operating costs.

A typical example of a reactive distillation process is methyl acetate production by the reaction of methanol with acetic acid.

$$CH_3OH + CH_3COOH \rightleftarrows CH_3COOCH_3 + H_2O$$

Production of methyl acetate is a slightly exothermic reversible reaction, and fractional conversion of methanol is limited by equilibrium. Usually, an acidic catalyst, such as sulfuric acid in a homogeneous system or an acidic ion-exchange resin for a heterogeneous system, is used in this process. Due to its exothermic nature, higher temperatures do not favor methyl acetate yield. Differences in boiling points of reactants and the products of this reaction (Table 13.1) are essential factors for the use of a reactive distillation process. Such a system may be proposed to eliminate the equilibrium limitations and to achieve very high methyl acetate yields.

As seen in Figure 13.4, a reactive distillation column has reactive and non-reactive zones. The trays in the reactive zone can be packed with a solid acid catalyst. Methanol should be fed to

Table 13.1 Normal boiling points of species involved in methyl acetate synthesis.

Species	Methyl acetate	Water	Methanol	Acetic acid
Boiling point, °C	57	100	65	118

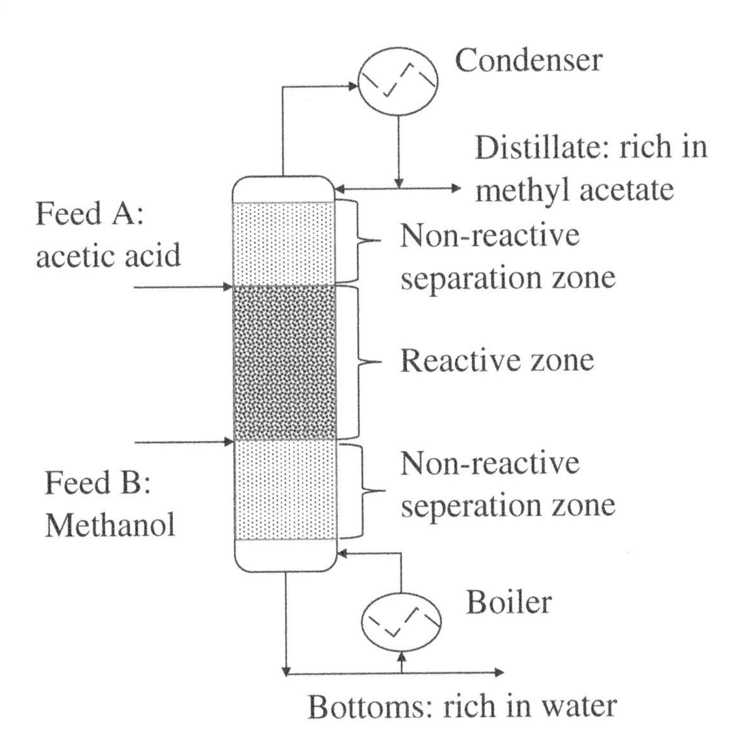

Figure 13.4 Schematic diagram of a reactive distillation column for methyl acetate synthesis.

the column from the lower end of the reaction zone, while acetic acid should be fed from the upper end of the reaction zone. Methanol, which has a boiling point of 65 °C, is expected to evaporate in the reaction zone and flow upwards through the column, while the acetic acid, which has a much higher boiling point (118 °C), will flow downwards in liquid form. As soon as the reaction takes place, the methyl acetate product, which has a boiling point of 57 °C, will transfer to the vapor phase, and it will immediately leave the reaction zone, flowing upward. On the other hand, the other product, namely water, leaves the reaction zone, flowing downward in liquid form. Immediate removal of the products from the reaction zone eliminates the equilibrium limitations in methyl acetate synthesis. Achievement of reaction and separation in the same unit not only enhances the methyl acetate yield but also causes significant improvements in the investment cost. It also improves the energy efficiency of the process by utilizing the heat of this reaction for the evaporation of volatile components within the reaction zone.

The feasibility of the synthesis of methanol in a reactive distillation system is illustrated in a recent publication [4]. Some other applications of reactive distillation are the synthesis of MTBE (methyl *tert*-butyl ether) from isobutene and methanol, alkylation of aromatic hydrocarbons, and hydration of ethylene oxide to produce ethylene glycol. Different approaches can be used for the modeling and design of reactive distillation processes [5–7].

13.2.1 Equilibrium-Stage Model

For continuous distillation, either a multi-stage column or a packed column can be used. In the case of a multi-stage column, an equilibrium-stage model can be used by writing the species conservation equations for the reactive component A around a stage containing no catalyst (separation zone) and also around a stage containing the catalyst (reactive zone) (Figure 13.5), as follows:

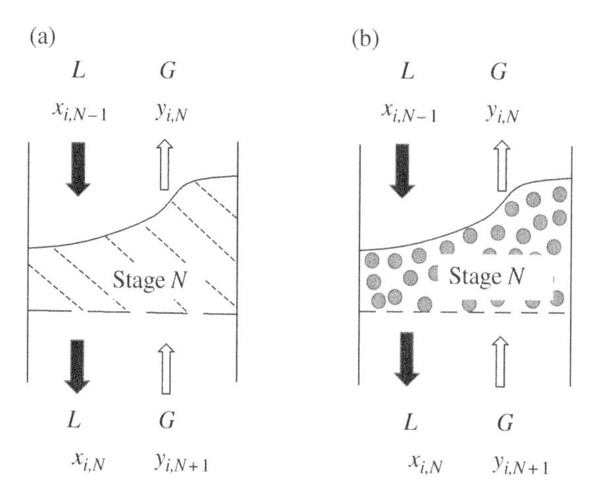

Figure 13.5 Schematic diagram of (a) non-reactive and (b) reactive stages in a reactive distillation-stage column.

Stage with no catalyst (separation zone):

$$\left(L_{N-1}x_{i,N-1} + G_{N+1}y_{i,N+1}\right) - \left(L_N x_{i,N} + G_N y_{i,N}\right) = 0 \tag{13.15}$$

$$y_{i,N} = K_{H,i}x_{i,N} \tag{13.16}$$

Equation (13.16) is the phase equilibrium relation between the vapor and the liquid mole fractions of species i leaving the stage N, namely $y_{i,N}$ and $x_{i,N}$, respectively, and $K_{H,i}$ is Henry's constant for species i.

Stage with the catalyst (reactive zone):

$$\left(L_{N-1}x_{i,N-1} + G_{N+1}y_{i,N+1}\right) - \left(L_N x_{i,N} + G_N y_{i,N}\right) + w_N\left(R_{i,N}\right)_w = 0 \tag{13.17}$$

Here, $(R_{i,N})_w$ is the reaction rate of species i in stage N (based on the unit mass of the catalyst). L_N, G_N are the molar flow rates of liquid and vapor streams, leaving stage N of the column, and w_N is the mass of the catalyst in stage N, respectively. In some cases, L_N and G_N are assumed as constant within the column. If the molar heats of vaporizations of the components are approximately the same, and if all of the heat effects are negligible, it is safe to make this assumption. In this case, for every mole of the liquid evaporated, one mole of vapor will condense. Note that the following conditions should also be satisfied for each stage:

$$\sum_{i=1}^{n_c} x_{i,N} = 1.0; \quad \sum_{i=1}^{n_c} y_{i,N} = 1.0 \tag{13.18}$$

Here, n_c is the total number of species in the reactive distillation column.

In the analysis of a multicomponent distillation column, generally, two components of the mixture are selected as the light-key and the heavy-key components. Any component lighter than the light-key (having higher relative volatility than the relative volatility of the light-key component) is generally assumed to appear only in the distillate, while any component having lower relative volatility than the heavy-key is assumed to appear only in the bottoms stream. Details of the stage-by-stage solution procedure of a multicomponent distillation column are beyond the scope of this book and available in the literature on separation processes.

If the heat effects in such an equilibrium-stage distillation column are not negligible, constant molar flow rate assumptions for the liquid and vapor phases within the column may fail, and an energy balance should also be written around each stage.

$$\left(L_{N-1}\hat{H}_{L,N-1} + G_{N+1}\hat{H}_{G,N+1}\right) - \left(L_N\hat{H}_{L,N} + G_N\hat{H}_{G,N}\right) = 0 \tag{13.19}$$

Here, $\hat{H}_{L,N}$ and $\hat{H}_{G,N}$ correspond to the average molar enthalpies of the liquid and the vapor streams leaving the stage N. $\hat{H}_{L,N}$ and $\hat{H}_{G,N}$ depend upon the compositions of the liquid and the vapor streams, respectively. They can be estimated by summing up the products of molar enthalpies with the mole fractions of the species in the respective streams at the temperature of the stage N. The selection of the feed location, the temperature of the reaction zone, and the reflux ratio are some of the critical parameters affecting the efficiency of the column, and the product yield.

13.2.2 A Rate-Based Model for a Continuous Reactive Distillation Column

Rate-based models can be used for the design of continuous packed-bed distillation columns. For the description of the mass transfer rate at the gas–liquid interface, different models can be used. Most conventional models proposed for the mass transfer rate between the phases are the two-film model, penetration model, etc. In the case of a two-film model, it is considered that the mass transfer rate between the phases is controlled by the resistances within the thin films adjacent to the liquid–gas interface (Figure 13.6).

The species conservation equations in the liquid and the vapor phases, within a differential height of Δz in a continuous distillation column (Figure 13.7), can be expressed as follows:

Section of the column with no reaction:

$$-\frac{d(\overline{L}x_i)}{A_c dz} + K_l a_{vl}\left(\frac{y_i}{K_{H,i}} - x_i\right) = 0 \tag{13.20}$$

$$-\frac{d(\overline{G}y_i)}{A_c dz} - K_l a_{vl}\left(\frac{y_i}{K_{H,i}} - x_i\right) = 0 \tag{13.21}$$

Here, K_l is the overall mass transfer coefficient and a_{vl} is the interfacial surface area between the liquid and the vapor streams, based on the unit volume of the column. $\overline{L}$ and $\overline{G}$ are the mean liquid- and the vapor-phase molar flow rate values, respectively. Assuming that the solid catalyst pellets within the column are thoroughly wetted by the liquid stream, and there is no fluid-phase reaction taking place in the vapor phase, the species conservation equation for the region of the bed containing the catalyst can be expressed as follows:

Section of the column with catalytic reaction:

$$-\frac{d(\overline{L}x_i)}{A_c dz} + K_l a_{vl}\left(\frac{y_i}{K_{H,i}} - x_i\right) + \eta R_{i,p}(1 - \varepsilon_b) = 0 \tag{13.22}$$

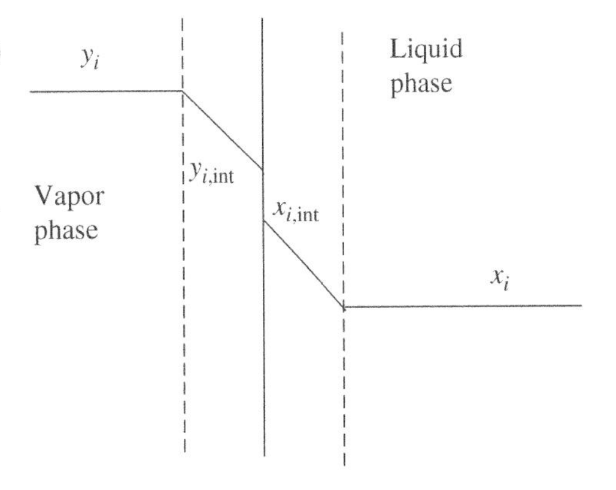

Figure 13.6 Schematic representation of two-film theory.

Here, $R_{i,p}$ is the reaction rate of species i based on catalyst volume, η is the effectiveness factor of the catalyst pellet, K_l is the overall mass transfer coefficient, $(1 - \varepsilon_b)$ is the ratio of catalyst volume to the total volume of the bed.

Instead of using the overall mass transfer coefficient in Eqs. (13.20) and (13.21), liquid or vapor side mass transfer coefficients can also be used. In that case, an equilibrium relation between the liquid-side and vapor-side mole fractions of

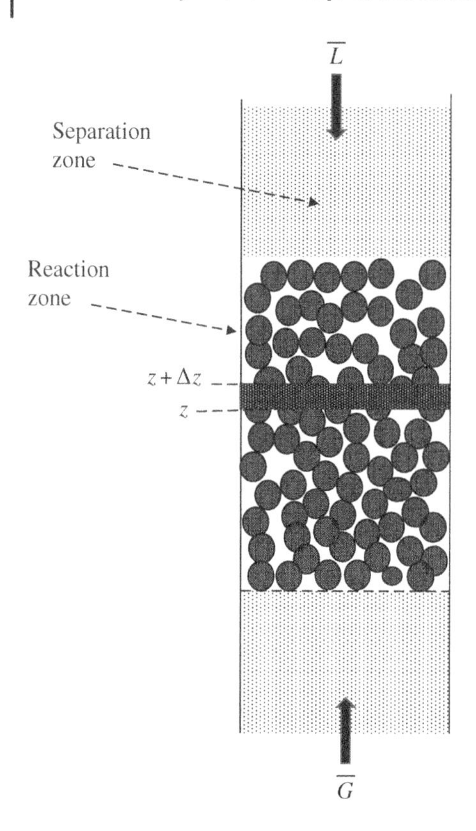

Figure 13.7 Schematic representation of reaction and separation zones in a fixed-bed catalytic reactive distillation column.

species i should be used at the interface of the vapor and liquid streams.

Species conservation equations for all of the other components can also be written following a similar procedure. Liquid and vapor flow rates, $\overline{L}$ and $\overline{G}$, can be taken as constant if heat effects are negligible. Otherwise, a differential energy balance should also be written and solved simultaneously with the species conservation equations [6].

13.3 Sorption-Enhanced Reaction Process

In some reversible reactions, removal of one of the products from the reaction zone through a sorption process may eliminate the equilibrium limitation and significantly enhances the conversion of the limiting reactant. A typical example of a sorption-enhanced reaction process is the production of hydrogen by steam reforming of methane, which is followed by the water–gas shift reaction.

$$CH_4 + H_2O \rightleftharpoons CO + 3H_2 \quad \text{SRM} \quad \Delta H^o_{298} = 206\ \text{kJ/mol}$$

$$CO + H_2O \rightleftharpoons CO_2 + H_2 \quad \text{WGSR} \quad \Delta H^o_{298} = -41\ \text{kJ/mol}$$

Both of these catalytic reactions are reversible, and equilibrium limitations may be quite significant. If one of the products of a reversible reaction is removed from the reaction zone, the reaction shifts to the product side. For instance, if CO_2 is removed from the reaction zone by a sorption process, this decreases the concentrations of both CO_2 and CO in the product stream. Both reactions shift to the product side, yielding purer hydrogen in the reactor exit stream.

Removal of CO_2 from the reaction zone can be achieved by mixing the catalyst with an adsorbent of CO_2. For this purpose, certain clays, hydrotalcites, or CaO can be used. For instance, if the catalyst is mixed with CaO particles, one would expect an in situ reaction of CO_2 with CaO, yielding $CaCO_3$ within the reactor [8–10].

$$CaO(s) + CO_2(g) \rightleftharpoons CaCO_3(s) \quad \Delta H^o_{298} = -179 \text{ kJ/mol}$$

This carbonation reaction is exothermic and reversible. The heat generated due to this reaction will contribute to the heat requirement of the endothermic SRM reaction. With an increase in temperature, the reverse of the sorption reaction becomes more significant. The following expression is given in the literature for the equilibrium volume fraction of carbon dioxide in the reactor [8, 10]:

$$v_{CO_2,eq} = 4.137 \times 10^7 \exp\left(-\frac{-20\,474}{T}\right) \tag{13.23}$$

Equilibrium volume fraction values of CO_2 estimated from this expression at 550 and 650 °C are 6.5×10^{-4} and 0.011, respectively. These estimations indicate that temperatures lower than 650 °C should be preferred in the sorption-enhanced reforming process using CaO as the sorbent. Removal of carbon dioxide, as soon as it is formed within the catalyst bed, will minimize the reverse reactions of SRM and WGSR and hence will increase the conversion of methane and the purity of hydrogen in the product stream. The overall reaction stoichiometry can then be written as follows:

$$CH_4 + 2H_2O + CaO \rightleftharpoons 4H_2 + CaCO_3$$

We would expect to have some CO and CO_2 in the product stream depending upon the reaction temperature, feed composition, and the catalyst to CaO ratio in the reactor.

Steam reforming of methanol is another example of a sorption-enhanced process

$$CH_3OH + H_2O \rightleftharpoons CO_2 + 3H_2$$

Usually, Ni-based or Co-based catalysts are used in such reforming processes. Packing of the tubular reactor with a mixture of the catalyst and a CO_2 adsorbent will allow a higher conversion of methanol. As a result of the sorption process, available sites of the adsorbent for the sorption of CO_2 will decrease as a function of reaction time. For instance, if we use CaO for the removal of CO_2 from the reaction zone, available CaO for the sorption step will decrease with reaction time as a result of the conversion of CaO to $CaCO_3$. Hence, we will expect to observe an S-shaped concentration variation curve of CO_2 as a function of time at the reactor outlet (Figure 13.8). This curve is called the breakthrough curve. Modeling of an adsorber and derivation of a breakthrough expression is also discussed in Chapter 15.

Many of the sorption processes are reversible, and the net rate of sorption decreases as the equilibrium is approached. For a reversible sorption process, the maximum degree of sorption is limited by equilibrium. Let us illustrate a reversible sorption-enhanced reaction process, in which one of the products of the reversible reaction (for instance, product C) is removed from the reaction zone as a result of its sorption on an adsorbent E.

$$A + B \rightleftharpoons C + D; \quad -R_{A,p} = R_{(1),p} = k_1 P_A P_B - k_{-1} P_C P_D \text{ (mol/l)/s} \tag{R.13.1}$$

$$C + E \text{ (solid reactant)} \rightleftharpoons CE \text{ (solid product)}; \quad R_{(2),p} \text{ (mol/l)/s} \tag{R.13.2}$$

Here, $R_{(1),p}$ and $R_{(2),p}$ are the rates of reactions (R.13.1) and (R.13.2), based on catalyst volume and the solid adsorbent volume, respectively. The pseudo-homogeneous species conservation equations for product C and reactant A in the fixed-bed reactor can then be written as follows:

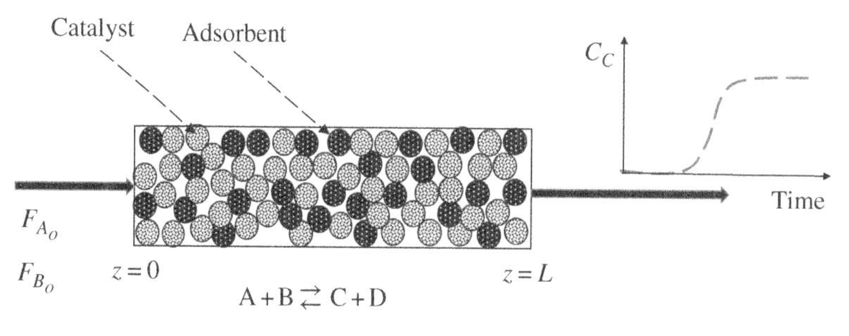

Figure 13.8 Schematic representation of a sorption-enhanced reaction process.

$$\varepsilon_b \frac{\partial C_C}{\partial t} = -\frac{\partial F_C}{\partial V} + (1 - \varepsilon_b - \varepsilon_a) R_{(1),p} - \varepsilon_a k_2 (P_C - P_{C_e})(x_{E_{max}} - x_E) \tag{13.24}$$

$$\varepsilon_b \frac{\partial C_A}{\partial t} = -\frac{\partial F_A}{\partial V} - (1 - \varepsilon_b - \varepsilon_a) R_{(1),p} \tag{13.25}$$

Here, ε_b is the void fraction of the bed and ε_a is the fraction of the bed occupied by the solid adsorbent. The third term on the right-hand side of Eq. (13.24) corresponds to the sorption rate of gaseous product C on the solid adsorbent E. In this model, the rate of sorption of the gaseous product C (CO_2 in the steam-reforming reaction) is considered as proportional to the difference of the partial pressures of C in the fluid phase and its equilibrium partial pressure P_{C_e}. As the concentration of C approaches to the equilibrium concentration, the rate of the sorption process slows down. The rate of the sorption process is also expected to be proportional to the number of available sites of the solid adsorbent. For instance, in the case of the removal of CO_2 by the reaction with CaO, the rate of carbonation reaction is expected to be proportional to the available CaO sites for the removal of CO_2. The term x_E in the sorption rate term of Eq. (13.24) corresponds to the fractional conversion of the solid reactant at an arbitrary reaction time. The complete conversion of the solid reactant is not possible in many cases due to equilibrium limitation and structural changes of the solid reactant during the reaction. For instance, in the case of the reaction of carbon dioxide with calcium oxide, the structural changes of the solid reactant during the reaction limit the value of the maximum fractional conversion of CaO to $CaCO_3$. During the reaction of CO_2 with porous CaO, some of the pores may get narrower, and pore mouth closure may take place due to the formation of product $CaCO_3$. The complete conversion of CaO to $CaCO_3$ is almost impossible in most cases. The maximum possible conversion of the solid reactant may be less than 1 in many cases. The term $x_{E_{max}}$, which appears in Eq. (13.24), corresponds to the maximum possible conversion of the solid reactant.

The rate of conversion of solid reactant (adsorbent) based on its volume may be expressed as

$$R_{(2),p} = \frac{\rho_E}{M_E} \frac{dx_E}{dt} = k_2 (P_C - P_{C_e})(x_{E_{max}} - x_E) \tag{13.26}$$

Here, the density and the molecular weight of the solid reactant are denoted as ρ_E and M_E, respectively [8–10].

In many cases, the order of magnitudes of the reaction rate and the convective transport rate terms in Eqs. (13.24) and (13.25) (the order of magnitudes of the terms on the right-hand sides of these equations) are much higher than the order of magnitude of the accumulation terms of these expressions. Hence, the pseudo-steady-state assumption is justified in most cases. In that case, Eqs. (13.24) and (13.25) reduce to

$$-\frac{dF_C}{dV}+(1-\varepsilon_b-\varepsilon_a)R_{(1),p}-\varepsilon_a k_2(P_C-P_{C_e})(x_{E_{max}}-x_E)=0 \tag{13.27}$$

$$-\frac{dF_A}{dV}-(1-\varepsilon_b-\varepsilon_a)R_{(1),p}=0 \tag{13.28}$$

The molar flow rates of the other species involved in reaction (R.13.1) may be expressed in terms of the molar flow rate of the limiting reactant A as follows:

$$F_D=F_{A_o}-F_A \tag{13.29}$$

$$F_B=F_{B_o}-(F_{A_o}-F_A) \tag{13.30}$$

The partial pressures of the species are then expressed in terms of their molar flow rates as follows:

$$P_A=P\frac{F_A}{F_T},\quad P_B=P\frac{F_B}{F_T},\quad P_C=P\frac{F_C}{F_T},\quad P_D=P\frac{F_D}{F_T} \tag{13.31}$$

Here, P and F_T are the total pressure and the total molar flow rate in the reactor, respectively.

$$F_T=F_A+F_B+F_C+F_D \tag{13.32}$$

The simultaneous solution of Eqs. (13.27)–(13.30) together with (13.26), (13.31), and (13.32) will yield the variation of flow rates of the species as a function of time and axial position in the reactor. The breakthrough curve of the product C at the exit of the reactor can then be predicted by the solution of these equations.

If the reaction rate expressions are given in terms of concentrations of the species instead of their partial pressures, all these equations may be expressed in terms of concentrations instead of molar flow rates. Also, a constant superficial velocity assumption may be made if the mole fraction of the limiting reactant is not high. These assumptions simplify the model expressions.

Unsteady species conservation equations (Eqs. (13.24) and (13.25)) may also be used instead of the equations expressed using the pseudo-steady state assumption. The use of a pseudo-steady-state assumption in gas–solid non-catalytic reactions is further discussed in Chapter 15. Another simplification may also be made by taking a pseudo-first-order rate expression for reaction (R.13.1) for the case of having an excess concentration of reactant B.

In a realistic model for the sorption-enhanced steam reforming, both steam-reforming and water–gas shift reactions should be considered. The rate expressions for the catalytic reactions may also be multiplied by an effectiveness factor if pore diffusion resistance within the catalyst pellets is significant. The model proposed here is written for an isothermal system. A non-isothermal model, including energy balance, is available in the literature for the sorption-enhanced steam-reforming reaction of methane [8, 9].

If the rate expressions are available in terms of the concentrations of the species involved in this process instead of their partial pressures, these species conservation equations may also be expressed in terms of concentrations. The molar flow rates of the species may be expressed as the products of their concentrations with total volumetric flow rate in the reactor. Total volumetric flow rate may be assumed constant if the mole fraction of the limiting gaseous reactant is low.

Another approach for modeling a sorption-enhanced reaction process may be the use deactivation model described in Section 11.4 for catalyst deactivation and in Chapter 15 for gas–solid reactions. The rate of sorption of the gaseous product C on an adsorbent E depends upon the concentration of product C and the number of active sites available per unit mass (or per unit volume) of the adsorbent.

$$C + E(\text{adsorbent}) \rightarrow CE(\text{adsorbed state}) \tag{R.13.3}$$

Several models are discussed in Chapter 15 for the rate of decrease of activity of solid reactants in gas–solid reactions, and also in Chapter 11 for the deactivation of catalysts as a result of sorption of poisons on the active sites. An activity term of the solid material (a_c) may be introduced into the rate expression for the sorption reaction (R.13.3) given above.

$$R_{\text{sorp}} = k_2 a_c P_C \tag{13.33}$$

Here, R_{sorp} is the rate of sorption based on the volume of the solid adsorbent. a_c corresponds to the activity factor of the adsorbent (or solid reactant), which is unity at the initial time. The rate of deactivation of the adsorbent (or the solid reactant) may be expressed as a function of the partial pressure of the adsorbing product (P_C) and the activity itself.

$$-\frac{da_c}{dt} = k_d a_c P_C^m \tag{13.34}$$

Here, k_d is the deactivation rate constant. The simplest form of the deactivation rate can be expressed by taking $m = 0$.

$$-\frac{da_c}{dt} = k_d a_c \tag{13.35}$$

For this case, the time dependence of the activity factor is expressed as follows:

$$a_c = \exp(-k_d t) \tag{13.36}$$

This corresponds to a concentration-independent deactivation process. The prediction of a fixed-bed catalytic reactor performance for the concentration-independent deactivation of the catalyst is illustrated in Section 11.4. Treatment of a concentration-dependent deactivation process is also discussed in Section 11.4, and also in Chapter 15, by taking $m = 1$.

The pseudo-homogeneous species conservation equations for product C and reactant A, can then be expressed considering the deactivation model as follows:

$$\varepsilon_b \frac{\partial C_C}{\partial t} = -\frac{\partial F_C}{\partial V} + (1 - \varepsilon_b - \varepsilon_a) R_{(1),p} - \varepsilon_a k_2 a_c P_C \tag{13.37}$$

$$\varepsilon_b \frac{\partial C_A}{\partial t} = -\frac{\partial F_A}{\partial V} - (1 - \varepsilon_b - \varepsilon_a) R_{(1),p} \tag{13.38}$$

Note that, $R_{(1),p}$ corresponds to the reversible rate expression for Reaction (R.13.1), based on the volume of the catalyst. With the pseudo-steady-state assumption, Eqs. (13.37) and (13.38) reduce to

$$-\frac{dF_C}{dV} + (1 - \varepsilon_b - \varepsilon_a) R_{(1),p} - \varepsilon_a k_2 a_c P_C = 0 \tag{13.39}$$

$$-\frac{dF_A}{dV} - (1 - \varepsilon_b - \varepsilon_a) R_{(1)} = 0 \tag{13.40}$$

These expressions are written with the plug-flow assumption. The simultaneous solution of the species conservation equations and the deactivation rate expression (Eq. (13.34)) is needed for the prediction of variations of reactant and product concentrations at the reactor outlet as a function of time. Partial pressures, which appear in the reaction rate expressions, should be expressed as the products of mole fractions with the total pressure. Note that mole fraction of a species in the reactor is equal to the ratio of its molar flow rate to the total molar flow rate.

The time required to reach a desired value of the concentration ratio of C to A (C_C/C_A) at the reactor outlet is called the breakthrough point. This ratio may be arbitrarily selected as 0.05 or lower, depending upon the specific application. The breakthrough curve of the gaseous product C at the reactor outlet is expected to become sharper as the ratio of the sorption rate constant to the forward reaction rate constant of the main reaction increases. A shift of the breakthrough point to longer times is also expected with an increase in this ratio.

Cyclic operation of carbonation and calcination reactors may be considered for a sorption-enhanced reforming process using CaO as the adsorbent. Regenerability of the adsorbent after its saturation with the adsorbing species is quite important for its repeated use. Structural changes and sintering of the adsorbent during the regeneration cycle may cause some decrease in the activity of the adsorbent in a multiple cycle process.

13.4 Monolithic and Microchannel Reactors

Catalytic monoliths found widespread application in catalytic converters used in automobiles. One major advantage of a monolithic catalyst block is its low-pressure drop as compared to a fixed-bed catalytic reactor. More recently, monolithic catalysts also found applications in non-automotive reactions. A monolithic catalyst is composed of parallel channels in a ceramic or metal block, through which the reaction mixture flows. These channels may have a square cross-section in most cases. Cell dimensions of these channels are usually in the order of 10^{-3} m. However, flow channels with circular or rectangular cross-section are also possible.

In most cases, the surfaces of these channels are coated with an active catalyst layer [11, 12]. In the case of automobile catalytic converters, the surfaces of the flow channels are wash-coated with a thin porous alumina layer, which also contains impregnated active metals, like Pt, Pd, Rh, etc. Oxidation of CO and unburned hydrocarbons in the exhaust stream, as well as reduction of nitrogen oxides, can be achieved in such bi-functional catalytic converters. An image of a monolithic block is shown in Figure 13.9. A schematic diagram of the cross-section of a monolithic block and cross-section of a unit cell is shown in Figure 13.10a,b, respectively.

Modeling of a monolithic reactor involves mass and heat transfer expressions between the fluid phase and the catalyst layer, as well as pore diffusion resistance within the catalyst layer. Also, the hydrodynamics of the flow within the channels has a significant influence on the performance of such a catalyst block. Simultaneous solution of the equation of motion within the channels, species conservation equations for the fluid phase in the channels and the catalyst layer, and energy balance equations are needed to predict the performance of a monolithic reactor. Here, a simple isothermal model is presented, assuming axially dispersed flow in the channels.

Assuming that the axial dispersion model is a good representation of flow within the channels, the following pseudo-homogeneous steady-state species conservation expression can be written for the fluid phase within the channels:

$$D_z \frac{d^2 C_A}{dz^2} - U_o \frac{dC_A}{dz} - \frac{2\varepsilon_m D_e}{b} \left. \frac{dC_{A_p}}{dy} \right|_{y=\delta} = 0 \tag{13.41}$$

Here, the flux of reactant A into the wash-coated catalyst layer is equal to the rate of consumption of A within this layer.

Figure 13.9 An image of monolithic block.

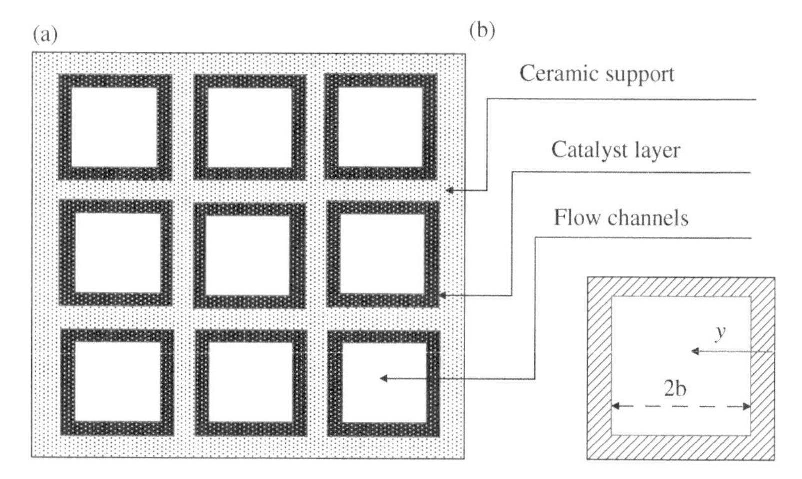

Figure 13.10 (a) Schematic representation of the cross-section of a monolith. (b) Cross-sectional view of a unit cell.

$$D_e \left. \frac{dC_{A_p}}{dy} \right|_{y=\delta} = -\delta\eta \left(R_{A,p}\right)_{y=\delta} \tag{13.42}$$

Here, η is the effectiveness factor and C_{A_p} is the concentration within the porous catalyst layer wash-coated on the surfaces of the monolithic channels, b is the half-cell size of a square channel in the monolithic block, and δ is the thickness of the catalyst layer. In this system, the void fraction ε_m is defined as the ratio of the volume of the flow channels to the total volume of the monolith, U_o is the superficial velocity within the monolithic block (volumetric flow rate divided by the total cross-sectional area of the monolithic block), and D_e is the effective diffusion coefficient within the porous catalyst layer. Equation (13.41) is written considering no fluid-phase (homogeneous) reaction taking place in the flow channels and also by considering a constant volumetric flow rate. Since the concentrations of reacting species in an automobile exhaust stream are quite low, the volume expansion due to changes in the total number of moles during the reaction is neglected in this derivation. Considering that the active catalyst layer thickness is very small, slab geometry is assumed for this layer.

Equation (13.41) should be solved simultaneously with the pseudo-homogeneous species conservation equation written for the catalyst layer. The direction of y is illustrated in Figure 13.10b.

$$D_e \frac{d^2 C_{A_p}}{dy^2} + R_{A,p} = 0 \tag{13.43}$$

For a single first-order catalytic reaction taking place within the wash-coated catalyst layer, Eq. (13.43) reduces to:

$$D_e \frac{d^2 C_{A_p}}{dy^2} - kC_{A_p} = 0 \tag{13.44}$$

Considering the possible contribution of film mass transfer resistance on the surface of the catalyst layer, the boundary conditions for Eq. (13.44) can be expressed as follows:

$$D_e \left.\frac{dC_{A_p}}{dy}\right|_{y=\delta} = k_m \left(C_A - \left.C_{A_p}\right|_{y=\delta}\right) \tag{13.45}$$

$$\left.\frac{dC_{A_p}}{dy}\right|_{y=0} = 0 \tag{13.46}$$

In the case of a multiple reaction system, species conservation equations should be written for all of the species having independent rates. Details of a model with the assumption of plug flow within the channels of the monolithic block are presented in Online Appendix O6.6.

Equations given above are written assuming an isothermal monolithic block. However, if the heat effects of the reactions are significant, one would expect temperature variations, both in the fluid phase and also in the catalyst layer. The temperatures of the fluid phase, as well as the catalyst phase, may change both in the axial and the y-directions across the catalyst layer. There may also be temperature variations in the monolith block. Evaluation of the physical parameters, such as effective diffusion coefficient within the washcoat layer, thermal conductivity values, and estimation of transfer coefficients, which are needed for the design of a monolithic reactor, are available in the literature [12, 13].

Another way of modeling a monolithic catalytic reactor can be done by assuming the fluid and the wash-coated catalyst zones as a pseudo-homogeneous phase within the channels of the monolith. That is, the fluid and the catalyst phases within a channel can be considered a single pseudo-homogeneous phase, and the model equations can be expressed for this phase. Let us consider a monolithic reactor in which a first-order pseudo-homogeneous reaction occurs in the channels of the reactor, which is considered a pseudo-homogeneous phase. A simplified model can be developed for this system. If the reaction is exothermic, the heat generated as a result of the reaction should be transferred to the solid phase, that is to the metallic (or ceramic) block of the monolith. If we consider axial temperature variations both in the channels and in the solid block of the monolith, one-dimensional model equations can be expressed as given by Eqs. (13.47)–(13.49). In writing these equations, plug flow is assumed within the monolith channels.

$$-\frac{U_o}{\varepsilon_m}\frac{dC_A}{dz} - kC_A = 0; \quad k = k_o \exp\left(-\frac{E_a}{R_g T}\right) \tag{13.47}$$

Here, kC_A is the first-order reaction rate based on unit volume of the channel, which is considered as a pseudo-homogeneous phase.

The energy balance equations for the pseudo-homogeneous phase and the solid block of the monolith may be expressed as

$$\sum_i N_{i_z} C_{p_i} \frac{dT}{dz} + k_o e^{-\frac{E_a}{R_g T}} C_A \Delta H^o = -\frac{2}{b} h(T - T_m) \tag{13.48}$$

$$\lambda_m \frac{d^2 T_m}{dz^2} = -\frac{2\varepsilon_m}{b(1-\varepsilon_m)} h(T - T_m) \tag{13.49}$$

Here, T_m and λ_m are the temperature and thermal conductivity of the monolithic block (solid phase), respectively. Axial heat conduction contribution and axial dispersion terms within the channels are neglected in writing these equations.

13.4.1 Microchannel Reactors

Significant advances were achieved in the microchannel reactor technology and modeling of such reactors in recent decades [14–18]. Microchannel reactors are similar to the monolithic reactors with smaller channel dimensions. In the production of a microchannel reactor, parallel channels with square or rectangular cross-sections and cell dimensions being as low as 10^{-4} m are incorporated into a metal substrate. The surfaces of the channels are generally covered with a thin layer of a catalyst. In some cases, some of the channels are used for the flow of cooling or the heating fluid, while the other channels are used for the chemical conversion. Heat transfer coefficients between the reaction fluid and the metal substrate were reported as being 10^4 times higher than the conventional packed-bed reactors. Heat and mass transfer limitations can be significantly reduced in the microchannel reactors. Hence, higher production yields can be achieved in a much smaller volume as compared to the conventional systems.

The design procedure of microchannel reactors is quite similar to that of monolithic reactors and conventional reactors. Important parameters to be considered are flow patterns in the microchannels and the transport parameters. An example of a microchannel reactor is illustrated for the steam reforming of ethanol to produce hydrogen [14]. Due to the endothermic nature of this reaction, the heat transfer rate to the reaction zone is quite important. While the ethanol steam mixture was flowing through a set of catalytic microchannels, hot flue gas may flow through the adjacent microchannels of the microreactor block. Due to the very high heat transfer rates achieved in such an arrangement, increased yield of the desired product can be achieved.

Microchannel reactors can also be used for coupling endothermic and exothermic reactions. In these cases, some of the channels can be used for an exothermic reaction, while an endothermic reaction may take place in the other channels. The heat generated by the exothermic reaction can be used by the endothermic reaction, and the energy efficiency of the overall process can be significantly enhanced by this arrangement. The modeling of a microchannel reactor and a micro heat-exchanger reactor involves the equation of motion, equation of continuity, species conservation equations, and equation of energy for the fluid phase and the wash-coated catalyst layer. An energy equation for the solid wall phase is also written in some cases. The performance of such reactors can be predicted by the simultaneous solution of these transport equations [16–18].

13.5 Chromatographic Reactors

Chromatography is a conventional analytical technique for the separation of species with different adsorption equilibrium constants on a solid phase. A typical example is a flow of a fluid (mobile) phase, containing a mixture of adsorbing species, through a fixed bed of adsorbent particles. If a pulse of a mixture containing the adsorbing species is injected into the mobile phase at the inlet of the chromatographic column, each component in the sample will spend different retention times in this column. This is due to the differences in the adsorption rate and equilibrium parameters of different components of the mixture on the solid phase. In the case of a chromatographic reactor, separation and the chemical reaction processes take place simultaneously in the fixed-bed reactor. Separation of products from the reactants in such a chromatographic reactor can enhance the yield of the desired product, especially for the equilibrium-limited reactions [19].

Unsteady-state species conservation equations for the reactant A and the product B can be written considering a first-order pseudo-homogeneous reversible reaction taking place in a chromatographic reactor.

$$\varepsilon_b \frac{\partial C_A}{\partial t} + (1-\varepsilon_b)\rho_p \frac{\partial n_{A,\text{ads}}}{\partial t} = -U_o \frac{\partial C_A}{\partial z} + D_z \frac{\partial^2 C_A}{\partial z^2} - k_1\left(C_A - \frac{1}{K} C_B\right) \tag{13.50}$$

Here, $n_{A,\text{ads}}$ is the adsorbed concentration of species A per unit mass of the adsorbent, and it may be related to the mobile-phase concentration assuming a linear adsorption isotherm relation;

$$n_{A,\text{ads}} = K_A C_A \tag{13.51}$$

Such a linear adsorption relation is justified for low concentrations of species A. An adsorption isotherm relation like Langmuir adsorption isotherm may be used at high concentrations, instead of this linear adsorption relation. Substitution of Eq. (13.51) into Eq. (13.50) gives the following partial differential equation for species A.

$$\left(\varepsilon_b + (1-\varepsilon_b)\rho_p K_A\right)\frac{\partial C_A}{\partial t} = -U_o \frac{\partial C_A}{\partial z} + D_z \frac{\partial^2 C_A}{\partial z^2} - k_1\left(C_A - \frac{1}{K} C_B\right) \tag{13.52}$$

A similar expression should be written for species B (Eq. (13.53)), and these equations should be solved simultaneously with the proper boundary conditions for the prediction of the performance of a chromatographic reactor.

$$\left(\varepsilon_b + (1-\varepsilon_b)\rho_p K_B\right)\frac{\partial C_B}{\partial t} = -U_o \frac{\partial C_B}{\partial z} + D_z \frac{\partial^2 C_B}{\partial z^2} + k_1\left(C_A - \frac{1}{K} C_B\right) \tag{13.53}$$

13.6 Alternative Energy Sources for Chemical Processing

The accelerated increase of global energy consumption led to significant progress in the development of the applications of alternative energy sources. The heating of chemical processing units is conventionally achieved by conductive and convective mechanisms from heat sources, such as a steam boiler. The use of alternative energy sources, like microwave, solar energy, ultrasound, etc., in chemical processing and the increase of thermal efficiency of these units gained importance in recent decades as part of the process intensification approach. The integration of such energy sources to chemical conversion units is still mainly in small and pilot scales for most applications.

Significant research is in progress for the development of the commercial scale implementation of these alternative energy sources in chemical reactors.

13.6.1 Microwave-Assisted Chemical Conversions

Microwaves have first been used in liquid-phase homogeneous reactions, such as for organic synthesis. More recently, significant advancement has been achieved in the area of microwave-enhanced solid-catalyzed reactions [20]. Electromagnetic radiations with wavelengths between 0.001 and 1 m are called microwaves. The corresponding frequency range of microwaves is between 300 and 0.3 GHz. In most household applications and chemical conversions, a microwave frequency of 2.45 GHz is used. This corresponds to a wavelength of about 0.12 m. A microwave frequency of 915 MHz (which corresponds to a wavelength of about 0.32 m) is also available, especially for large-scale microwave reactor applications [21, 22].

The heating mechanism of a liquid or solid material by microwave radiation is quite different than the conduction–convection type mechanism of conventional reaction systems. In the case of liquids, molecules having high dipole moments start to rotate when exposed to microwave radiation. Due to the friction between these rotating molecules, mechanical energy is converted to heat. Polar molecules and ions are the ones that are most affected by microwave radiation. For instance, water can be quickly heated by microwaves. However, gases are least affected by microwave radiation and can be taken as microwave transparent. In the case of solids, microwave heating depends upon their dielectric properties.

The heating capacity of a material is related to its dielectric loss tangent [21, 22].

It is reported in the related literature that the materials having dielectric loss tangent values of 0.1 and higher are good microwave absorbers. Air has a loss tangent value of zero, while the loss tangent values of polar liquids, like water and ethanol, are reported as 0.123 and 0.941, respectively. These polar liquids are excellent microwave absorbers. Pyrex and quartz are reported to have loss tangent values of 0.026 and less than 0.001, respectively. Hence, these materials can be considered as microwave transparent. Loss tangent values of activated carbon and silicon carbide are 0.907 and 0.58, respectively, indicating excellent microwave absorption capacities of these materials [21, 22]. On the other hand, conventional catalyst support materials, like alumina and mesoporous silica, have relatively low microwave absorption capacities.

When microwave interacts with a solid, it is either reflected, absorbed, or permeated through the material without being absorbed. Most of the metals cause reflection of the microwave from the surface. These materials are called conductors. However, it is reported that the metal particle size has an important effect on the microwave absorptivity of these materials. Metal particles being smaller than the microwave penetration depth were reported to be heated rapidly to high temperatures by microwave radiation. For the conventionally used microwave frequency of 2.45 GHz, metal particles being smaller than 1–10 μm are reported as good microwave absorbers [21]. This indicates that active metal clusters (Ni, Pt, etc.) within alumina and silicate-based supported catalyst may be rapidly heated by microwave irradiation. Hence, some temperature differences may develop between the metal nanoparticles and the support. This is claimed as one of the reasons for having higher reaction rate values in the microwave-heated catalytic reaction systems, as compared to the conventionally heated reactors.

In the case of microwave-assisted gas–solid catalytic reactors, selective heating of the catalyst pellets can be achieved while the gas phase is at a much lower temperature. This can be achieved by using catalyst support materials having high microwave absorptivity. For instance, carbon-based porous materials or silicon carbide, which have very high loss tangent values, can be used as the catalyst support material [23]. Another way of heating the catalyst zone of the reactor by microwave irradiation is by mixing the catalyst pellets having low microwave absorption capacity with a material having relatively high dielectric loss tangent value, such as activated carbon. This is illustrated in microwave-assisted reforming of ethanol over mesoporous alumina and SBA-15-based supported catalysts [24, 25].

In the conventionally heated tubular catalytic reactors, significant temperature gradients may occur in the radial and axial directions. For instance, for highly endothermic reactions, like reforming reactions, the temperature at the centerline of the fixed-bed reactor may be considerably less than the reactor wall temperature. Such temperature variations may cause significant changes in the reaction rate constant, as well as in the occurrence of undesired side reactions and coke formation. On the contrary, microwave heating is generally considered as volumetric heating, minimizing such temperature gradients. In the case of endothermic reactions heated by microwave irradiation, the temperature at the centerline of the tubular reactor may even become higher than the wall temperature. More importantly, the formation of hot-spots (microplasmas) is reported in the microwave-heated catalytic reactors. The formation of such hot-spots is believed to be mainly due to the differences in microwave absorptivities of metal or metal oxide clusters and the support materials, and due to the loss tangent differences of different phases within the reactor. The temperatures of the hot spots are reported as being at least 100–200 K higher than the bulk temperature in the reactor [21, 26]. Achievement of more uniform temperature due to volumetric heating of the reaction medium is an important advantage of microwave heating for homogeneous liquid-phase reactions. Some important characteristics and advantages of microwave-heated fixed-bed catalytic reactors can be listed as

- Selective heating of the catalyst bed without heating the reactants
- Higher reaction rates and higher yields due to the formation of local hot-spots
- Improved selectivities of the desired products; optimization of product distribution
- Low coke formation
- High heating rate to reach the desired temperature, and easy control of the heating process
- Less energy input for heating the active sites of the catalyst bed.

Microwave-enhanced catalytic reactions were investigated for a variety of chemical reactions. Steam and dry reforming of methane, steam reforming of alcohols, pyrolysis of methane to produce hydrogen, oxidation of carbon monoxide, dehydrogenation reactions, and fuel-cell-grade hydrogen production by decomposition of ammonia are some of these catalytic reactions. Coke minimization is reported in reforming reactions carried out in microwave-heated reactors. Since carbon has a very high microwave absorption potential, any coke formed during the reaction causes local hot spots at carbon-containing places, which may facilitate the gasification of the produced coke. Also, coke formation due to exothermic Bouduard reaction is expected to lose its significance due to the formation of such hot spots.

Comparison of the performances of conventionally heated and microwave-heated fixed-bed reactors in ammonia decomposition and steam-reforming reactions showed a significant increase in the fractional conversion of the limiting reactant in the microwave system at the measured temperature. For instance, in the ammonia decomposition reaction investigated with a Mo-incorporated mesoporous carbon catalyst, almost complete conversion of ammonia was achieved at 400 °C in the microwave-heated reactor [23]. On the other hand, only about 50% conversion was achieved with the same catalyst and at the same space-time at a much higher temperature of 600 °C. This is explained by the occurrence of hot-spots in the microwave system [23]. The measurement of the actual temperatures of these hot-spots is quite challenging.

Microwave-heated batch or flow reactors of different configurations may be used for chemical conversions [21, 22]. Most microwave-heated batch processes were performed in multimode microwave applicators. Typical examples of these applicators are similar to domestic microwave ovens. The reaction vessel is placed into these ovens. Reflections of the microwave in such an oven cause the occurrence of several resonance modes, and hence the reaction medium is highly nonuniformly heated. Fixed-bed flow reactors may also be placed into such microwave ovens. Another type of microwave-heated reactor is called a monomode applicator, in which a stable standing microwave is generated inside the cavity by the magnetron, and it reaches the reaction vessel (batch or flow) through the waveguide [20, 21]. A higher electromagnetic field was reported to be achieved in these monomode applicators. Another type of microwave-heated reactor configuration is reported as a coaxial traveling-wave microwave reaction system. These reactors are proposed for more uniform heating of the catalyst bed, better utilization of microwave energy, and easier scale-up potential for industrial applications. In this type of reactor configuration, microwave travels along the reactor in the same direction as the gaseous reactants. Monolithic or other types of catalysts may be placed in the annular space between the coaxial cylindrical inner and outer conductors. Details of the design challenges of microwave-heated reactors are discussed in the literature [20, 21, 27].

13.6.2 Ultrasound Reactors

Integration of ultrasound with chemical reactors has promising advantages in terms of increased selectivity of the desired product, increased rate of the reaction, and hence reduction in reaction time and lower energy consumption for a given production rate of the desired product in a batch reactor.

The effect of ultrasound on the rates and mechanisms of catalytic reactions is due to both physical and chemical factors [28, 29]. Application of ultrasound to a liquid-phase reaction mixture creates cavitation of microbubbles, which contain the vapors of the liquid mixture. These bubbles may grow in size to a bubble size of about a couple of hundred microns. Then, they collapse in an extremely short time, in the order of microseconds. This collapse of microbubbles leads to a significant amount of energy release, causing the formation of local hot spots with very high temperatures. The collapse of these microbubbles in an extremely short time also produces microjets, shockwaves, and turbulence in the liquid. Such intense motion of the fluid in the reaction medium and near the liquid–solid interface causes a substantial increase in the mass transfer rate, eliminating the interpellet transport limitations on the observed rate of the catalytic reaction. In addition to the intensification effect of the ultrasound by enhancing the physical rate processes in a catalytic reactor, local hot-spots formed due to the cavitation process help to increase the reaction rate, as well as cause the formation of short life-time reactive radicals [28, 29]. Different types of ultrasound-assisted reactors are designed for different applications. Significant enhancement of the heat and mass transfer rates are reported in the ultrasound-enhanced microchannel reactors.

13.6.3 Solar Energy for Chemical Conversion

Synthesis of chemicals and fuels by the use of solar energy can be achieved through thermochemical, photocatalytic, or electrochemical processes. Advances in electrocatalytic and photocatalytic processes opened new avenues for the efficient synthesis of valuable chemicals [30, 31]. Production of hydrogen via water splitting or from hydrocarbons by the action of TiO_2-, ZnO-, etc. based photocatalysts, oxidation of organic pollutants in wastewater, and artificial photosynthesis are some of the photocatalytic reactions, which have been extensively investigated. The use of solar energy for photocatalytic conversion processes is the main objective of most of these investigations. Advances in the use of solar energy also opened new avenues in electrocatalytic processes for the synthesis of chemicals and fuels. Solar electrolysis of water is a possible route for hydrogen production.

Solar energy is used as the power source for producing hydrogen and syngas in solar thermochemical processes. Steam reforming of methane, gasification of carbonaceous materials, and solar cracking of methane are some of the possible routes for solar hydrogen production. Solar energy is concentrated in these systems containing parabolic mirrors, solar furnaces, etc. to achieve very high temperatures for chemical conversions. This energy is then used in the solar reactors for chemical conversion by a solar thermochemical process. Catalytic membrane reactors, fluidized-bed, or packed-bed reactors may be used as solar-energy-heated reaction systems. These reactors are either directly heated by the concentrated solar energy or indirectly heated by a heat-absorbing material, which is heated by the solar radiation, and then it transfers heat to the reactor. In the case of directly heated solar reactors, the reactor is placed at the focal point of the solar energy concentrator. Different types of reactors are designed for both directly and indirectly heated systems [32, 33].

References

1 Stankiewicz, A. and Moulijn, J.A. (2003). *Re-engineering the Chemical Processing: Plant Process Intensification*. New York: Marcel Deccer Inc.

2 Keil, F.J. (2007). *Modeling Process Intensification*. Weinheim: Wiley-VCH.

3 Basile, A., Di Paola, L., Hai, F.I., and Piemonte, V. (eds.) (2015). *Membrane Reactors for Energy Applications and Basic Chemical Production*. Amsterdam: Woodhead Publishers/Elsevier.

4 Ghosh, S. and Sreethamraju, S. (2019). *Chem. Eng. Process. Process Intensif.* 145: 107673.

5 Mallaiah, M.K., Kishore, A., and Reddy, G.V. (2017). *Chem. Biochem. Eng. Q.* 31: 293–302.

6 Noeres, E., Kenig, C.Y., and Gorak, A. (2003). *Chem. Eng. Process.* 42: 157–178.

7 Avami, A., Marquardt, W., Saboohi, Y. et al. (2012). *Chem. Eng. Sci.* 71: 166–177.

8 Fernandez, J.R., Abanades, J.C., and Murillo, R. (2012). *Chem. Eng. Sci.* 84: 1–11.

9 Abbas, S.Z. and Dupont, V. (2017). *Int. J. Hydrogen Energy* 42: 18910–18921.

10 Baker, E.H. (1962). *J. Chem. Soc.*: 464–470.

11 Meille, V. (2006). *Appl. Catal., A* 315: 1–17.

12 Leclerc, J.P. and Schweich, D. (1992). Modeling Catalytic Monoliths for Automobile Emission Control. In: *Chemical Reactor Technology for Environmentally Safe Reactors and Products*, NATO ASI Series 225 (eds. H. de Lasa, G. Dogu and A. Ravella), 547–576. Dordrecht: Kluwer Acadenic Publishers.

13 Votruba, J., Mikus, O., Nguen, K. et al. (1975). *Chem. Eng. Sci.* 30: 201–206.

14 Roychowdhury, S., Sundararajan, T., and Das, S.K. (2019). *Int. J. Heat Mass Transfer* 139: 660–674.

15 Rodriguez, M.L., Cadus, L.E., and Borio, D.O. (2019). *Chem. Eng. J.* 377: 120139.

16 Ehrenfeld, W., Hessel, V., and LOve, H. (2000). *Microreactors: New Technology for Modern Chemistry*. Mainz: Wiley-VCH.

17 Venkateswaran, S., Kravaris, C., and Wilhite, B. (2019). *Chem. Eng. J.* 377: 120501.

18 Bac, S. and Avcı, A.K. (2019). *Chem. Eng. J.* 377: 120104.

19 Qamar, S., Bibi, S., Khan, F.U. et al. (2014). *Ind. Eng. Chem. Res.* 53: 2461–2472.

20 Stankiewicz, A., Sarabi, F.E., Baubaid, A. et al. (2019). *Chem. Rec.* 19: 40–50.

21 Stefanidis, G.D., Munoz, A.N., Guido, S.J. et al. (2014). *Rev. Chem. Eng.* 30: 233–259.

22 Pham, T.T.P., Ro, K.S., Chen, L. et al. (2020). *Environ. Chem. Lett.* https://doi.org./10.1007/s10311-020-01055-0.

23 Guler, M., Dogu, T., and Varisli, D. (2017). *Appl. Catal., B* 219: 173–182.

24 Sarıyer, M., Bozdağ, A.A., Sezgi, N.A. et al. (2019). *Chem. Eng. J.* 377: 1201–1208.
25 Gunduz, S. and Dogu, T. (2015). *Appl. Catal., B* 168–169: 497–508.
26 Zhang, X.L., Hayward, D.O., and Mingos, D.M.P. (1999). *Chem. Commun.*: 975–976.
27 Sarabi, F.E., Ghorbani, M., Stankiewicz, A. et al. (2020). *Chem. Eng. Res. Des.* 153: 677–683.
28 Sancheti, S.V. and Gogate, P.R. (2017). *Ultrason. Sonochem.* 36: 527–543.
29 Dong, Z., Delacour, C., Mc Carogher, K. et al. (2020). *Materials* 13: 344.
30 Dogu, D., Meyer, K.E., Fuller, A. et al. (2018). *Appl. Catal., B* 227: 90–101.
31 De Lasa, H., Serrano, B., and Salaices, M. (2005). *Photocatalytic Reaction Engineering*. New York: Springer.
32 Villafan-Vidales, H.I., Arancibia-Bulnes, C.A., Riveros-Rosas, D. et al. (2017). *Renewable Sustainable Energy Rev.* 75: 894–908.
33 Murmura, M.A. and Annesini, M.C. (2020). *Processes* 8: 308. https://doi.org/103390/pr8030308.

14

Multiphase Reactors

Besides the conventional two-phase fixed-bed catalytic reactor systems, multiphase reactors involving gas, liquid, and solid phases are used in several chemical processes. Among these, three-phase reaction systems, such as slurry reactors, trickle-bed reactors, gas sparged bubble columns are widely used in the production of chemicals and fuels. Generally, these reactors involve volatile and non-volatile reactants or sometimes a liquid solvent with gaseous reactants and solid-phase catalysts. Some typical processes conducted in three-phase reactors are hydrogenation of oils over a solid catalyst, polymerization of ethylene in a slurry of solid catalyst particles, hydrotreating of petroleum residues, oxidation of liquid hydrocarbons, some fermentation reactions, biological waste treatment, flue gas desulfurization, aliphatic and aromatic chlorination reactions, etc. Hydrodynamics of such three-phase systems dictates the most appropriate model to be used for design purposes.

Two-phase (gas–solid or liquid–solid) fluidized-bed reactors are also widely used in the chemical and petrochemical industry. Fluidized Catalytic Cracking (FCC) is a very well known application of such reactors in the petroleum industry. Hydrodynamics of the fluidized beds and transport resistances between the phases are essential factors in setting up the models for the design and analysis of such reactors.

In reactors involving more than a single phase, transport of reactant molecules from one phase to another should be considered. The actual reaction may take place in one phase only. However, if one of the reactants is in a phase where the reaction does not occur, it should be transported to the phase where the reaction takes place. Mass transfer resistance at the interphase may become the rate-controlling step in many cases. In this chapter, some examples of modeling of slurry, trickle-bed, and fluidized-bed reactors are presented.

14.1 Slurry Reactors

A catalytic slurry reactor is a mixed vessel containing a solid catalyst which is suspended in the liquid phase. The slurry is generally mixed either mechanically or by gas-induced agitation. Gas is generally sparged from the bottom of the stirred tank. The liquid phase may be continuously flowing or sometimes in batch mode. A typical schematic diagram of a well-mixed slurry reactor is shown in Figure 14.1.

A slurry reactor generally contains gas and liquid phase reactants. The complexity of such reactors is due to the presence of gas–liquid, liquid–particle mass transfer, and surface chemical

Fundamentals of Chemical Reactor Engineering: A Multi-Scale Approach, First Edition. Timur Doğu and Gülşen Doğu.

Companion website: www.wiley.com/go/dogu/chemreacengin

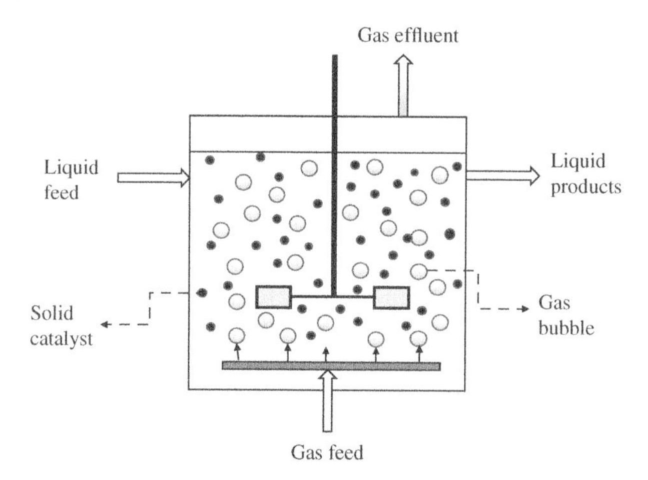

Figure 14.1 Schematic diagram of a slurry reactor.

reaction within the pores of the solid catalyst. Mixing in the reactor and mass transfer are affected by the presence of various phases within the reactor [1–3]. Flow patterns in such a reactor strongly depend upon the type and speed of the impeller.

Mass transfer and surface reaction steps in a slurry reactor can be listed as follows [3]:

- Gas-phase diffusion of reactants within a bubble to the gas–liquid interface
- Mass transfer of reactant from the gas–liquid interface to the bulk liquid
- Transport of the reactant from the bulk liquid to the catalyst external surface
- Diffusion of reactant within the pores of the catalyst
- Adsorption of the reactant on the active surface sites of the catalyst
- Surface reaction on the pore surfaces of the catalyst
- Desorption of products from the catalyst surface
- Diffusion of products in the pores of the catalyst to its external surface
- Transport of products from the external surface of the catalyst to the bulk liquid phase
- Transport of gaseous products from bulk liquid to the gas bubbles

Catalyst particles used in slurry reactors are usually quite small, and hence, intraparticle transport effects might be negligible in most cases. However, if the particle size is 100 μm or larger, the effectiveness factor values smaller than unity are possible for porous catalysts. This is mainly due to much smaller diffusivity values of reactants in the liquid-filled pores than in the gas-filled pores. The external mass transfer resistance between the bulk liquid and the external surface of the catalyst is also usually quite important. Small particles tend to move together with the liquid.

The main assumption of modeling such a reactor is for the hydrodynamics of the slurry. The simplest model that can be made is the perfect mixing of the liquid phase so that concentrations are uniform in the liquid phase within the reactor. Several other models were proposed in the literature for the hydrodynamics of a slurry reactor [1, 2]. Concentration profiles of reactant A in a slurry reactor and schematic representation of mass transfer resistances are shown in Figure 14.2, considering a film-theory approach.

Let us consider a second-order catalytic gas–liquid reaction between the gaseous reactant A and the liquid reactant B.

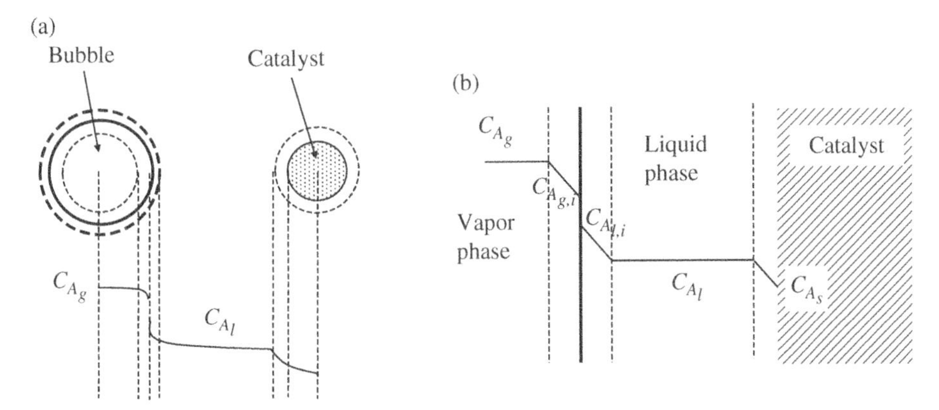

Figure 14.2 (a) Concentration profiles in a slurry reactor. (b) Schematic diagram of mass transfer resistances in a slurry system.

$$-R_A = a_e k_s C_{A_s} C_B \tag{14.1}$$

Here, $k_s C_{A_s} C_B$ is the reaction rate on the catalyst surface per unit area of the external surface, and a_e is the external surface area of the catalyst per unit volume of the bubble-free liquid.

a_e: (external surface area of catalyst particles)/(volume of liquid in the reactor)

If pore diffusion resistance within the catalyst pellet is significant, this rate expression should also be multiplied by the effectiveness factor.

In many cases, the concentration of the liquid reactant B is much higher than the surface concentration of reactant A, which is absorbed in the liquid phase. With this assumption, the reaction rate can be considered as a pseudo-first-order expression with respect to reactant A.

$$\text{If } C_B \gg C_{A_s}; \quad -R_A = \eta a_e (k_s C_B) C_{A_s} = \eta a_e k_s' C_{A_s} \tag{14.2}$$

Here, k_s' is the pseudo-first-order surface reaction rate constant.

In this system, the rate of surface reaction is expected to be equal to the rate of transport of reactant A to the external surface of the catalyst, which is also equal to its rate of transport from the gas bubbles to the liquid phase.

$$-R_A = k_c a_e (C_{A_l} - C_{A_s}) \tag{14.3}$$

$$-R_A = k_l a_b \left(C_{A_{l,i}} - C_{A_l}\right) \tag{14.4}$$

$$-R_A = k_g a_b \left(C_{A_g} - C_{A_{g,i}}\right) \tag{14.5}$$

Here,

k_c is the mass transfer coefficient from bulk liquid to catalyst external surface
k_l is the liquid side mass transfer coefficient at the gas–liquid interface
k_g is the gas side mass transfer coefficient at the gas–liquid interface
a_b interfacial area of the bubbles per unit volume of liquid in the reactor

In these relations, C_{A_g} and C_{A_l} are the gas phase and absorbed concentration of A in the bulk liquid, respectively. Gas and the liquid side concentrations of reactant A can be assumed to be in equilibrium at the gas–liquid interphase.

$$C_{A_{g,i}} = K_{H,i} C_{A_{l,i}} \tag{14.6}$$

Here, $K_{H,i}$ is Henry's law equilibrium constant.

Rearrangement of Eqs. (14.2)–(14.5) together with Eq. (14.6) gives the following relations:

$$-\frac{R_A}{\left(\frac{\eta a_e k'_s}{K_{H,i}}\right)} = K_{H,i} C_{A_s} \tag{14.7}$$

$$-\frac{R_A}{\left(\frac{k_c a_e}{K_{H,i}}\right)} = K_{H,i}(C_{A_l} - C_{A_s}) \tag{14.8}$$

$$-\frac{R_A}{\left(\frac{k_l a_b}{K_{H,i}}\right)} = K_{H,i}\left(C_{A_{l,i}} - C_{A_l}\right) \tag{14.9}$$

$$-\frac{R_A}{(k_g a_b)} = \left(C_{A_g} - C_{A_{g,i}}\right) \tag{14.10}$$

Then, the global rate in terms of gas-phase concentration of reactant A can be expressed by the side-by-side summation of Eqs. (14.7)–(14.10).

$$-\frac{R_A}{\left(\frac{\eta a_e k'_s}{K_{H,i}}\right)} - \frac{R_A}{\left(\frac{k_c a_e}{K_{H,i}}\right)} - \frac{R_A}{\left(\frac{k_l a_b}{K_{H,i}}\right)} - \frac{R_A}{(k_g a_b)} = C_{A_g} \tag{14.11}$$

$$-R_A = \frac{C_{A_g}}{\frac{K_{H,i}}{\eta k'_s a_e} + \frac{K_{H,i}}{k_c a_e} + \frac{K_{H,i}}{k_l a_b} + \frac{1}{k_g a_b}} = k_{obs} C_{A_g} \tag{14.12}$$

Here, the observed rate constant k_{obs} is inversely proportional to the summation of surface reaction and mass transfer resistances at the catalyst surface and the gas–liquid interface (gas and liquid side resistances).

$$k_{obs} = \frac{1}{\frac{K_{H,i}}{\eta k'_s a_e} + \frac{K_{H,i}}{k_c a_e} + \frac{K_{H,i}}{k_l a_b} + \frac{1}{k_g a_b}} \tag{14.13}$$

Depending upon their magnitudes, one of these resistances may control the reaction rate. For instance, if the slowest step is the mass transfer rate of reactant A from the interfacial surface of the bubbles to the bulk liquid (if the highest resistance is the one on the liquid side of the gas–liquid interface), the observed reaction rate can be expressed as follows:

$$-R_A = \left(\frac{k_l a_b}{K_{H,i}}\right) C_{A_g} \tag{14.14}$$

In this case, knowledge about the liquid side mass transfer coefficient is needed to design the reactor.

The specific surface area of the bubbles should be evaluated from the average bubble diameter, which can be estimated from the correlations available in the literature. The relation between the bubble diameter (d_b) and the specific bubble surface area can be expressed assuming spherical bubbles as follows:

$$a_b = \frac{\pi d_b^2}{\pi \frac{d_b^3}{6}} \varepsilon_{bubble} \tag{14.15}$$

Here, ε_{bubble} is the volume fraction of the bubbles within the slurry (gas holdup) and can be estimated using the correlations given in the literature [2]. The specific surface area of the catalyst particles (a_e) can be estimated knowing the external surface area of a single catalyst particle per unit mass and mass of the catalyst present in the reactor per unit volume of the slurry. For a stirred tank reactor containing a relatively small fraction of solids, the mass transfer coefficient from gas to liquid phase can be evaluated using the correlations proposed in the literature [3, 4]. As for the mass transfer coefficient from liquid to a solid surface is concerned, several correlations are also available in the literature [3–5].

If we assume that the slurry within the reactor is well mixed, a CSTR model can be used for the design of a slurry reactor.

$$\frac{V}{F_{A_o}} = \frac{x_{A_f}}{k_{obs} C_{A_g}} \tag{14.16}$$

In the case of bubble-column slurry reactors, the height to diameter ratio of the reactor is longer than the conventional well-mixed slurry reactors. In this case, there is no mechanical mixer in the column. Depending upon the catalyst particle size, catalyst loading, column diameter, and gas velocity, different flow regimes are expected to occur in a bubble-column slurry reactor. In the case of gas velocities smaller than 5 cm/s and for small catalyst loadings homogeneous bubble flow regime is expected. With an increase in gas velocity, heterogeneous turbulent flow and slug flow regimes may be obtained. Details of the bubble column reactors are available in the literature [1–4].

14.2 Trickle-Bed Reactors

Trickle-bed reactors are co-current downflow packed columns. Gas and liquid phases flow co-currently over a fixed bed of solid catalyst pellets in most cases. In some cases, we may also have a counter-current flow of gas and liquid streams. In these reactors, pores of the catalyst are generally filled by the liquid, and the size of the catalyst pellets is much larger than the size of catalyst pellets in the slurry reactors. Hence, a significant effect of intrapellet transport resistance is expected on the observed rates.

In these reactors, the gas phase is considered as continuous, while the liquid phase trickles down over the solid pellets (Figure 14.3). Liquid holdup, axial mixing, mass transfer from gas to liquid, from liquid to solid, and wetting efficiency of the solid pellets are some of the critical parameters affecting the efficiency of these reactors. Some critical factors in the design and operation of trickle-bed reactors are; uniform distribution of the liquid phase over the cross-section of the bed, complete wetting of the catalyst particles, particle size to reactor diameter ratio, elimination of bypass, etc. [3, 6].

Differential species conservation equations for the reactant A within the gas phase, liquid phase, and within the porous solid can be expressed as in Eqs. (14.17), (14.18), and (14.19), respectively. In writing these equations, the gas phase is assumed as plug flow, the catalytic reaction is taken as first-order, and the axial dispersion model is considered for the liquid phase.

$$-U_g \frac{dC_{A_g}}{dz} = K_l a_b \left(\frac{C_{A_g}}{K_H} - C_{A_l} \right) \tag{14.17}$$

$$-U_l \frac{dC_{A_l}}{dz} + D_z \frac{d^2 C_{A_l}}{dz^2} + K_l a_b \left(\frac{C_{A_g}}{K_H} - C_{A_l} \right) - k_c a_c \left(C_{A_l} - C_{A_p}\big|_{r=r_p} \right) = 0 \tag{14.18}$$

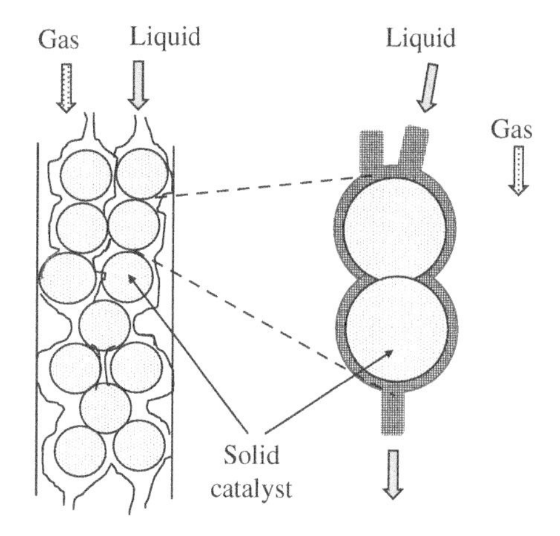

Figure 14.3 Liquid and gas flow in a trickle-bed reactor.

$$\frac{D_e}{r^2}\left(\frac{d}{dr}\left(r^2\frac{dC_{A_p}}{dr}\right)\right) - kC_{A_p} = 0 \tag{14.19}$$

Here, U_g and U_l are the superficial velocity of the gas and the liquid phases and K_l is the overall mass transfer coefficient at the gas–liquid interface. This system of differential equations can be solved with the following boundary conditions:

$$\text{At } z = 0; \quad C_{A_g} = C_{A_{g,o}} \tag{14.20}$$

$$\text{At } z = 0; \quad U_l C_{A_{l,o}} = U_l C_{A_l}\big|_{z=0} - D_z \frac{dC_{A_l}}{dz}\bigg|_{z=0} \tag{14.21}$$

$$\text{At } z = L; \quad \frac{dC_{A_l}}{dz} = 0 \tag{14.22}$$

$$\text{At } r = 0; \quad \frac{dC_{A_p}}{dr} = 0 \tag{14.23}$$

$$\text{At } r = r_o; \quad D_e \frac{dC_{A_p}}{dr}\bigg|_{r=r_p} = k_c\left(C_{A_l} - C_{A_p}\big|_{r=r_p}\right) \tag{14.24}$$

Several correlations are available for the system parameters of a trickle bed reactor [3, 6, 7].

14.3 Fluidized-Bed Reactors

Let us consider a bed of particles on a horizontal distributor within a column, through which there is an upward flow of gas or liquid (Figure 14.4). Pressure drop due to friction between the flowing fluid and the solid particles within this bed can be estimated using the Ergun equation.

$$\frac{(P_o - P_L)}{L} = 150\frac{\mu(1-\varepsilon_b)^2}{d_p^2\varepsilon_b^3}U_o + 1.75\frac{\rho(1-\varepsilon_b)}{d_p\varepsilon_b^3}U_o^2 \tag{14.25}$$

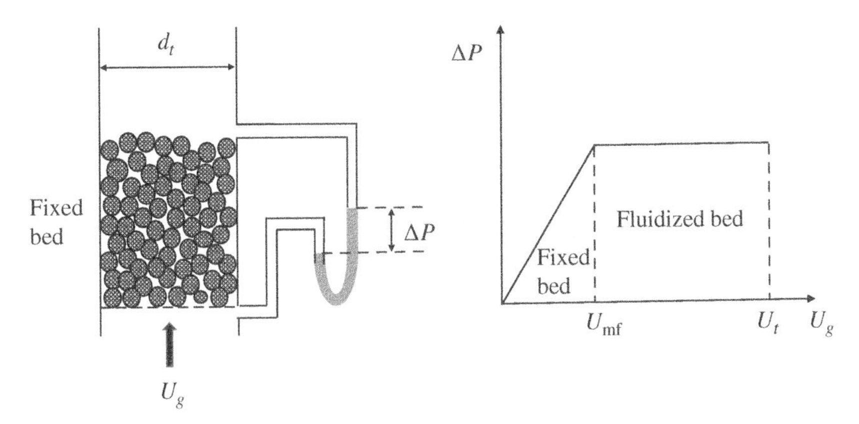

Figure 14.4 Pressure drop in a fixed bed and a fluidized bed.

The second term on the right-hand side of this equation becomes negligible at low particle Reynolds numbers. In a fixed bed of particles, the pressure drop is expected to increase with an increase in superficial fluid velocity (Figure 14.4). When the velocity reaches the minimum fluidization velocity (U_{mf}), the drag force applied by the flow on the fixed bed of particles becomes sufficient to support the particles, and fluidization starts (Eq. (14.26a)). This phenomenon is called incipient fluidization.

$$\left(\rho_s - \rho_g\right)g(1-\varepsilon_{mf}) = \frac{(P_o - P_L)}{L} \tag{14.26a}$$

$$\left(\rho_s - \rho_g\right)g(1-\varepsilon_{mf}) \cong 150\frac{\mu(1-\varepsilon_{mf})^2}{d_p^2\varepsilon_{mf}^3}U_{mf} \tag{14.26b}$$

$$U_{mf} \cong \frac{\left(\rho_s - \rho_g\right)g d_p^2 \varepsilon_{mf}^3}{150\mu(1-\varepsilon_{mf})} \tag{14.26c}$$

Here, ε_{mf} is the void fraction of the bed at minimum fluidization. For superficial velocities of fluid being higher than the minimum fluidization velocity, the bed expands by increasing its total void fraction (Figure 14.5). It is considered that gas bubbles are formed at velocities over the minimum fluidization velocity. Bed expansion continues until the terminal velocity is reached. For velocities over the terminal velocity, a two-phase flow of the gas–solid mixture is expected. In the case of gas fluidized beds, agitation of the solid particles together with the formation of bubbles occur. This is called aggregative fluidization. For this kind of fluidization, excess flow above the value needed for minimum fluidization flows through the bubbles. However, for liquid fluidized-bed systems, the bed expands smoothly, and bubble formation does not take place. This phenomenon is called particulate fluidization.

We expect a good contact of gas and solid phases within a gas fluidized-bed. Good mixing cause close to the uniform temperature within the fluidized-bed reactors. This allows easy temperature control. Fluidized beds are frequently used as catalytic and non-catalytic reactors for reactions involving gas and solid phases [7–11]. Analysis of complex interactions between physical and chemical processes within a fluidized-bed reactor involves understanding of flow characteristics of such systems. In gas fluidized beds, bubbles are expected to form at an unpredictable rate. These bubbles grow, coalesce, and break-up along the bed. Such bubbling beds are generally considered as a two-

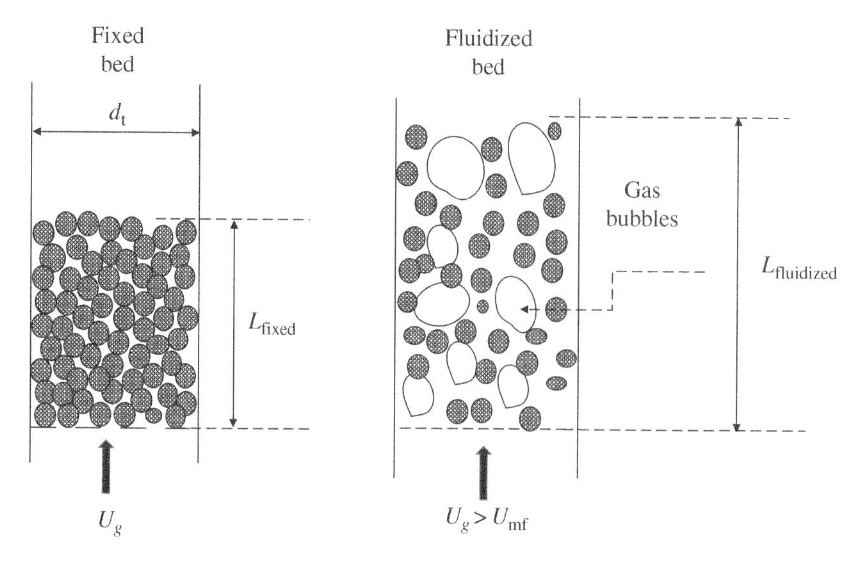

Figure 14.5 Schematic representation of a gas fluidized bed.

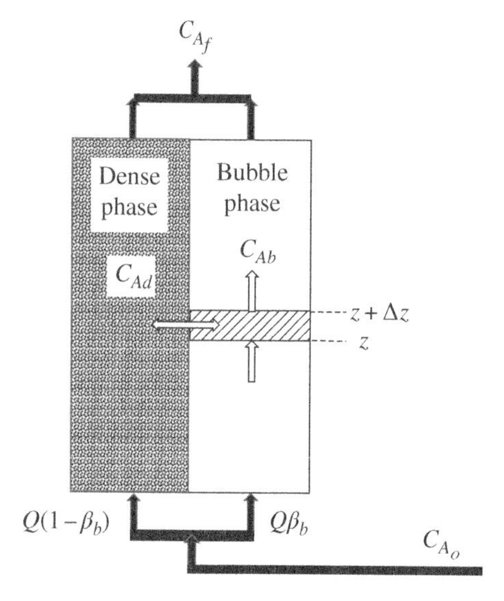

Figure 14.6 Schematic diagram of the two-phase model for a gas fluidized bed reactor.

phase system, namely the bubble phase and dense phase, which is also called the emulsion phase. A fraction of the inlet gas flows through the bubbles, while another fraction flows through the dense phase (D-phase), which is composed of gas and solid particles. Interphase mass transfer is expected between these two phases. Various models were proposed to analyze and design fluidized-bed reactors, which are extensively reviewed in the literature. Here, a macroscopic model, namely a simple two-phase model, is illustrated as an example.

A schematic diagram for the two-phase model considered here is illustrated in Figure 14.6. In this model, the bubble phase (B-phase), which is assumed to be free of solid particles, is considered as a continuous phase within the reactor. It is also assumed that the void fraction in the dense phase (emulsion phase) is constant and equal to the void fraction at incipient fluidization. The hydrodynamic assumption made for this phase is plug flow in the model described below. However, models considering mixing in the bubble phase are also available in the literature. There are also some models considering the cloud-wake phase of the bubbles as a separate phase [7].

As for the hydrodynamics of the dense phase (D-phase), perfect mixing (CSTR) or dispersed flow assumptions can be made. The dense phase is composed of solid catalyst particles dispersed in the reacting gas mixture. Wakes of the bubbles are the main reason for particle circulation in this system. CSTR assumption is made for the D-phase considering good mixing in this phase. The model equations are written based on this assumption, and the catalytic reaction was assumed to take place only within the D-phase.

Species conservation equations for the bubble and the dense phases in the two-phase model described above are as follows:

Bubble phase (assuming plug flow with no catalyst particles):

$$-U_o\beta_b\frac{dC_{Ab}}{dz} - K_l a_b(C_{Ab} - C_{Ad}) = 0 \tag{14.27}$$

Dense phase (assuming perfect mixing with a solid catalyzed reaction):

$$U_o(1-\beta_b)(C_{A_o}) - U_o(1-\beta_b)(C_{Ad}) + \int_0^L K_l a_b(C_{Ab} - C_{Ad})dz + R_{A,w}\rho_p(1-\varepsilon_{\mathrm{mf}})L(1-\phi_b) = 0 \tag{14.28}$$

Here, β_b and ϕ_b are the fraction of the total volumetric flow rate flowing through the bubble phase and volume fraction of the bubble phase within the reactor, respectively. $\varepsilon_{\mathrm{mf}}$ corresponds to the volume fraction of the gas phase within the dense phase, which is equal to the void fraction of the bed at minimum fluidization velocity. The velocity in the dense phase is generally assumed to be equal to the minimum fluidization velocity. Integration of Eq. (14.27) gives

$$C_{Ab} = (C_{A_o} - C_{Ad})\exp\left(-\frac{K_l a_b}{U_o\beta_b}z\right) + C_{Ad} \tag{14.29}$$

Simultaneous solution of Eqs. (14.28) and (14.29) yields the following relations for C_{Ab} and C_{Ad} at the exit of the reactor:

$$C_{Ad} = C_{A_o} + \frac{R_{A,w}\rho_p(1-\varepsilon_{\mathrm{mf}})L(1-\phi_b)}{U_o(1-\beta_b)\left[1-\frac{\beta_b}{(1-\beta_b)}\left(\exp\left(-\frac{K_l a_b}{U_o\beta_b}L\right)-1\right)\right]} \tag{14.30}$$

$$C_{Ab}|_{z=L} = C_{A_o} + \frac{R_{A,w}\rho_p(1-\varepsilon_{\mathrm{mf}})L(1-\phi_b)}{U_o(1-\beta_b)\left[1-\frac{\beta_b}{(1-\beta_b)}\left(\exp\left(-\frac{K_l a_b}{U_o\beta_b}L\right)-1\right)\right]}\left(1-\exp\left(-\frac{K_l a_b}{U_o\beta_b}L\right)\right) \tag{14.31}$$

In the derivation of these equations, the superficial velocity of the gas within the fluidized bed $\left(U_o = \frac{Q}{A_c}\right)$ is assumed constant. The estimation of C_{Ad} and C_{Ab} from Eqs. (14.30) and (14.31) needs information about the rate expression. For instance, for a first-order catalytic reaction in the dense phase

$$-R_{A,w} = k_r C_{Ad} \tag{14.32}$$

the expressions given in Eqs. (14.30) and (14.31) can be written for C_{Ad} and C_{Ab} as follows:

$$C_{Ad} = C_{A_o}\left\{\frac{U_o(1-\beta_b)\left[1-\frac{\beta}{(1-\beta)}\left(\exp\left(-\frac{K_l a_b}{U_o\beta_b}L\right)-1\right)\right]}{U_o(1-\beta_b)\left[1-\frac{\beta}{(1-\beta)}\left(\exp\left(-\frac{K_l a_b}{U_o\beta_b}L\right)-1\right)\right] + k_r\rho_p(1-\varepsilon_{\mathrm{mf}})L(1-\phi_b)}\right\} \tag{14.33}$$

$$C_{Ab}|_{z=L} = C_{A_o}\left\{1 - \frac{k_r\rho_p(1-\varepsilon_{mf})L(1-\phi_b)\left(1-\exp\left(-\frac{K_l a_b}{U_0\beta_b}L\right)\right)}{U_o(1-\beta_b)\left[1-\frac{\beta}{(1-\beta)}\left(\exp\left(-\frac{K_l a_b}{U_o\beta_b}L\right)-1\right)\right]+k_r\rho_p(1-\varepsilon_{mf})L(1-\phi_b)}\right\} \quad (14.34)$$

The concentration of reactant A at the exit of the fluidized bed reactor, and hence its fractional conversion can then be expressed as follows:

$$C_{A_f} = C_{A_o}\left(1-x_{A_f}\right) = \beta_b C_{Ab}|_{z=L} + (1-\beta_b)C_{Ad} \quad (14.35)$$

As can be concluded from this analysis, the design of such a fluidized bed reactor needs quantitative information for the parameters β_b, ϕ_b, ε_{mf}, mass transfer coefficient between the phases $K_l a_b$, as well as the observed reaction rate constant of the catalytic reaction. Several correlations were reported in the literature for the estimation of the hydrodynamic and transport parameters. The two-phase model presented here is only one of the possible alternative models that can be used for the modeling of a fluidized bed reactor. There are several other more complex models in the literature, which consider bubble dimensions, bubble velocity, and cloud-wake formation in the fluidized-bed reactors.

References

1 Chaudhari, R.V. and Shah, Y.T. (1986). Recent advances in slurry reactors. In: *Concepts and Design of Chemical Reactors* (eds. S. Whitaker and A.E. Cassano), 242–287. Gordon Breach Science Publishers.

2 Dudukovic, M.P. and Devanathan, N. (1992). Bubble column reactors: some recent developments. In: *Chemical Reactor Technology for Environmentally Safe Reactors and Products*, NATO ASI Series 225 (eds. H.I. de Lasa, G. Doğu and A. Ravella), 353–377. Dordrecht: Kluwer Academic Publishers.

3 Levec, J. and Goto, S. (1986). Mass transfer and kinetics in three-phase reactors. In: *Handbook of Heat and Mass Transfer*, vol. 2 (ed. N.P. Cheremisinoff), 869–902. Houston, TX: Gulf Publ. Co.

4 Yagi, H. and Yoshida, F. (1975). *IEC Process Des. Dev.* 14: 488–493.

5 Boon-Long, S., Laguereie, C., and Couderc, J.P. (1978). *Chem. Eng. Sci.* 33: 813–819.

6 Hanika, J. and Stanek, V. (1986). Operation and design of trickle-bed reactors. In: *Handbook of Heat and Mass Transfer*, vol. 2 (ed. N.P. Cheremisinoff), 1029–1080. Houston, TX: Gulf Publ. Co.

7 Davidson, J.F., Harrison, D., Darton, R.C. et al. (1977). The two-phase theory of fluidization and its applications to chemical reactors. In: *Chemical Reactor Theory: A Review* (eds. L. Lapidus and N.A. Amundson), 583–685. London: Prentice-Hall.

8 Kunii, D. and Levenspiel, O. (1991). *Fluidization Engineering*. Stoneham, MA: Butterworth Heinemann.

9 Bukur, D., Caram, H.S., and Amundson, N.A. (1977). Some models of fluidized bed reactors. In: *Chemical Reactor Theory: A Review* (eds. L. Lapidus and N.A. Amundson), 686–757. London: Prentice-Hall.

10 Pyle, D.L. (1974). Fluidized bed reactors: a review. In: *Chemical Reaction Engineering* (ed. B. Bischoff), 106–140. ACS.

11 Sozen, Z.Z. and Dogu, G. (1985). *Chem. Eng. J.* 31: 145.

15

Kinetics and Modeling of Non-catalytic Gas–Solid Reactions

Gas–solid non-catalytic reaction systems are involved in numerous industrial processes. The general form of a gas–solid reaction can be expressed as follows:

$$aA(g) + bB(s) \rightarrow cC(g) + dD(s) \tag{15.1}$$

A solid reactant decomposes to give gas and solid products in some cases.

$$CaCO_3(s) \rightarrow CaO(s) + CO_2(g)$$

In most cases, a solid reactant reacts with a gas, such as in the oxidation, reduction, and gasification reactions. Some examples are

Combustion:

$$C(s) + O_2(g) \rightarrow CO_2(g)$$

The reaction of CaO with SO_2:

$$CaO(s) + SO_2(g) + \frac{1}{2}O_2(g) \rightarrow CaSO_4(s)$$

Gasification of coal and biowaste:

$$C_xH_y(s) + xH_2O(g) \rightarrow xCO(g) + \left(x + \frac{y}{2}\right)H_2(g)$$

There is another group of gas-phase reactions that take place on the surface of a solid, yielding solid and gaseous products, which are chemical vapor deposition (CVD) reactions. Some examples of CVD reactions are

Boron deposition by CVD:

$$BCl_3(g) + \frac{3}{2}H_2(g) \rightarrow B(s) + 3HCl(g)$$

Silicon deposition by CVD:

$$SiCl_4(g) + 2H_2(g) \rightarrow Si(s) + 4HCl$$

Gas–solid non-catalytic reactions are quite complex processes involving changes in the mass, shape, and pore structure of the solid reactant. If a solid product is formed as a result of a

Fundamentals of Chemical Reactor Engineering: A Multi-Scale Approach, First Edition. Timur Doğu and Gülşen Doğu.

Companion website: www.wiley.com/go/dogu/chemreacengin

non-catalytic gas–solid reaction, it may deposit over the solid reactant. The formation of a solid product layer over the solid reactant may create additional resistance for the surface reaction between the solid and gaseous reactants. The observed reaction rate and mechanism of such reactions strongly depend upon whether the solid reactant is porous or not. If it is porous, the gaseous reactant should penetrate the pores before the surface reaction. Pore and product layer diffusion resistances, changes in the pore structure, and the active surface area during the reaction are critical factors to be considered in modeling such reactions [1, 2].

15.1 Unreacted-Core Model

The simplest and the most frequently used model for a gas–solid reaction is the unreacted-core model. This model is also called the shrinking-core model. In this model, the reaction is assumed to take place on the external surface of the solid reactant, which is assumed to shrink as the reaction proceeds. The solid product is deposited as a shell over the solid reactant, which is in the core section. This reaction model is illustrated in Figure 15.1. Let us consider a gas–solid reaction between a gaseous reactant A and a solid reactant B.

$$A(g) + B(s) \rightarrow C(g) + D(s)$$

The observed rate of this reaction will be affected by the mass transfer resistances for the transport of reactant A from the bulk gas phase to the external surface of the solid reactant, diffusion resistance through the product layer formed over the solid reactant, and the surface reaction resistance at the surface of the solid reactant core. Model equations for the unreacted-core model can be expressed considering the following assumptions:

- spherical particle
- the reaction takes place only at the interface of reactant core and product shell sections (no reaction within the shell and the core sections)
- isothermal system
- no change in particle size during the reaction (r_p constant)
- pseudo-steady-state.

Pseudo-steady-state species conservation equation for the gaseous reactant A within the product layer (shell section) can be expressed as follows:

$$\left(4\pi r^2 N_{A_r}\right)_r - \left(4\pi r^2 N_{A_r}\right)_{r+\Delta r} = 0 \tag{15.2}$$

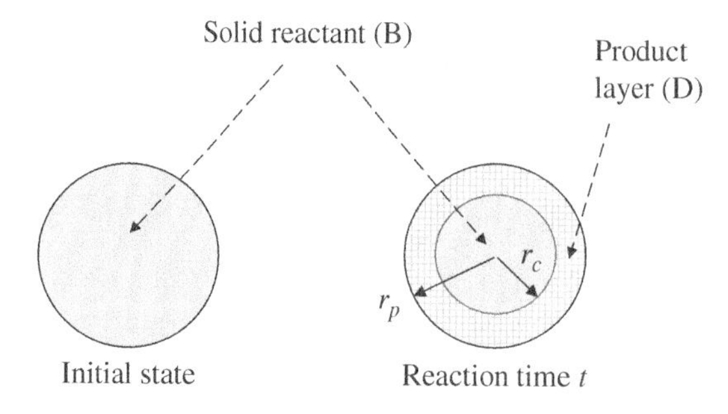

Figure 15.1 Schematic diagram of the unreacted core model.

This relation can be expressed in differential form as given in Eq. (15.3).

$$\frac{1}{r^2}\frac{d}{dr}\left(r^2 N_{A_r}\right) = 0 \tag{15.3}$$

Here, N_{A_r} is the effective diffusion flux of reactant A in the product layer. Using Fick's law of diffusion expression for the transport of reactant A through the shell section, the species conservation equation for this reactant can be written as follows:

$$\frac{D_e}{r^2}\frac{d}{dr}\left(r^2 \frac{dC_{A_p}}{dr}\right) = 0 \tag{15.4}$$

In the derivation of this expression, a pseudo-steady-state assumption is made for the concentration of reactant A in the shell section. This assumption is generally justified considering that the shrinking rate of the core radius is much less than the rate of diffusion of reactant A through the product layer.

Integration of Eq. (15.4) gives

$$C_{A_p} = -\frac{C_1}{r} + C_2 \tag{15.5}$$

Evaluation of the integration constants C_1 and C_2 needs two boundary conditions. The concentration of the reactant A at the external surface of the solid reactant can be assumed to be the same as its bulk concentration if the mass transfer resistance at the external surface of the particle is negligible.

$$\text{B.C.1: at } r = r_p; \quad C_{A_p} = C_A \tag{15.6}$$

The second boundary condition can be written at the core radius r_c, by equating the diffusion flux to the surface reaction rate.

$$\text{B.C.2: at } r = r_c; \quad D_e \left.\frac{dC_{A_p}}{dr}\right|_{r=r_c} = k_s C_{A_p}\big|_{r=r_c} \tag{15.7}$$

Evaluation of integration constants from the boundary conditions and using Eq. (15.5), the concentration profile of reactant A within the product layer can be expressed as follows:

$$\frac{C_{A_p}}{C_A} = \frac{1 + \frac{r_c k_s}{D_e}\left(1 - \frac{r_c}{r}\right)}{1 + \frac{r_c k_s}{D_e}\left(1 - \frac{r_c}{r_p}\right)} \tag{15.8}$$

Surface reaction rate at the interface between the core and shell sections can then be expressed as follows:

$$-R_{A_s} = -R_{B_s} = k_s C_{A_p}\big|_{r=r_c} = C_A \frac{k_s}{1 + \left(\frac{k_s r_c}{D_e}\right)\left(1 - \frac{r_c}{r_p}\right)} \quad (\text{mol/m}^2\ \text{s}) \tag{15.9}$$

The rate of conversion of solid reactant B per unit time is then expressed by multiplying Eq. (15.9) with $4\pi r_c^2$.

$$-\frac{dn_B}{dt} = 4\pi r_c^2 \frac{k_s C_A}{1 + \left(\frac{k_s r_c}{D_e}\right)\left(1 - \frac{r_c}{r_p}\right)} \tag{15.10}$$

The rate of change of the number of moles of B (n_B) with reaction time can also be expressed in terms of the rate of change of core radius with respect to time.

$$\frac{dn_B}{dt} = \frac{\rho_B}{M_B} 4\pi r_c^2 \frac{dr_c}{dt} \tag{15.11}$$

Here, ρ_B and M_B are the density and the molecular weight of solid reactant B, respectively. Combining Eqs. (15.10) and (15.11), the following differential equation can be written for the variation of core radius with respect to time.

$$-\frac{\rho_B}{M_B}\frac{dr_c}{dt} = \frac{k_s C_A}{\left(1 + \left(\frac{k_s r_c}{D_e}\right)\left(1 - \frac{r_c}{r_p}\right)\right)} \tag{15.12}$$

Integration of Eq. (15.12) with the initial condition of $r_c|_{t=0} = r_p$ gives the following relation for the time dependence of core radius.

$$\left(\frac{M_B}{\rho_B}\right)\left(\frac{k_s C_A}{r_p}\right) t = \left(1 - \frac{r_c}{r_p}\right)\left[1 + \frac{k_s r_p}{6D_e}\left(1 + \frac{r_c}{r_p} - 2\left(\frac{r_c}{r_p}\right)^2\right)\right] \tag{15.13}$$

Fractional conversion of solid reactant B can be expressed in terms of core radius, as follows:

$$x_B = \left(\frac{\left(\frac{4}{3}\pi r_p^3\right)\rho_B - \left(\frac{4}{3}\pi r_c^3\right)\rho_B}{\left(\frac{4}{3}\pi r_p^3\right)\rho_B}\right) = 1 - \left(\frac{r_c}{r_p}\right)^3 \tag{15.14}$$

The following expression is then obtained for the relation between the fractional conversion of solid reactant B and reaction time:

$$t = \left(\frac{\rho_B}{M_B}\right)\left(\frac{r_p}{k_s C_A}\right)\left(1 - (1 - x_B)^{1/3}\right)\left[1 + \frac{k_s r_p}{6D_e}\left(1 + (1 - x_B)^{1/3} - 2(1 - x_B)^{2/3}\right)\right] \tag{15.15}$$

We can consider two limiting cases, namely diffusion control and surface reaction control limits of this equation. These limiting forms can be expressed for the cases of

$$\frac{k_s r_p}{D_e} \ll 1 \text{ (surface reaction control, negligible diffusion resistance)}$$

$$\frac{k_s r_p}{D_e} \gg 1 \text{ (product layer diffusion control, surface reaction very fast)}$$

The dimensional group $\frac{k_s r_p}{D_e}$ is proportional to the ratio of product-layer diffusion to the surface reaction resistances. As the value of this dimensionless group becomes very large, product-layer diffusion becomes the controlling mechanism and the concentration of A at $r = r_c$ approaches zero. The relations between fractional conversion and time for these two limiting cases are as follows:

$$t = \left(\frac{\rho_B}{M_B}\right)\left(\frac{r_p}{k_s C_A}\right)\left(1 - (1 - x_B)^{1/3}\right) \text{ (surface reaction control)} \tag{15.16}$$

$$t = \left(\frac{\rho_B}{M_B}\right)\left(\frac{r_p^2}{6D_e C_A}\right)\left(1 - 3(1 - x_B)^{2/3} + 2(1 - x_B)\right) \text{ (diffusion control)} \tag{15.17}$$

Expressions for the time required to reach complete conversion are also different for these two cases:

$$T = t|_{x_B = 1} = \left(\frac{\rho_B}{M_B}\right)\left(\frac{r_p}{k_s C_A}\right); \text{ (surface reaction control)} \tag{15.18}$$

$$T = t|_{x_B = 1} = \left(\frac{\rho_B}{M_B}\right)\left(\frac{r_p{}^2}{6D_e C_A}\right); \text{ (diffusion control)} \tag{15.19}$$

As can be predicted from Eqs. (15.18) and (15.19), the time required to reach the complete conversion of a solid reactant is proportional to the particle size for the surface reaction control case. However, it is proportional to the square of the particle size in the case of product layer diffusion control.

The analysis shown above was made by neglecting the possible contribution of external mass transfer resistance over the solid particles. Suppose that the external mass transfer resistance for the transport of gaseous reactant A over the solid reactant particles is also significant. In that case, the boundary condition written by Eq. (15.6) at $r = r_p$ should be replaced by a flux boundary condition.

$$k_m\left(C_A - C_{A_p}|_{r = r_p}\right) = D_e \frac{dC_{A_p}}{dr}\bigg|_{r = r_p} = k_s C_{A_c}|_{r = r_c}\left(\frac{4\pi r_c^2}{4\pi r_p^2}\right) \tag{15.20}$$

If the rate of transport of reactant A to the external surface of the solid reactant is much slower than its diffusion rate through the product layer and the surface reaction rate at the core surface, film mass transfer resistance over the external surface of the catalyst pellet becomes the rate-determining step. In this case, the concentration of reactant A at the external surface of the solid reactant approaches zero, and the rate of conversion of solid reactant B can be expressed as

$$-\frac{dn_B}{dt} = 4\pi r_p^2 k_m\left(C_A - C_{A_p}|_{r = r_p}\right) \cong 4\pi r_p^2 k_m (C_A) \tag{15.21}$$

$$-\frac{dn_B}{dt} = 4\pi r_p^2 k_m (C_A) = -\frac{\rho_B}{M_B} 4\pi r_c^2 \frac{dr_c}{dt} \tag{15.22}$$

Integration of this relation yields the following expression for the variation of fractional conversion of B as a function of reaction time:

$$t = \left(\frac{\rho_B}{M_B}\right)\left(\frac{r_p}{3k_m C_A}\right) x_B; \text{ (external mass transfer control)} \tag{15.23}$$

The relation between the fractional conversion of solid reactant and the reaction time is linear for the external mass transfer control case. In this case, the expression for the time required for the complete conversion of B becomes

$$T = t|_{x_B = 1} = \left(\frac{\rho_B}{M_B}\right)\left(\frac{r_p}{3k_m C_A}\right) \tag{15.24}$$

15.2 Deactivation and Structural Models for Gas–Solid Reactions

The unreacted-core model assumes a non-porous solid reactant and does not consider the penetration of the reactant into its pores. Instead, it is assumed that the reaction takes place only at the interface of the solid reactant (core section) and the product layer (shell section). However, in many cases, the solid reactant may be porous, and hence the gaseous reactant may diffuse into the solid reactant, and reaction may take place throughout the particle. Of course, we would expect a decrease in the concentration of the gaseous reactant from the external surface toward the center of the pellet.

As a result of the gas–solid reaction, changes are expected in the pore structure of the solid reactant. Solid product may deposit on the surfaces of the pores, causing changes in pore radius, even causing pore-mouth closure (Figure 15.2). In such cases, diffusion resistance for the transport of the gaseous reactant into the pores, as well as the diffusion resistance in the product layer, should be considered. Structural changes may even lead to changes in the particle size of the solid reactant with the reaction extent.

Several models were considered in the literature to describe the complex behavior of such gas–solid non-catalytic reactions. Grain models and pore models are two of the most used approaches for these complex systems [3, 4]. The schematic representations of the pore and the grain models are shown in Figures 15.2 and 15.3, respectively. Details of these models, their assumptions, model equations, and solution procedures are available in the literature.

Reactivity of the solid reactant strongly depends upon the pore or grain size distributions. Prediction of the observed rate values from the pore or grain models depends strongly upon the availability of detailed experimental information for the pore structure. Another approach used to predict the observed rate, and conversion–time relation is deactivation model [3, 5, 6]. In this

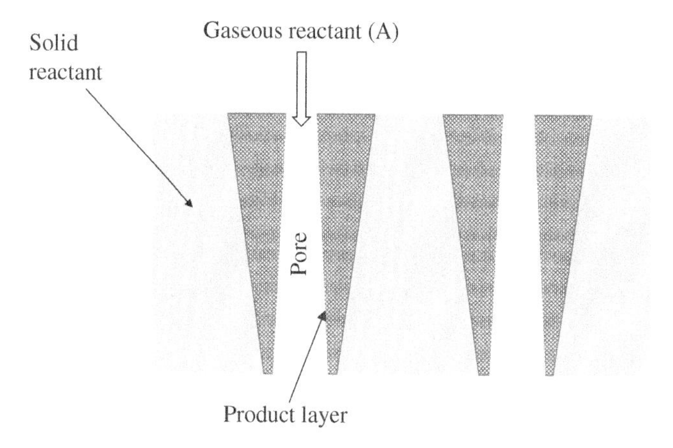

Figure 15.2 Schematic representation of pore-mouth closure as a result of a gas–solid non-catalytic reaction.

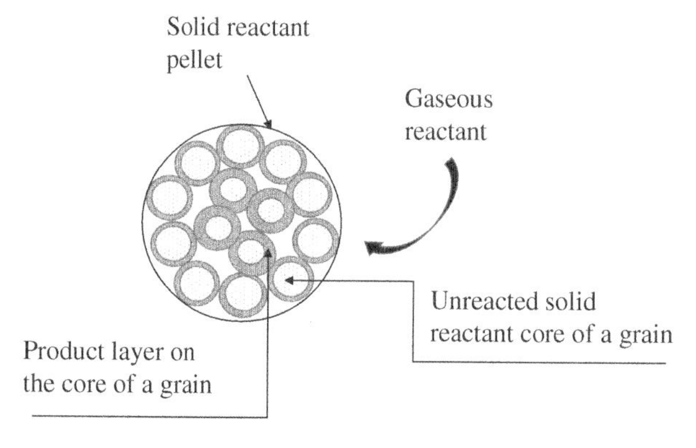

Figure 15.3 Schematic representation of the grain model.

model, an activity factor is defined for the solid reactant. The pore structure and the diffusion resistance in the pores of the solid reactant, as well as diffusion resistance through the product layer on the pore surfaces, influence the value of this activity factor. The rate of a gas–solid non-catalytic reaction is then expressed as follows:

$$-R_{A,p} = \left(S_g^o \rho_p^o k_s\right) a_c C_A = k^o a_c C_A = \frac{\rho_p^o}{M_B}\frac{dx_B}{dt} \quad (\text{mol/m}^3\ \text{s}) \tag{15.25}$$

Here, parameters S_g^o, ρ_p^o, and k^o are the initial values of the surface area (m^2/kg), density (kg/m^3), and the initial observed rate constant of the gas–solid reaction, respectively. The parameter a_c is the activity factor of the solid reactant, and it decreases with time due to the changes in surface area, pore structure, and also due to the formation of product layer over the reactive surfaces of the pores. At time zero, the activity factor a_c is equal to unity.

Following a procedure similar to catalyst deactivation, the rate of change in activity with time can be modeled as follows:

$$-\frac{da_c}{dt} = k_d a_c^n C_A^m \tag{15.26}$$

In this equation, the parameter k_d is the deactivation rate constant. The zeroth solution of the deactivation model can be obtained by taking $n = 1$ and $m = 0$. In this case, the integration of Eq. (15.26) gives the following expression for the activity factor:

$$a_c = \exp(-k_d t) \tag{15.27}$$

Incorporation of Eq. (15.27) into Eq. (15.25) and integration of that equation give the following expression for the case of a constant concentration of the gas-phase reactant:

$$x_B = \frac{M_B k^o}{\rho_p^o k_d} C_A[1 - \exp(-k_d t)] \tag{15.28}$$

The deactivation model is a two-parameter model, containing the initial reaction rate constant and the deactivation rate constant as the model parameters.

The use of this model in a fixed-bed adsorber can also be illustrated to predict the breakthrough curve of the adsorbing gas. Consider a fixed-bed adsorber packed with adsorbent particles. Suppose a step input of adsorbing gas is introduced into the inlet of the fixed-bed adsorber (Figure 15.4). In that case, an S-shaped breakthrough curve is expected for the variation of concentration of adsorbing gas at the exit of the adsorber. While the adsorbing species is flowing along the fixed bed of adsorbent particles, it is removed from the fluid phase as a result of sorption on the adsorbent.

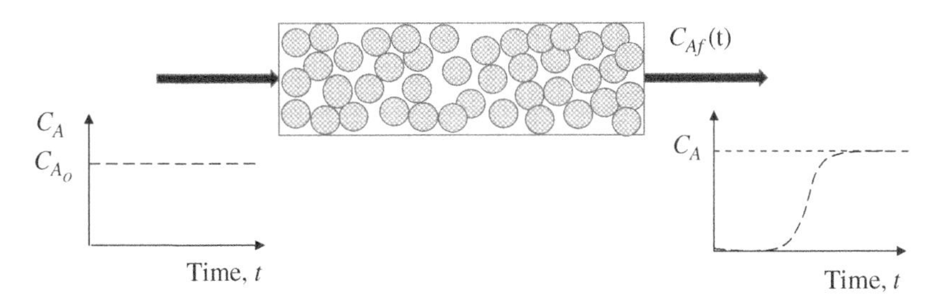

Figure 15.4 The breakthrough curve for the adsorbing gas in a fixed-bed adsorber.

However, the adsorbent will be saturated by the adsorbed molecules as time progresses, and hence its activity for further removal of the adsorbing species will decrease.

Assuming plug flow in the fixed-bed adsorber and making a pseudo-steady-state assumption for the system, the species conservation equation for the adsorbing gas A can be expressed as follows:

$$-Q\frac{dC_A}{dW} - k_w^o a_c C_A = 0 \tag{15.29}$$

The rate of change of concentration with respect to time at a given location in the fixed-bed column is much less than the rate of sorption, and the rate of transport terms of the adsorbent, in most cases. Hence, the pseudo-steady-state assumption is justified in many cases. Here, W is the mass of the adsorbing particles in the adsorber. In the case of concentration-independent deactivation rate, the integration of this equation with the activity relation given in Eq. (15.27) gives the breakthrough expression for the concentration of the adsorbing species at the exit of the adsorber.

$$C_A = C_{A_o} \exp\left[-\frac{k_w^o W}{Q}\exp(-k_d t)\right] \tag{15.30}$$

The parameters of this adsorbing system, namely k_w^o and k_d, can be evaluated by the analysis of the experimental breakthrough data.

If we consider the concentration dependence of the deactivation process, we can take $n = 1$ and $m = 1$ in Eq. (15.26). In this case, a numerical solution of Eq. (15.29), together with (15.26), predicts the breakthrough curve. An approximate solution of Eq. (15.29) was reported for this case as [6]

$$C_A = C_{A_o} \exp\left\{\frac{\left[1 - \exp\left(\frac{k_w^o W}{Q}(1 - \exp(-k_d t))\right)\right]}{(1 - \exp(-k_d t))}\exp(-k_d t)\right\} \tag{15.31}$$

15.3 Chemical Vapor Deposition Reactors

CVD is a versatile method for the production of materials from gas-phase reactants. Reactant species are in the vapor phase in a CVD reactor. A solid product is formed and deposited on a substrate as a result of a chemical reaction. During this process, deposition of the solid on a substrate surface may involve the gas phase and the surface reactions, forming a solid film. Gas-phase reactions may yield some other gas-phase species, which may then react, yielding the solid product. CVD is a quite common process in materials processing. A few examples of CVD reactors are illustrated in this section.

Analysis of a CVD process should include heat and mass transfer processes, as well as surface and gas-phase reactions. One of the examples of a CVD process is the deposition of boron films on a substrate. Boron fiber is a corrosion-resistant, and high-tensile-strength fibrous material, which may be used to produce refractory coatings, electronic components, and the production of composite materials. Boron-coated surfaces and boron fiber can be produced through a CVD process. CVD of boron can be illustrated by the reaction of boron chloride with hydrogen, as given by the following stoichiometry:

$$BCl_3(g) + \frac{3}{2}H_2(g) \rightarrow B(s) + 3HCl(g)$$

CVD of boron through the reaction of boron chloride with hydrogen may involve a gas-phase reaction, followed by the deposition of solid boron.

$$BCl_3(g) + H_2(g) \rightarrow BHCl_2(g) + HCl(g)$$

$$BHCl_2(g) + \frac{1}{2}H_2(g) \rightarrow B(s) + 2HCl(g)$$

CVD processes are usually quite complex, and modeling and design of a CVD reactor require the simultaneous solution of the continuity equation, species conservation equation, energy balance equations, together with the equation of motion. In many of these systems, the substrate on which CVD takes place is hot, while the reactants are at a much lower temperature. Radiation heat transfer between the heated substrate and the reactants flowing in the reactor should also be taken into account for the modeling of such a reactor. The geometry of a CVD reactor and hydrodynamics within this reactor are important factors for setting the model equations and design of a CVD reactor.

The CVD process can also be used to produce fibrous materials, such as SiC fiber and boron fiber. An example of a CVD reactor, which will be discussed here, is a boron fiber production system, which is shown in Figure 15.5 [7, 8]. In this CVD reactor, a cold reactant gas mixture, namely BCl_3 and hydrogen, enters the flow system and flows parallel to the tungsten filament on which boron deposition takes place. This filament is heated directly by applying a DC voltage between the two ends of the filament. The temperature of the filament is over 1000 °C. The deposition of the boron on the hot filament takes place through a CVD process. Tungsten filament (thin wire) continuously moves from one end of the CVD reactor to the other by the rotation of the feeding and the collection spools (Figure 15.5). Boron-coated filament is then continuously produced and collected with the collecting spool. Hydrodynamics of the species within this reactor, diffusion of

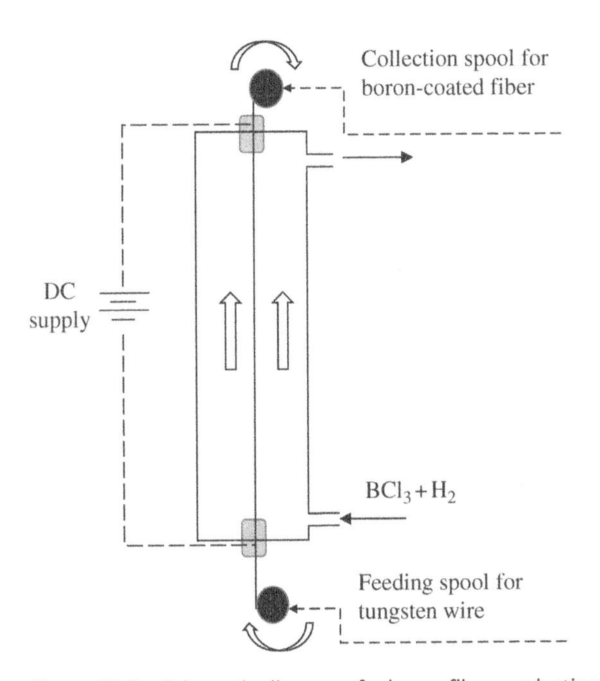

Figure 15.5 Schematic diagram of a boron fiber production reactor.

boron chloride to the substrate surface, and temperature distributions are critical parameters for modeling such a CVD reactor.

Equation of motion should be written for the prediction of the hydrodynamics of such a CVD reactor. Variation of the concentration of the gaseous reactant in a CVD reactor, which is illustrated in Figure 15.5, can be represented by the species conservation equation

$$-v_z\frac{\partial C_A}{\partial z} + D_{AB}\frac{\partial^2 C_A}{\partial z^2} + \frac{D_{AB}}{r}\frac{\partial}{\partial r}\left(r\frac{\partial C_A}{\partial r}\right) = 0 \tag{15.32}$$

This expression is written for a case with no fluid-phase reaction taking place within the reactor and also considering a constant value of the diffusion coefficient. The CVD reaction term on the substrate surface appears in the boundary condition at the surface of the hot wire ($r = r_1$). The boundary conditions can be expressed as follows for this system:

$$D_{AB}\left.\frac{\partial C_A}{\partial r}\right|_{r=r_1} = k_s C_A\big|_{r=r_1} \text{ (on the surface of the hot wire)} \tag{15.33}$$

$$\left.\frac{\partial C_A}{\partial r}\right|_{r=r_2} = 0 \text{ (on the surface of the reactor tube)} \tag{15.34}$$

Here, r_1 and r_2 are the radii of the tungsten wire and the cylindrical reactor, respectively. Equation of motion should be used for the prediction of the variation of velocity (v_z) in the radial direction in the reactor. The order of magnitude of the diffusion term in the axial direction is generally much less than the order of magnitude of the convective transport term in Eq. (15.32) and may be neglected.

The temperature of the fiber can be assumed as constant along the fiber length. However, significant temperature variations are expected in the fluid phase, both in the axial and the radial directions. Radiative heat transfer, as well as convection and conduction of heat from the fiber to the fluid, should be considered in writing the energy equation. In a more realistic model, possible variations of density and viscosity on the hydrodynamics of the system should be considered. In that case, the equation of continuity, equation of motion (momentum balance equation), species conservation equation, and the energy equation should be solved simultaneously for the two-dimensional modeling of such a reactor.

Another typical example of a CVD system is the formation of silicon or silicon dioxide films on a substrate for electronic applications. In these reactors, homogeneous gas-phase reactions also take place for the production of CVD precursors. For instance, gas-phase pyrolysis of silane may occur by the removal of one hydrogen molecule [9].

$$SiH_4(g) \rightarrow SiH_2(g) + H_2$$

CVD of silicon may then take place by the adsorption of SiH_2 molecules on the substrate, which may then be followed by a deposition reaction on the surface.

$$SiH_2(ads) \rightarrow Si(s) + H_2$$

The rate expression for silicon deposition on the substrate can be expressed as follows:

$$R_{Si} = \frac{kC_{SiH_2}}{1 + K_A C_{SiH_2}} \tag{15.35}$$

Here, K_A is the adsorption equilibrium constant of SiH_2 on the substrate surface.

The formation of silicon dioxide on the wafers by the reaction of silane with an excess amount of oxygen is another CVD system. This reaction is generally considered as a first-order reaction in terms of SiH_4.

$$SiH_4(g) + O_2(g) \rightarrow SiO_2(s) + 2H_2(g)$$

The reaction of silicon chloride with hydrogen is another example of a CVD of silicon on a substrate [9]. We also expect a set of gas-phase reactions, together with the deposition reaction of silicon on the substrate surface.

$$SiCl_4(g) + H_2(g) \rightarrow SiHCl_3(g) + HCl(g)$$
$$SiHCl_3(g) + H_2(g) \rightarrow Si(s) + 3HCl(g)$$

Many different reactor configurations were used for CVD processes. Among them, impinging-jet, barrel, horizontal-flow, rotating disc, pancake, fluidized-bed-type reactors can be mentioned. Also, while some of these reactors operate at atmospheric pressure, some of them work in a vacuum. There are also cold-wall and hot-wall CVD reactors. Modeling of these CVD reactors involves the simultaneous solution of the equation of continuity, equation of motion (momentum equation), equation of species conservation, and energy equation, considering the geometry of the reaction system.

Exercises

15.1 High-temperature removal of H_2S from process gases can be achieved by using oxide sorbents. Using the breakthrough data reported by Yasyerli et al. [6] for the removal of H_2S on an oxide sorbent, and assuming that the unreacted core model is a good representation of this reaction, predict conversion vs. time relations for the surface reaction control and the external film mass transfer control cases. Also, obtain the rate parameters and obtain an expression for the time required to reach 10% of the feed concentration (breakthrough point) at the reactor outlet.

15.2 The surface reaction rate constant for the reduction of Fe_2O_3 to Fe is given as 3 cm/s at 1100 K.

$$Fe_3O_4 + 4H_2 \rightarrow 3Fe + 4H_2O$$

Assuming that the unreacted-core model can be used for the prediction of conversion-time relation for this reaction, estimate the time required to reach the complete reduction of Fe_3O_4 using the given data. Neglect external film mass transfer resistance on the observed rate of this reaction, and discuss whether diffusion resistance of hydrogen in the product layer (shell section) or the surface reaction resistance on the surface of Fe_2O_4 core is more significant for this reaction.

Data: Particle size of Fe_2O_4: 0.5 cm
The effective diffusion coefficient in the product layer: 0.02 cm^2/s.

15.3 The reaction of a non-porous solid reactant B with a gaseous reactant A yields solid and gaseous products:

$$A(g) + B(s) \rightarrow C(s) + D(g)$$

Assuming that the unreacted-core model can explain the conversion–time relation for this reaction, answer the following questions:

(a) Reaction time needed for the complete conversion of solid reactant increased from 60 to 240 minutes by increasing the particle size from 0.5 to 1.0 cm. Discuss whether pore diffusion resistance in the product layer or the surface reaction resistance controls this reaction rate.

(b) This reaction is carried out in a continuous-flow reactor, in which the solids move from one end to the exit of the reactor with a velocity of 0.5 cm/s without mixing. The concentration of the gaseous reactant is kept constant in the reactor. Estimate the length of the reactor to reach a 95% conversion of a solid reactant.

15.4 The formation of a silicon dioxide layer by the oxidation of a silicon wafer is illustrated in Figure 15.6. This oxidation process takes place by the following reaction:

$$Si(s) + O_2(g) \rightarrow SiO_2(s)$$

The silicon dioxide layer grows over the silicon surface with time, and the diffusion resistance of oxygen within this layer is quite significant. Experiments have also shown that the surface reaction on the silicon surface is very fast, and the film mass transfer resistance at the gas–solid (SiO_2) surface is also significant.

(a) Derive equations for the variation of oxygen concentration in the SiO_2 layer and the relation between z_f and time. The solubility of oxygen in the silicon dioxide at the gas–solid interface can be predicted by assuming a linear equilibrium relation, similar to Henry's law expression.

(b) If the initial thickness of the silicon wafer is L, obtain an expression for the time required to reach complete oxidation of silicon.

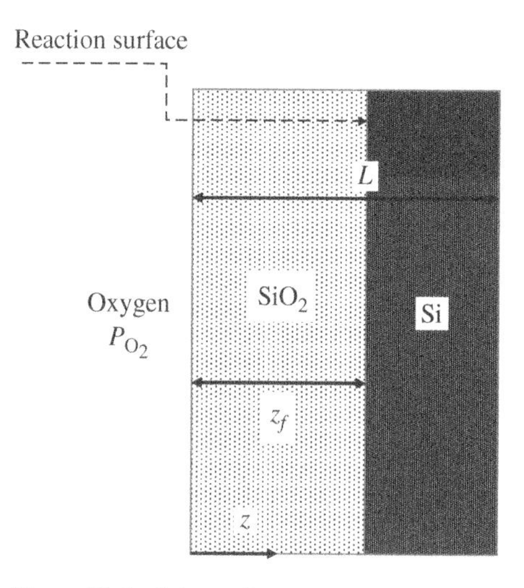

Figure 15.6 Schematic representation of the system for Exercise 15.4.

References

1 Levenspiel, O. (1999). *Chemical Reaction Engineering*. New York: Wiley.
2 Smith, J.M. (1981). *Chemical Engineering Kinetics*. Singapore: McGraw Hill.
3 Dogu, G. and Dogu, T. (1992). Kinetics of capture of sulfur dioxide and applications to flue gas desulfurization. In: *Chemical Reactor Theory for Environmentally Safe Reactors and Products*, NATO ASI Series 225 (eds. H.I. de Lasa, G. Dogu and A. Ravella), 467–498. Dordrecht: Kluwer Academic Publishers, Kluver Publ.
4 Ramachandran, P.A. and Smith, J.M. (1977). *AIChE J.* 23: 353–360.
5 Balcı, S., Dogu, G., and Dogu, T. (1987). *Ind. Eng. Chem. Res.* 26: 1454–1458.
6 Yasyerli, S., Dogu, G., Ar, I., and Dogu, T. (2001). *Ind. Eng. Chem. Res.* 40: 5206–5214.
7 Sezgi, N.A., Dogu, T., and Ozbelge, H.O. (1997). *Ind. Eng. Chem. Res.* 36: 5537–5540.
8 Scholtz, J.H. and Hlavacek, V. (1990). *J. Electrochem. Soc.* 137: 3459–3468.
9 Lee, H.H. (1990). *Fundamentals of Microelectronics Processing*. New York: McGraw Hill Co.

Appendix A

Some Constants of Nature

Gas constant, R_g	1.987	cal/mol K
Gas constant, R_g	1.987	BTU/lbmol R
Gas constant, R_g	8.314	J/mol K
Gas constant, R_g	0.083 14	l bar/mol K
Gas constant, R_g	82.056×10^{-6}	m^3 atm/mol K
Boltzmann constant	$1.380\,649 \times 10^{-23}$	J/K
Planck's constant	6.626×10^{-34}	J s
Avogadro number	6.022×10^{23}	mol^{-1}

Fundamentals of Chemical Reactor Engineering: A Multi-Scale Approach, First Edition. Timur Doğu and Gülşen Doğu.

Companion website: www.wiley.com/go/dogu/chemreacengin

Appendix B

Conversion Factors

To convert from	To	Multiply by
atm	Pa	1.01325×10^5
atm	lbf/in.^{2}	14.696
bar	Pa	1×10^5
Btu	J	1055.06
Btu	cal	251.99
Btu/lb	J/kg	2.326×10^3
Btu/lb R	J/kg K	4.184×10^3
cal	J	4.184
cal/g	J/kg	4.184×10^3
cm	in.	0.3937
cm mercury (at 0 °C)	Pa	1.33×10^3
cP (centipoise)	kg/m s	1×10^{-3}
dyne	N	1×10^{-5}
erg	J	1×10^{-7}
foot	m	0.3048
ft lbf	J	1.355 82
g/cm^3	kg/m^3	1×10^3
gal (US)	m^3	3.785×10^{-3}
hp	kW	0.746
in.	m	2.54×10^{-2}
kg	lb	2.2046
kWh	J	3.6×10^6
l	m^3	1×10^{-3}
lb	kg	0.4536
lb/ft^3	kg/m^3	16.02
Lbf/in.^{2} (psi)	Pa	6.895×10^3
N	lbf	0.224 81
Torr (mm Hg)	Pa	1.33×10^2

Fundamentals of Chemical Reactor Engineering: A Multi-Scale Approach, First Edition. Timur Doğu and Gülşen Doğu.

Companion website: www.wiley.com/go/dogu/chemreacengin

Appendix C

Dimensionless Groups and Parameters

Name	Symbol	Definition
Arrhenius group	γ	$\frac{E_a}{R_g T}$
Biot number for mass transfer	Bi_m	$\frac{k_m(V_p/A_e)}{D_e}$
Biot number for heat transfer	Bi_h	$\frac{h(V_p/A_e)}{D_e}$
Damköhler number	Da	$\frac{V_p}{A_e}\left(\frac{k}{k_m}\right)$
J factor for mass transfer on a catalyst pellet	J	$J_D = \frac{k_m}{U_o}\varepsilon_b(Sc)^{2/3} = \frac{1.15}{Re_p^{1/2}}$
Observable modulus	Φ	$\Phi = \eta\varphi^2 = \frac{(V_p/A_e)^2 R_{p_{obs}}}{D_e C_{A_s}}$
Parameter defined by Eq. (10.93)	α	$\alpha = 3(1-\varepsilon_a)\frac{D_i}{D_a}\left(\frac{r_p}{r_g}\right)^2$
Peclet number (axial)	Pe_z	$\frac{U_o L}{D_z}$
Peclet number based on pellet diameter	Pe_p	$\frac{U_o d_p}{D_z}$
Prater number	β	$\frac{(-\Delta H_R^o) D_e C_{A_s}}{\lambda_e T_s}$
Reynolds number	Re	$\frac{d_t U_o \rho}{\mu}$
Schmidt number	Sc	$\frac{\mu}{\rho D_{AB}} = \frac{\nu}{D_{AB}}$
Sherwood number	Sh	$\frac{k_m(V_p/A_e)}{D_{AB}}$
Thiele modulus (shape generalized, nth-order reaction)	φ_n	$\frac{V_p}{A_e}\left(\frac{k C_A^{n-1}}{D_e}\right)^{1/2}$

Fundamentals of Chemical Reactor Engineering: A Multi-Scale Approach, First Edition. Timur Doğu and Gülşen Doğu.

Companion website: www.wiley.com/go/dogu/chemreacengin

Index

Fundamentals of Chemical Reactor Engineering: A Multi-Scale Approach, First Edition. Timur Doğu and Gülşen Doğu.

Companion website: www.wiley.com/go/dogu/chemreacengin

d

t

Printed and bound by CPI Group (UK) Ltd, Croydon, CR0 4YY